Health Monitoring and Personalized Feedback using Multimedia Data

Alexia Briassouli • Jenny Benois-Pineau
Alexander Hauptmann
Editors

Health Monitoring and Personalized Feedback using Multimedia Data

Editors
Alexia Briassouli
Centre for Research and Technology, Hellas
Information Technologies Institute
Thermi, Thessaloniki, Greece

Jenny Benois-Pineau
Laboratoire Bordelais de Recherche
en Informatique/University Bordeaux
Talence, France

Alexander Hauptmann
Computer Science Department
Carnegie Mellon University
Pittsburgh, PA, USA

ISBN 978-3-319-35783-6 ISBN 978-3-319-17963-6 (eBook)
DOI 10.1007/978-3-319-17963-6

Springer Cham Heidelberg New York Dordrecht London

Softcover reprint of the hardcover 1st edition 2015

Printed on acid-free paper

Springer International Publishing AG Switzerland is part of Springer Science+Business Media (www.springer.com)

Preface

As the world population ages and the care ratio (ratio of healthy young citizens to elderly citizens) is in decline, monitoring and care of individuals with continuous recording of medical information in electronic form, remotely or during in-site medical visits, is becoming more and more incorporated in daily life. Multimodal technologies are constantly being developed to assist people in their daily life, with a vast range of wearable sensors now available for monitoring health parameters (e.g., blood pressure, sweat, body temperature, heart rate, etc.), lifestyle (e.g., monitoring utility use, levels of activity, sleep quantity and quality, etc.), and a person's ability to carry out activities of daily living. At the same time, health professionals have integrated new technologies into their workflow, for example by using various types of medical imagery to facilitate and support their clinical practice and diagnosis, and also by examining data from sensors and home medical devices, which allow them to remotely care for their patients. Health records and databases are now enriched with digital multimodal data on the patients, for which new methods need to be developed for accurate and fast access and retrieval.

One of the starting points for this book was the EU funded project Dem@Care. In Dem@Care, a multi-modal sensing platform was developed for the remote multimodal monitoring and care of people with dementia. Intelligent fusion and decision support made its outcomes immediately useable by clinicians, informal caregivers and also the people with dementia, via appropriate interfaces. This book covers, in its three parts, the above-mentioned aspects on the presence of multimodal data in healthcare, both for individuals (patients, family, and friends) and for medical professionals.

The first part entitled "Multimedia and Multimodal Pattern Recognition for Healthcare Applications" examines the analysis of different kinds of medical images and video data for diagnosis, evaluation of the health status of an individual, and their ability to live alone. From medical image analysis to multimodal video analytics, this part presents new multimedia technologies from healthcare applications.

The second part "Multimedia Analysis and Feedback in Medicine" presents current solutions to provide feedback to patients, for support, training, and rehabilitation purposes, or for monitoring them for safety and providing feedback if needed. All its chapters present tools which directly mainly communicate with patients in an efficient recommendation or self-assessment protocols.

Finally, the third part covers "Multimedia and Technology in Medicine," where aspects of accessing multimodal health-related data (e.g., electronic health records, medical images) are presented.

We hope that this book will give medical practitioners and general scientists a good overview of the most advanced practices and tools of multimedia data analysis for health and will help them in their everyday practice.

Thermi, Thessaloniki, Greece — Alexia Briassouli
Talence, France — Jenny Benois-Pineau
Pittsburgh, PA, USA — Alexander Hauptmann

Contents

Part II Multimedia Event and Activity Detection and Recognition for Health-Related Monitoring

Part III Multimedia-Based Personalized Health Feedback Solutions

Overview of Multimedia in Healthcare

Alexia Briassouli, Jenny Benois-Pineau, and Alexander Hauptmann

As the world population ages and the care ratio (ratio of healthy young citizens to elderly citizens) is in decline, monitoring and care of individuals with continuous recording of medical information in electronic form remotely or during in-site medical visits is becoming more and more incorporated in daily life. Multimodal technologies are constantly being developed to assist people in their daily life, with a vast range of wearable sensors now available for monitoring health parameters (e.g. blood pressure, sweat, body temperature, heart rate etc.), lifestyle (e.g. monitoring utility use, levels of activity, sleep quantity and quality etc.), a person's ability to carry out activities of daily living. At the same time, health professionals have integrated new technologies into their workflow, for example by using various types of medical imagery to facilitate and support their clinical practice and diagnosis, and also by examining data from sensors and home medical devices, which allow them to remotely care for their patients. Health records and databases are now enriched with digital multimodal data on the patients, for which new methods need to be developed for accurate and fast access and retrieval.

A. Briassouli (✉)
Centre for Research and Technology, Hellas, Information Technologies Institute,
Thermi, Thessaloniki, Greece
e-mail: abria@iti.gr

J. Benois-Pineau
Laboratoire Bordelais de Recherche en Informatique/University Bordeaux,
Talence, France
e-mail: jenny.benois@labri.fr

A. Hauptmann
Computer Science Department, Carnegie Mellon University, Pittsburgh, PA, USA
e-mail: alex@cs.cmu.edu

A. Briassouli et al. (eds.), *Health Monitoring and Personalized Feedback using Multimedia Data*, DOI 10.1007/978-3-319-17963-6_1

One of the starting points for this book was the EU funded project Dem@Care. In Dem@Care, a multi-modal sensing platform was developed for the remote multi-modal monitoring and care of people with dementia. Intelligent fusion and decision support made its outcomes immediately useable by clinicians, informal caregivers and also the people with dementia, via appropriate interfaces. This book covers, in its three parts, the abovementioned aspects on the presence of multimodal data in healthcare. The book targets a wide audience, including both individuals (patients, family, friends), as well as medical professionals. The goal of this book is to provide a comprehensive overview, together with research contributions, on the use of multimedia signals and derived knowledge nowadays for efficient computer aided diagnosis, monitoring and patient feedback.

The first part entitled "Multimedia Data Analysis" examines the analysis of different kinds of medical data, ranging from medical images and video data to health records.

The second part, "Multimedia Event and Activity Detection and Recognition for Health-Related Monitoring" presents methods for detecting and classifying events in long-duration recordings. This is particularly relevant to real-world applications, where events and activities need to be detected in long, otherwise uneventful, recordings.

The third part, "Multimedia Based Personalized Health Feedback Solutions" presents current solutions to provide feedback to patients, for support, training and rehabilitation purposes, or for monitoring them for safety and providing feedback if needed. The tools that are presented focus on direct and personalized.

1 Types of Multimedia Data and Types of Health-Related Applications

Healthcare is considered one of the most important testbeds for new multimedia applications, which have a central role in modern medicine. All types of multimedia data are involved in health-related applications, ranging from simple physiological signals to four-dimensional medical images and high resolution CAT scans. Modern day diagnostic equipment is in many cases digital, producing specialized medical measurements. For example, digital imagery is used in numerous healthcare applications, e.g. in MRI, CAT scans, and other forms of medical imaging such as mammography, neuroimaging, optoacoustic imaging, thermography (infrared imaging), ultrasound etc.

Together with specific imaging modalities mentioned above which are being developed for imaging in vivo, classical image/video acquisition techniques are increasingly targeting specific, healthcare-related applications. The unobtrusive observation of patients by ambient video cameras allows for objective assessment of capacities of patients with dementia and help in diagnostics and disease monitoring (http://www.grafton.org/smart-home-technology-a-tool-to-improve-quality-of-services/). As an existing example, non-intrusive ambient video observation allows for the assessment of gait in Parkinson's disease [3, 4], while it is extended in this work for the recognition of Activities of Daily Living in the assessment of individuals with dementia [1].

Numerous wearable sensors are also being developed (http://www.pcmag.com/article2/0,2817,2470806,00.asp; http://www.wearable-technologies.com/2014/01/the-new-wave-of-wristbands/), for daily monitoring of the health status of individuals, measuring blood pressure, blood oxygen, glucose levels, sleep patterns, heart activity etc., resulting in great amounts of digital data. At the same time, Electronic Healthcare Records (EHR) are being linked with multimodal data for a richer picture of the condition of patients (http://www.nibib.nih.gov/news-events/newsroom/linking-multimedia-electronic-health-records-improve-health-care). Mobile health-related data is also becoming increasingly widespread as mHealth technologies advance, e.g. via apps providing information about physical activity levels, physiological responses and other health-related apps, for which consumer demand is high (http://obssr.od.nih.gov/scientific_areas/methodology/mhealth/). The vast, continuously increasing, amounts of digital multimedia data related to healthcare, have led to the advancement of methods for analyzing "Big Data" in healthcare today (http://www.technologyreview.com/news/529011/can-technology-fix-medicine/). McKinsey is predicting that the business of analyzing health-related digital data may be worth $300–450 billion, while large IT companies are investing in all aspects of the domain, ranging from low-cost apps to expensive analytical medical systems. Thus, the extensive use of digital medical data and the variety of issues that arise in parallel with it is now driving innovation in digital health technologies with the creation of numerous health-related apps, gadgets and startups. Concluding, the proliferation of digital multimodal data in health is only increasing, necessitating consistent efforts for its analysis and seamless integration in daily life and the workflow of clinicians, which can be addressed by startups, but also established companies (http://hitconsultant.net/2014/10/02/philips-telehealth-apps-receives-510k-clearance-for-new-digital-health-platform/; http://www.intel.com/content/www/us/en/healthcare-it/healthcare-overview.html; http://www-935.ibm.com/industries/healthcare/; http://www.forbes.com/sites/dandiamond/2014/08/22/apple-google-are-jumping-into-health-care-is-amazon-next/) who are investing heavily in the healthcare sector.

2 Clinical Applications: Monitoring and Feedback

Health-related clinical applications include analyzing medical data and providing related feedback to clinicians. Monitoring of a person's health condition can take place in a hospital even when the nurse or physician is not present, to provide a complete picture of their physiological status (e.g. vitals over time) and to notify in case of alarm. The field of telehealth applications, devices and companies that have been created to address the needs of clinical monitoring and feedback is vast and continually evolving. Health or biomedical informatics analyze data for improved and automated monitoring of diseases and their evolution. This field alone ranges from clinical informatics to pharmacy, public, community, even consumer health informatics, among others. These fields automatically analyze medical data to provide analysis results on a person's health status, on the progression of an epidemic, on the effectiveness of certain interventions etc. Feedback can be in the form of paging emergency notifications to carers, or may involve more sophisticated analyses of multimodal medical data resulting in recommendations for lifestyle changes, and new medical knowledge.

3 Assisted Living Applications: Monitoring and Feedback

Remote health monitoring can also help people to live independently for a longer period of time, e.g. in smarthome environments [2] (www.demcare.eu; http://www.scientificamerican.com/article/smart-home-sensors-could-help-aging-population-stay-independent/; http://www.grafton.org/smart-home-technology-a-tool-to-improve-quality-of-services/). The EU project Dem@Care involved the creation of a multimodal sensing platform for comprehensive monitoring and intelligent decision support, aiming at better care for people with dementia. The multimodal nature of the sensing involved led to reliable assessments of the person's condition, rapid alerts in case of emergency and personalized care. Feedback was provided via appropriately designed interfaces, both for the person with dementia, and for their formal and informal caregivers. Health-related multimodal data can be used to provide support for chronic health management at home via remote monitoring and feedback. Emergencies can be detected quickly thanks to digital monitoring devices, as well as more subtle health-related deteriorations that can be detected from lifestyle changes, such as reduction in mobility or poor quality of sleep. The monitoring can take place using either ambient or wearable sensors, in most cases via a combination of both. Ambient sensors that are very commonly used include motion and contact sensors, to monitor mobility in the home, which rooms are used most often, where the person spends most of their time etc. Utility usage can also be tracked, e.g. the use of electricity and water, both for preventing emergencies, and for monitoring a person's lifestyle. Wearable sensors can be physiological sensors, measuring for example blood pressure, galvanic skin response, temperature, sleep quantity and quality, and even audiovisual [3, 4] in more advanced monitoring systems.

Feedback can be provided at numerous levels and in various forms. Clinical experts are provided with detailed medical feedback, to provide them with a complete and detailed clinical picture of the person's overall health status, either in the hospital or in their daily life. Individuals being monitored can receive feedback that is tailored to their condition and personal needs. For example, a person suffering from dementia can benefit from memory-enhancing feedback, such as visual diaries and reminders, while a person with a heart condition can benefit from motivational feedback on healthier lifestyle choices. The form of feedback can range from simple smartphone reminders, in text, audio or visual form, to environmental modifications, such as discreet lights blinking, changes in lighting and temperature, even the diffusion of scents.

4 User Considerations (Target Groups, Unobtrusiveness, Integration in Daily Life and Workcycle, Ethics and Privacy Issues)

The monitoring a person's health and lifestyle can provide valuable insights on their individual health status and its evolution, due to the detailed and varied outcomes of multimodal sensing. However, these advantages of multimodal monitoring also entail privacy and ethics concerns that need to be addressed before the deployment of such solutions. Each country has ethics committees that need to approve new protocols, such as those involving multisensory monitoring. An important part of the approval includes the informed consent of the individual to be monitored and their carers. Once ethics and privacy concerns are addressed, the monitoring solutions need to be integrated in daily life, to ensure consistent use and compliance of the end users: for example, patients need to wear the monitoring device and carers need to check the new, digital interface providing them with the analysis results of the sensor measurements from their patients. Ambient, environmental sensors may be easier to integrate in a person's daily life, as they are usually unobtrusive. Wearable sensors may be forgotten by the end user or may not be easy to use. For this reason, all new designs aim at familiar form factors, e.g. watches, pendants or sensing modalities integrated in smartphones, which are easy to use and understand. Similarly, the feedback that is provided to end users is tailored to them, so as to ensure that they understand it and do not find it distracting in their daily life or workflow (in the case of physicians). Participatory user centered design and the experience gained from ongoing advances in these technologies are continually improving their usability, while end users are becoming rapidly familiarized with them. It is therefore to be expected that multimodal monitoring and feedback will soon be an integral part of medicine and home care.

5 Ending

We hope that the book will give medical practitioners and general scientists a good overview of the most advanced practices and tools of multimedia data analysis for health and will help them in their everyday practice.

Acknowledgements The editors of the book would like to thank the French National Research network GDR CNRS ISIS and the U.S. National Science Foundation under Grant No. IIS-1251187 for their support in the preparation of this book.

They would also like to acknowledge the FP7 European Integrated Project Dem@care (www.demcare.eu) under grant agreement FP7-288199 (2011–2015), which served as the basis and motivation for this book.

References

1. Avgerinakis, K., Briassouli, A., & Kompatsiaris, I. (2013). Activity detection and recognition of daily living events. In *1st ACM MM workshop on multimedia indexing and information retrieval for healthcare (MIIRH), in conjunction with ACM MM 2013*. Barcelona, Spain, pp 3–10, October 2013.
2. Helal, A., Mokhtari, M., & Abdulrazak, B. (2007). *The engineering handbook on smart technology for aging, disability and independence*. Hoboken: John Wiley and Sons.
3. Mulin, E., Joumier, V., Leroi, I., Lee, J. H., Piano, J., Bordone, N., et al. (2012). Functional dementia assessment using a video monitoring system: Proof of concept. *Gerontechnology, 10*(4), 244–247.
4. Chen, S. W., Lin, S. H., Liao, L. D., Lai, H. Y., Pei, Y. C., Kuo, T. S., et al. (2011). Quantification and recognition of Parkinsonian gait from monocular video imaging using kernel-based principal component analysis. *Biomedical Engineering Online, 10*, 99. doi:10.1186/1475-925X-10-99.

Part I
Multimedia Data Analysis

Craniofacial Image Analysis

Ezgi Mercan, Indriyati Atmosukarto, Jia Wu, Shu Liang, and Linda G. Shapiro

1 Introduction to Craniofacial Analysis

Craniofacial research focuses on the study and treatment of certain congenital malformations or injuries of the head and face. It has become a multi-disciplinary area of expertise in which the players consist of not only oral and maxillofacial or plastic surgeons, but also craniofacial researchers including a large array of professionals from various backgrounds: basic scientists, geneticists, epidemiologists, developmental biologists, and recently computer scientists.

It is important to represent the shape of the human face in a standard way that facilitates modeling the abnormal and the normal. Morphometrics, the study of shape, has been a crucial toolbox for craniofacial research. Classical morphometrics-based craniofacial analyses use *anthropometric landmarks* and require taking physical measurements directly on the human face. These measurements are then used in a numerical analysis that compares the patient's measurements with the normal population to detect and quantify the deformation. Another technique for measuring the severity of shape deformations involves having clinical experts qualitatively match the shape of the patient's head to a set of templates. Template-matching is a common method in clinical practice, but it heavily depends on human judgment.

E. Mercan (✉) • S. Liang • L.G. Shapiro
Computer Science and Engineering, University of Washington, Seattle, WA, USA
e-mail: ezgi@cs.washington.edu; liangshu@cs.washington.edu; shapiro@cs.washington.edu

I. Atmosukarto
Singapore Institute of Technology, Singapore, Singapore
e-mail: indria@cs.washington.edu

J. Wu
Electrical Engineering, University of Washington, Seattle, WA, USA
e-mail: jiawu@uw.edu

A. Briassouli et al. (eds.), *Health Monitoring and Personalized Feedback using Multimedia Data*, DOI 10.1007/978-3-319-17963-6_2

As the field of computer vision has progressed, its techniques have become increasingly useful for medical applications. Advancements in 3D imaging technologies led craniofacial researchers to use computational methods for the analysis of the human head and face. Computational techniques aim to automate and improve established craniofacial analysis methods that are time consuming and prone to human error and innovate new approaches using the information that has become available through digital data.

This paper describes a set of computational techniques for craniofacial analysis developed by the University of Washington Multimedia Group. Section 2 describes the craniofacial syndromes whose analyses we have performed. Section 3 summarizes our previous work in craniofacial research and describes our image analysis pipeline, including preprocessing, feature extraction, quantification, and classification. Section 4 introduces our new work on the comparison of different features in similarity-based retrieval, and Sect. 5 concludes the paper.

2 Craniofacial Syndromes

We will describe three relevant syndromes: deformational plagiocephaly, 22q11.2 deletion syndrome, and cleft lip and palate.

2.1 *Deformational Plagiocephaly*

Deformational plagiocephaly can be defined as abnormal head shape (parallelogram shaped skull, asymmetric flattening, misalignment of the ears) due to external pressure on the infant's skull [17]. Figure 1 shows photographs of several infants' heads from the top view, with and without plagiocephaly. Although considered a minor cosmetic condition by many clinicians, if left untreated, children with plagiocephaly may experience a number of medical issues, ranging from social problems due to abnormal appearance to delayed neurocognitive development.

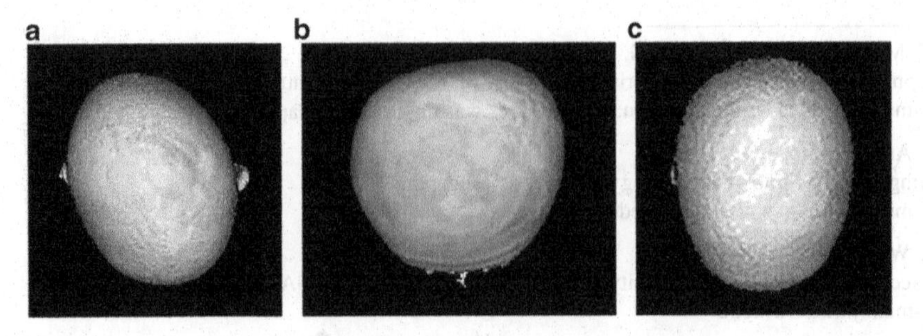

Fig. 1 *Deformational Plagiocephaly*—**(a,b)** *Top views* of heads of children with deformational plagiocephaly. **(c)** *Top view* of a child's head without deformational plagiocephaly

The severity of plagiocephaly ranges from mild flattening to severe asymmetry along a wide spectrum that is difficult to quantify. Clinical practices to diagnose and quantify plagiocephaly involve identifying anthropometric landmark points and taking measurements between the points. In one approach, the clinician determines the areas with the greatest prominence on the right and left sides of the head and measures diagonally the distances from these sites to the back of the head. The smaller length is subtracted from the larger resulting in an asymmetry number called the *transcranial diameter difference* [10]. Another technique compares the infant's skull shape to four templates: normal skull [score 0], mild shape deformation [score 1], moderate shape deformation [score 2], and severe shape deformation [score 3].

As an alternative to taking physical measurements directly on the infant's head, a technique called *HeadsUp* developed by Hutchinson et al. performs automated analysis of 2D digital photographs of infant heads fitted with an elastic head circumference band that has adjustable color markers to identify landmarks [13]. Although this semi-automatic approach is less intrusive and faster, it is still subjective, and the analysis is only 2D. There are some recently proposed techniques that use 3D surface data: Plank et al. [22] use a laser shape digitizer to obtain the 3D surface of the head, but still require manual identification of the landmarks. Lanche et al. [14] use a stereo-camera system to obtain a 3D model of the head and propose a method to compare the infant's head to an *ideal* head template.

2.2 22q11.2 Deletion Syndrome

22q11.2 deletion syndrome (22q11.2DS) is a disorder caused by a 1.5–3 MB deletion on chromosome 22 and occurs in 1 of every 4,000 individuals [21]. Over 180 phenotypic features are associated with this condition, including well-described craniofacial features such as asymmetric face shape, hooded eyes, bulbous nasal tip, tubular appearance to the nose, retrusive chin, prominent nasal root, small nasal alae, small mouth, open mouth and downturned mouth, among others. Some manifestations of facial features are very subtle, and even craniofacial experts find them difficult to identify without measurements and analysis. Figure 2 shows example manifestations of the syndrome on the face.

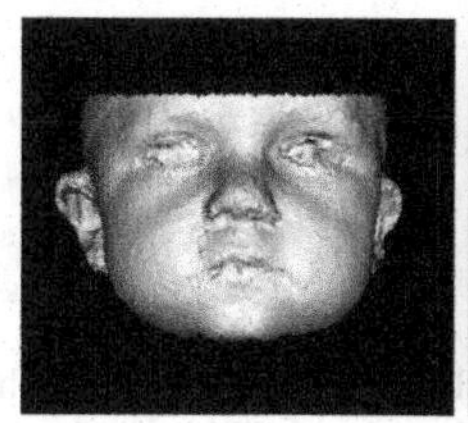
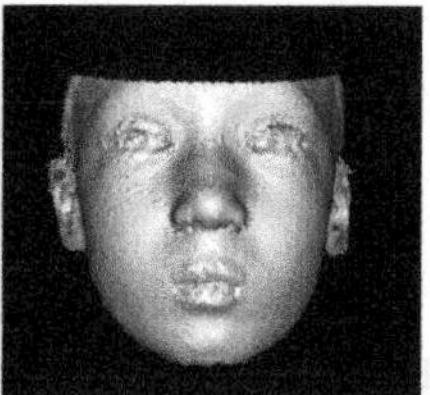

Fig. 2 *22q11.2DS craniofacial features*—Example 3D face mesh data of children with 22q11.2 deletion syndrome

Early detection of 22q11.2DS is important, because the condition is known to be associated with cardiac anomalies, mild-to-moderate immune deficiencies and learning disabilities. Similar to the detection of deformational plagiocephaly, the assessment of 22q11.2DS has commonly been through physical examination and craniofacial anthropometric measurements. After identification of the symptoms, genetic tests can be conducted to confirm and complete the diagnosis.

There has been little effort to automate the diagnosis and analysis of 22q11.2DS. Boehringer et al. [7] used Gabor wavelets to transform 2D photographs of the individuals and PCA to classify the dataset. However, the method requires manual placement of anthropometric landmarks on the face. Hammond et al. [12] proposed a *dense surface model* method followed by the application of PCA on 3D surface mesh data, which also requires manually placed landmarks to align the meshes.

2.3 Cleft Lip and Palate

Cleft lip is a birth defect that occurs in approximately 1 in every 1,000 newborns and can be associated with cleft palate [2]. The deformity is thought to be a result of the failure of fusion *in utero* and may be associated with underdevelopment of tissues. Cleft lip and palate can range from multiple deep severe clefts in the palate to a single incomplete or hardly noticeable cleft in the lip. Figure 3 shows examples of deformations on infants' faces caused by cleft lip and/or palate. The condition can be treated with surgery and the treatment can produce a dramatic change in appearance of the lip depending on the severity of the cleft. Since the potential results and treatment options depend on the severity of the cleft, it is important to have an objective assessment of the deformity.

The assessment of cleft deformities relies on a clinical description that can be subjective and landmark-based measurements that can be time consuming and difficult to perform on young infants. Additionally, there is no "gold standard" for evaluation and the correlation between the scores given by different medical experts can be very low [26].

There has not been much computational work done towards the quantification of the cleft lip using the face shape. Nonetheless, some work has been done on face symmetry, which can be used in cleft assessment, in computer vision. Although not

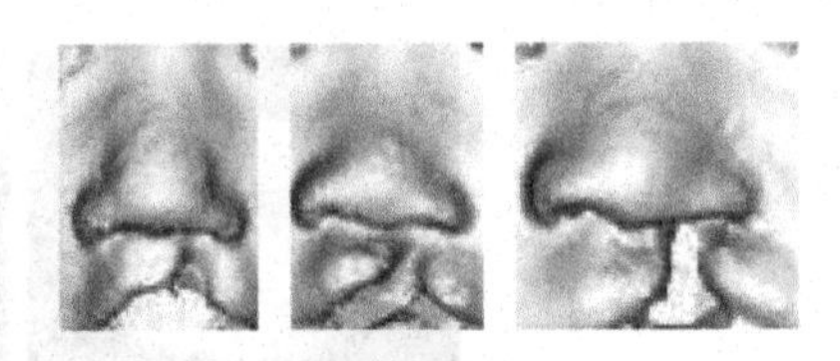

Fig. 3 *Cleft lip*—Example 3D face mesh data of children with cleft lip. Cleft of the lip can cause a wide range of deformations from mild to severe

applied to cleft assessment, Benz et al. introduced a method for 3D facial symmetry analysis using the iterative closest point algorithm to register the mirrored mesh to the original [5]. This method is reliable when the data is properly aligned and heavily depends on the choice of the initial plane about which the data is mirrored.

3 Craniofacial Image Analysis Pipeline

Our pipeline consists of data acquisition and preprocessing, feature extraction, and high-level operations of quantification, classification and content-based image retrieval. Figure 4 summarizes the steps of the pipeline and their relationships, as discussed in this section.

3.1 Data Acquisition and Preprocessing

With the developments in 3D imaging technologies, the use of 3D information has become widespread in research and applications. Craniofacial research greatly benefits from these developments due to the lower cost and higher accuracy of new imaging technologies like laser scanners and stereo-photography, in comparison to traditional methods such as direct measurements on the patient or 2D image-based techniques. Stereo imaging systems are popular among medical researchers, since they make it possible to collect large amounts of data in a non-invasive and convenient way.

The 3dMD® system is a commercial stereo-photography system commonly used for medical research. It uses texture information to produce a 3D mesh of the human face that consists of points and triangles. Figure 5 shows an example 3dMD setup where multiple pods of cameras are placed around a chair and simultaneously obtain photographs of the patient from different angles. Since the resulting mesh

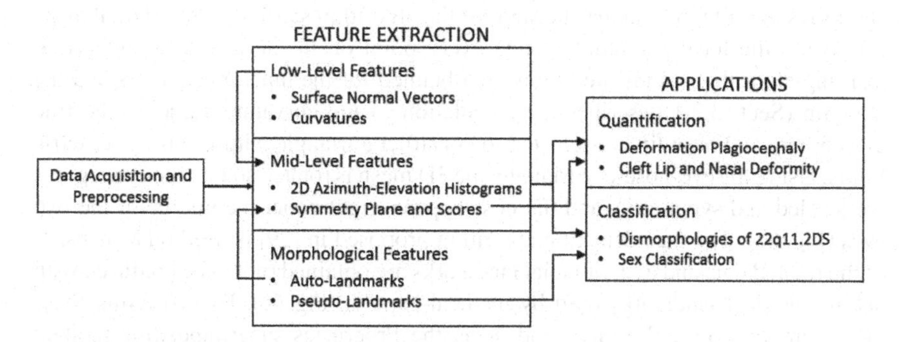

Fig. 4 The craniofacial image analysis pipeline overview

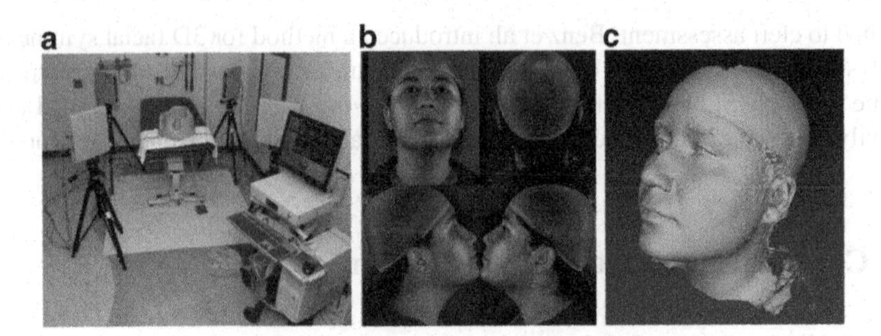

Fig. 5 *3D Face Mesh Acquisition*—(**a**) 3dMD® system with multiple cameras, (**b**) The texture images from four different cameras on front, back, left and right, (**c**) The 3D mesh produced with the stereophotography technique

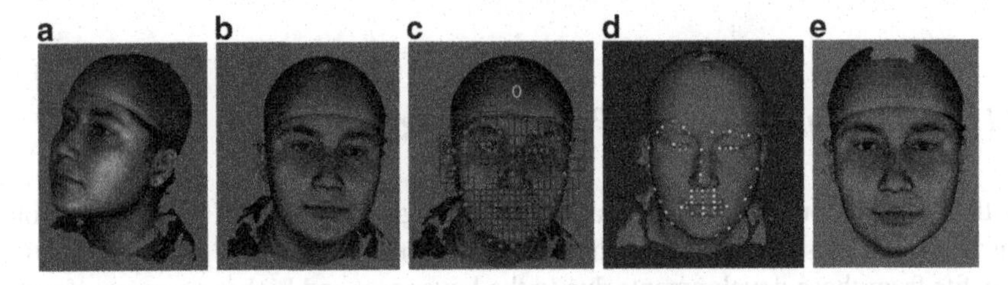

Fig. 6 *Automated mesh cleaning*—(**a**) Original mesh acquired by 3DMD®, (**b**) Frontal snapshot of the rotated mesh, (**c**) Facial landmarks detected by 2D face detection algorithm, (**d**) Pose normalized mesh using landmarks and Procrustes superimposition, (**e**) Extracted face

is not aligned and contains pieces of clothing, it needs to be processed before further analysis. It is usually an expert who cleans the mesh to obtain the face and normalizes the pose, so that the head faces directly front. The accuracy and efficiency of this step is crucial to any analysis conducted on the data.

Wu et al. proposed a method for automated face extraction and pose-normalization from raw 3D data [25]. The method makes use of established face detection algorithms for 2D images using multiple photographs produced by the stereo system. Figure 6 shows the steps of the algorithm starting from original mesh (a). Using the local curvature around every point on the surface in a supervised learning algorithm, candidate points are obtained for the inner eye corners and the nose tip (Sect. 3.2.1 describes the calculation of the curvature values). The true eye corners and nose tip are selected to construct a triangle within some geometric limits. Using the eye-nose-eye triangle, the 3D mesh is rotated so that the eye regions are leveled and symmetric, and the nose appears right under the middle of the two eyes (Fig. 6b). The face detection algorithm proposed in [29] is used on a snapshot of the rotated data and a set of facial landmarks are obtained (not to be confused with anthropological landmarks used by medical experts) Fig. 6c). By projecting these 2D landmarks to the 3D mesh and using the Procrustes superimposition method [11], the mesh is rotated so that the distance between the landmarks of the head and the average landmarks of the aligned data is minimal (Fig. 6d). After alignment, the bounding box for the 3D surface is used to cut the clothing and obtain the face region. Additionally, surface normal vectors are used to eliminate neck and shoulders from images where the bounding box is not small enough to capture only the face (Fig. 6e). The automation of mesh cleaning and pose normalization is an important step for processing large amounts of data with computational methods.

3.2 *Feature Extraction*

We describe both low-level and mid-level features used in craniofacial analysis.

3.2.1 Low-Level Features

A 3D mesh consists of a set of points, identified by their coordinates (x, y, z) in 3D space, and the connections between them. A cell of a mesh is a polygon defined by a set of points that form its boundary. The low-level operators capture local properties of the shape by computing a numeric value for every point or cell on the mesh surface. Low-level features can be averaged over local patches, aggregated into histograms as frequency representations and convoluted with a Gaussian filter to remove noise and smooth the values.

Surface Normal Vectors

Since the human head is roughly a sphere, the normal vectors can be quantified with a spherical coordinate system. Given the normal vector $n(n_x, n_y, n_z)$ at a 3D point, the azimuth angle θ is the angle between the positive x axis and the projection n' of n to the xz plane. The elevation angle ϕ is the angle between the x axis and the vector n.

$$\theta = \arctan\left(\frac{n_z}{n_x}\right) \qquad \phi = \arctan\left(\frac{n_y}{\sqrt{(n_x^2 + n_z^2)}}\right) \tag{1}$$

where $\theta \in [-\pi, \pi]$ and $\phi \in [-\frac{\pi}{2}, \frac{\pi}{2}]$.

Curvature

There are different measures of curvature of a surface: The *mean curvature* H at a point p is the weighted average over the edges between each pair of cells meeting at p.

$$H(p) = 1/|E(p)| \sum_{e \in E(p)} length(e) * angle(e) \tag{2}$$

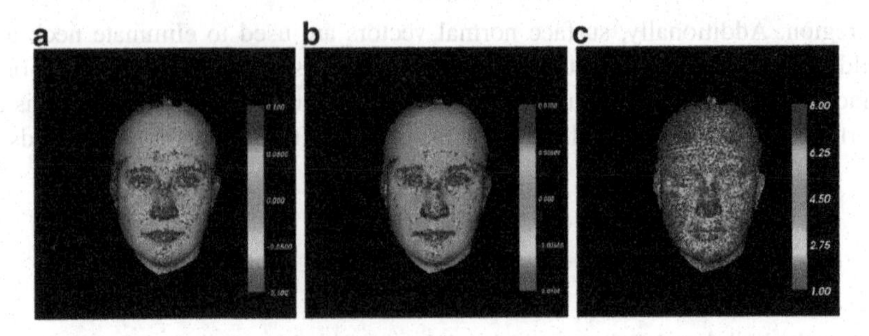

Fig. 7 *Curvature measures*—(**a**) Mean curvature, (**b**) Gaussian curvature and (**c**) Besl–Jain curvature visualized. Higher values are represented by cool (*blue*) colors while lower values are represented by warm (*red*) colors (Color figure online)

where $E(p)$ is the set of all the edges meeting at point p, and $angle(e)$ is the angle of edge e at point p. The contribution of every edge is weighted by $length(e)$. The *Gaussian curvature* K at point p is the weighted sum of interior angles of the cells meeting at point p.

$$K(p) = 2\pi - \sum_{f \in F(p)} area(f)/3 * interiorangle(f) \tag{3}$$

where $F(p)$ is the set of all the neighboring cells of point p, and $interiorangle(f)$ is the angle of cell f at point p. The contribution of every cell is weighted by $area(f)/3$.

Besl and Jain [6] suggested a surface characterization of a point p using the sign of the mean curvature H and the Gaussian curvature K at point p. Their characterization includes eight categories: peak surface, ridge surface, saddle ridge surface, plane surface, minimal surface, saddle valley, valley surface and cupped surface. Figure 7 illustrates mean, Gaussian and Besl–Jain curvature on a head mesh.

3.2.2 Mid-Level Features

Mid-level features are built upon low-level features to interpret global or local shape properties that are difficult to capture with low-level features.

2D Azimuth-Elevation Histograms

Azimuth and elevation angles, together, can define any unit vector in 3D space. Using a 2D histogram, it is possible to represent the frequency of cells according to their orientation on the surface. On relatively flat surfaces of the head, all surface normal vectors point in the same direction. In this case, all vectors fall into the same bin creating a strong signal in some bins of the 2D histogram. Figure 8 shows the visualization of an example 8×8 histogram.

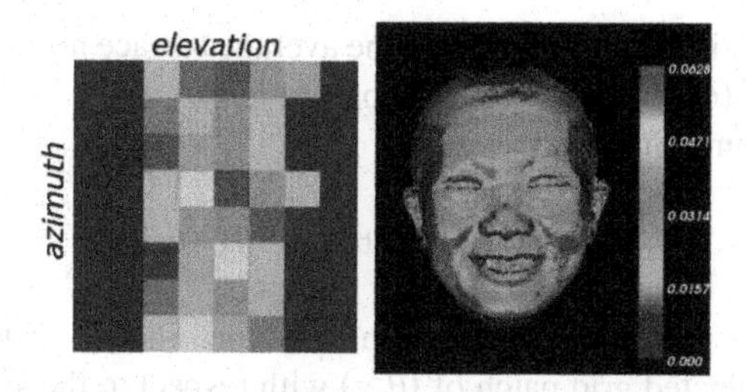

Fig. 8 Visualization of an 8 × 8 2D azimuth-elevation histogram. The histogram bins (*left*) with high values are shown with warmer colors (*red, yellow*). The image on *right* shows the localization of high-valued bins where the areas corresponding to bins are colored in a similar shade (Color figure online)

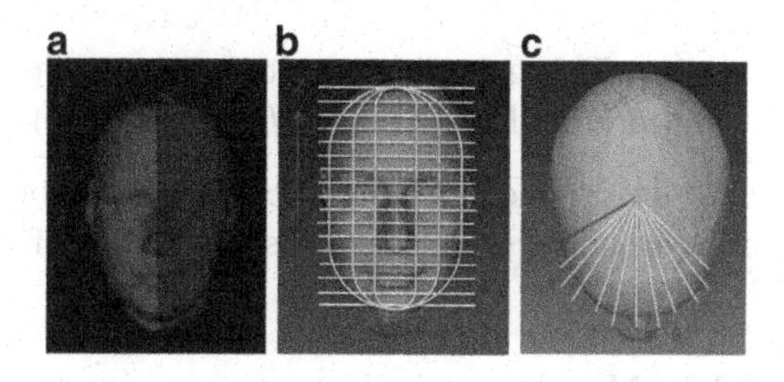

Fig. 9 (**a**) Symmetry plane, (**b**) *Front view* of grid showing θ and z for indexing the patches, (**c**) *Top view* of the grid showing the radius r and angle θ

Symmetry Plane and Related Symmetry Scores

Researchers from computer vision and craniofacial study share an interest in the computation of human face symmetry. Symmetry analyses have been used for studying facial attractiveness, quantification of degree of asymmetry in individuals with craniofacial birth defects (before and after corrective surgery), and analysis of facial expression for human identification.

Wu et al. developed a two-step approach for quantifying the symmetry of the face [24]. The first step is to detect the plane of symmetry. Wu described several methods for symmetry plane detection and proposed two methods: learning the plane by using point-feature-based region detection and calculating the mid-sagittal plane using automatically-detected landmark points (Sect. 3.2.3). After detecting the plane, the second step is to calculate the shape difference between two parts of the face. Wu proposes four features based on a grid laid out on the face (Fig. 9):

1. Radius difference:

$$RD(\theta, z) = |r(\theta, z) - r(-\theta, z)| \tag{4}$$

   where $r(\theta, z)$ is the average radius value in the grid patch(θ, z), and $(-\theta, z)$ is the reflected grid patch of (θ, z) with respect to the symmetry plane.
2. Angle difference:

$$AD(\theta, z) = cos(\beta_{(\theta,z),(-\theta,z)}) \tag{5}$$

where $\beta_{(\theta,z),(-\theta,z)}$ is the angle between the average surface normal vectors of each mesh grid patch (θ, z) and its reflected pair $(-\theta, z)$.

3. Gaussian curvature difference:

$$CD(\theta, z) = |K(\theta, z) - K(-\theta, z)| \tag{6}$$

where $K(\theta, z)$ is the average Gaussian curvature in the grid patch(θ, z), and $(-\theta, z)$ is the reflected grid patch of (θ, z) with respect to the symmetry plane.

4. Shape angle difference:

$$ED(\theta, z) = \left| \frac{\#points(\theta, z) > Th}{\#points(\theta, z)} - \frac{\#points(-\theta, z) > Th}{\#points(-\theta, z)} \right| \tag{7}$$

where Th is a threshold angle, and $\#points(\theta, z)$ is the total number of points with dihedral angle larger than Th in patch (θ, z).

These symmetry features produce a vector of length $M \times N$ where M is the number of horizontal grid cells and N is the number of vertical grid cells.

3.2.3 Morphometric Features

Most of the work on morphometrics in the craniofacial research community uses manually-marked landmarks to characterize the data. Usually, the data are aligned via these landmarks using the well-known Procrustes algorithm and can then be compared using the related Procrustes distance from the mean or between individuals [11]. Figure 10a shows a sample set of landmarks. Each landmark is placed by a medical expert using anatomical cues.

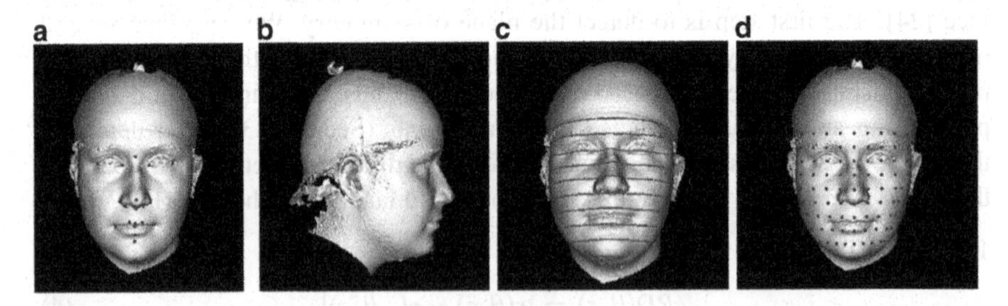

Fig. 10 (**a**) Twenty four anthropometric landmarks marked by human experts. (**b**) Sellion and tip of the chin are detected. (**c**) Two parallel planes go through chin tip and sellion, six parallel planes constructed between chin and sellion, and two above sellion. (**d**) On each plane, nine points are sampled with equal distances, placing the middle point on the bi-lateral symmetry plane. Ninety pseudo-landmark points calculated with ten planes and nine points

Auto-Landmarks

Traditional direct anthropometry using calipers is time consuming and invasive; it requires training of the expert and is prone to human error. The invasiveness of the method was overcome with the development of cost-effective 3D surface imaging technologies when experts started using digital human head data to obtain measurements . However, manual landmarking still presents a bottleneck during the analysis of large databases.

Liang et al. presented a method to automatically detect landmarks from 3D face surfaces [15]. The auto-landmarking method starts with computing an initial set of landmarks on each mesh using only the geometric information. Starting from a pose-normalized mesh, the geometric method finds 17 landmark points automatically, including 7 nose points, 4 eye points, 2 mouth points and 4 ear points. The geometric information used includes the local optima points like the tip of the nose (pronasale) or the sharp edges like the corners of the eyes. The sharp edges are calculated using the angle between the two surface normal vectors of two cells sharing an edge. The geometric method also uses information about the 17 landmarks and the human face such as the relative position of landmarks with respect to each other and the anatomical structures on the human face.

The initial landmark set is used for registering a template, which also has initial landmarks calculated, to each mesh using a deformable registration method [1]. The 17 landmark points (Fig. 11) provide a correspondence for the transformation of each face. When the template is deformed to the target mesh, the distance between the mesh and the deformed template is very small and every landmark point on the template can be transferred to the mesh. The average distance between the initial points generated by the geometric method and the expert points is 3.12 mm. This distance is reduced to 2.64 mm after deformable registration making the method very reliable. The method has no constraint on the number of landmarks that are marked on the template and transferred to each mesh. This provides a flexible work-flow for craniofacial experts who want to calculate a specific set of landmarks on large databases.

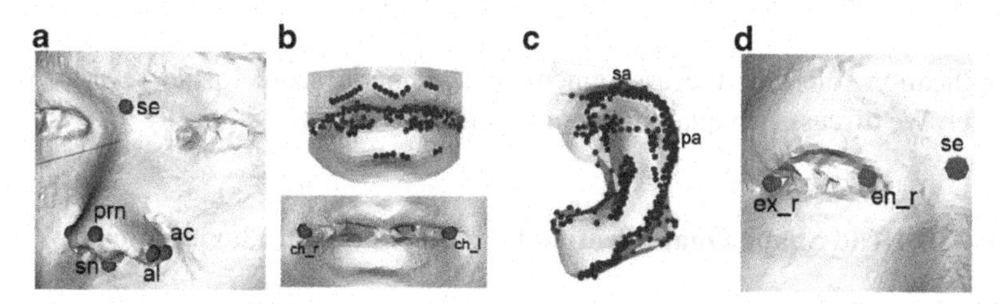

Fig. 11 Initial landmarks detected by the geometric method. (**a**) Nose points. (**b**) Mouth points. (**c**) Ear points. (**d**) Eye points

Pseudo-Landmarks

For large databases, hand landmarking is a very tedious and time consuming process that the auto-landmarking method tries to automate. Moreover, anthropometric landmarks cover only a small part of the face surface, and soft tissue like cheeks or forehead do not have landmarks on them. This makes pseudo-landmarks an attractive alternative. Hammond proposed a dense correspondence approach using anthropometric landmarks [12]. The dense correspondence is obtained by warping each surface mesh to average landmarks and finding the corresponding points using an iterative closest point algorithm. At the end, each mesh has the same number of points with the same connectivity and all points can be used as pseudo-landmarks. Claes et al. proposed a method called the *anthropometric mask* [8], which is a set of uniformly distributed points calculated on an average face from a healthy population and deformed to fit onto a 3D face mesh. Both methods required manual landmarking to initialize the process.

Motivated by the skull analysis work of Ruiz-Correa et al. [23], Lin et al. [16] and Yang et al. [27], Mercan et al. proposed a very simple, but effective, method that computes pseudo-landmarks by cutting through each 3D head mesh with a set of horizontal planes and extracting a set of points from each plane [18]. Correspondences among heads are not required, and the user does no hand marking. The method starts with 3D head meshes that have been pose-normalized to face front. It computes two landmark points, the sellion and chin tip, and constructs horizontal planes through these points. Using these two planes as base planes, it constructs m parallel planes through the head and from each of them samples a set of n points, where the parameters n and m are selected by the user. Figure 10 shows 90 pseudo-landmarks calculated with 10 planes and 9 points on a sample 3D mesh. Mercan et al. show in [18] that pseudo-landmarks work as well as dense surface or anthropometric mask methods, but they can be calculated without human input and from any region of the face surface.

3.3 *Quantification*

Quantification refers to the assignment of a numeric score to the severity of a disorder. We discuss two quantification experiments.

3.3.1 3D Head Shape Quantification for Deformational Plagiocephaly

Atmosukarto et al. used 2D histograms of azimuth-elevation angles to quantify the severity of deformational plagiocephaly [4]. On relatively flat surfaces of the head, normal vectors point in the same direction, and thus have similar azimuth and elevation angles. By definition, infants with flat surfaces have larger flat areas on their skulls causing peaks in 2D histograms of azimuth and elevation angles.

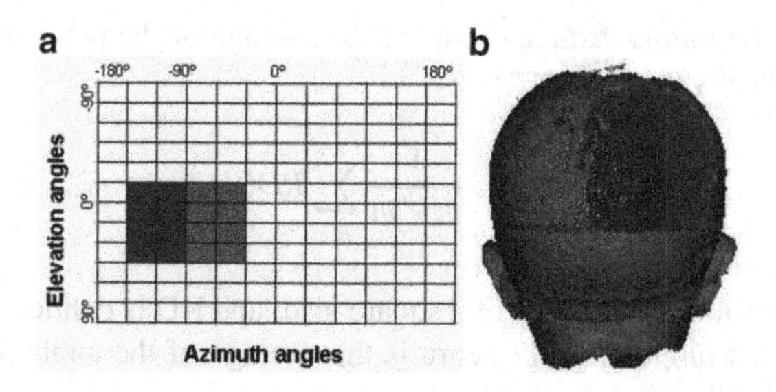

Fig. 12 On a 12 × 12 histogram (**a**), Left Posterior Flatness Score (*red*) and Right Posterior Flatness Scores (*blue*) are calculated by summing the relevant histogram bins. Selected bins correspond to points on the skull (**b**) that are relevant to the plagiocephaly (Color figure online)

Using a histogram with 12 × 12 bins, the method defines the sum of histogram bins corresponding to the combination of azimuth angles ranging from −90° to −30° and elevation angles ranging from −15° to 45° as the *Left Posterior Flatness Score* (LPFS). Similarly, the sum of histogram bins corresponding to the combination of azimuth angles ranging from −150° to −90° and elevation angles ranging from −15° to 45° gives the *Right Posterior Flatness Score* (RPFS). Figure 12 shows the selected bins and their projections on the back of the infant's head. The *asymmetry score* is defined as the difference between RPFS and LPFS. The asymmetry score measures the shape difference between two sides of the head, and the sign of the asymmetry score indicates which side is flatter.

The absolute value of the calculated asymmetry score was found to be correlated with experts' severity scores and the score calculated by Hutchinson's *HeadsUp* method [13] that uses anthropometric landmarks. Furthermore, the average flatness scores for left posterior flattening, right posterior flattening and control groups shows clear separation, providing a set of thresholds for distinguishing the cases.

3.3.2 Quantifying the Severity of Cleft Lip and Nasal Deformity

Quantifying the severity of a cleft is a hard problem even for medical experts. Wu et al. proposed a methodology based on symmetry features [26]. The method suggests that the asymmetry score is correlated with the severity of the cleft. It compares the scores with the severity of clefts assessed by surgeons before and after reconstruction surgery. Wu et al. proposed three measures based on asymmetry:

1. The *point-based distance* score is the average of the distances between points that are reflected around the symmetry plane:

$$PD_a = \frac{1}{n} \sum_{p} distance(p_s, q) \tag{8}$$

where n is the number of points and q is the reflection of point p.

2. The *grid-based radius distance* score is the average of the radius distance (RD) over the grid cells:

$$RD_a = \frac{1}{m \times m} \sum_{\theta,z} RD(\theta, z) \tag{9}$$

where m is the number of cells of a square grid, and RD is defined in (4).

3. The *grid-based angle distance* score is the average of the angle distance (AD) over the grid cells:

$$AD_a = \frac{1}{m \times m} \sum_{\theta,z} AD(\theta, z) \tag{10}$$

where m is the number of cells of a square grid, and AD is defined in (5).

Three distances are calculated for infants with clefts before and after surgery and compared with the rankings of the surgeons. The asymmetry scores indicate a significant improvement after the surgery and a strong correlation with surgeons' rankings. Figure 13 shows the visualization of RD_a scores for clefts with three different severity classes given by surgeons and the comparison of before and after surgery scores.

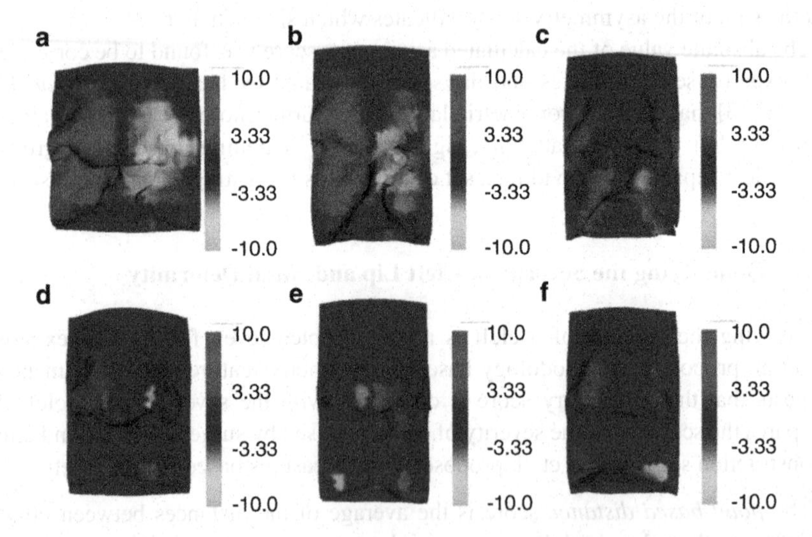

Fig. 13 RD_a reduction after the surgery for three cases. The *red* and *green* colors show the big difference between the *left* and *right* sides. *Red* means higher and green means lower. *Blue* means small difference between the two sides, (**a**) severe case pre-op $RD_a = 3.28$ mm, (**b**) moderate case pre-op $RD_a = 2.72$ mm, (**c**) mild case pre-op $RD_a = 1.64$ mm, (**d**) severe case post-op $RD_a = 1.03$ mm, (**e**) moderate case post-op $RD_a = 0.95$ mm, (**f**) mild case post-op $RD_a = 1.22$ mm (Color figure online)

3.4 Classification

We describe two classification experiments.

3.4.1 Classifying the Dismorphologies Associated with 22q11.2DS

The craniofacial features associated with 22q11.2 deletion syndrome are well-described and critical for detection in the clinical setting. Atmosukarto et al. proposed a method based on machine learning to classify and quantify some of these craniofacial features [3]. The method makes use of 2D histograms of azimuth and elevation angles of the surface normal vectors calculated from different regions of the face, but it uses machine learning instead of manually selecting histogram bins.

Using a visualization of the 2D azimuth-elevation angles histogram, Atmosukarto pointed out that certain bins in the histogram correspond to certain regions on the face, and the values in these bins are indicative of different face shapes. An example of different midface shapes is given in Fig. 14. Using this insight, a method based on sophisticated machine learning techniques was developed in order to learn the bins that are indicators of different craniofacial features.

In order to determine the histogram bins that are most discriminative in classification of craniofacial features, Adaboost learning was used to select the bins that give the highest classification performance of a certain craniofacial feature against others. The Adaboost algorithm is a strong classifier that combines a set of weak classifiers, in this case, decision stumps [9]. Different bins are selected for different craniofacial abnormalities. Note that the bins selected for each condition cover areas where the condition causes shape deformation.

After selecting discriminative histogram bins with Adaboost, a genetic programming approach [28] was used to combine the features. Genetic programming imitates human evolution by changing the mathematical expression over the selected histogram bins used for quantifying the facial abnormalities. The method aims to maximize a fitness function, which is selected as the F-measure in this work. The F-measure is commonly used in information retrieval and is defined as follows:

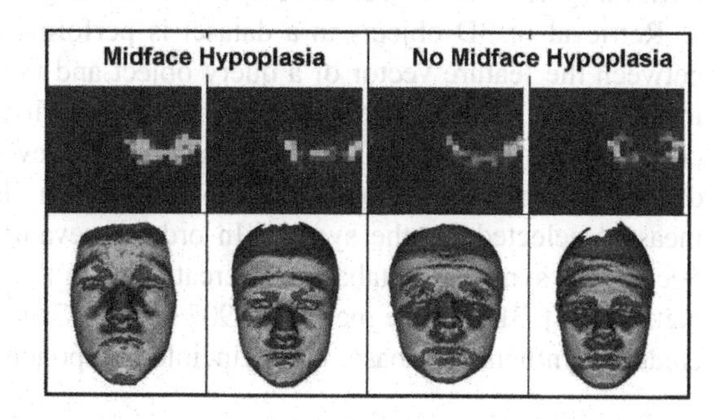

Fig. 14 Projections of 2D histograms of azimuth and elevation angles to the face. The projection shows discriminating patterns between individuals with and without midface hypoplasia

$$F(prec, rec) = 2 \times \frac{(prec \times rec)}{(prec + rec)} \tag{11}$$

where *prec* is the precision and *rec* is the recall metric. The mathematical expression with the highest F-measure is selected through cross-validation tests.

3.4.2 Sex Classification Using Pseudo-Landmarks

What makes a female face different from a male face has been an interest for computer vision and craniofacial research communities for quite some time. A great deal of previous work on sex classification in the computer vision literature uses 2D color or gray tone photographs rather than 3D meshes.

Mercan et al. used pseudo-landmarks in a classification setting to show their efficiency and the representation power over anthropometric landmarks [18]. L_1-regularized logistic regression was used in a binary classification setting where the features were simply the x, y and z coordinates of the landmark points. In a comparative study where several methods from the literature are compared in a sex classification experiment, it was shown that pseudo-landmarks (95.3 % accuracy) and dense surface models (95.6 % accuracy) perform better than anthropometric landmarks (92.5 % accuracy) but pseudo-landmarks are more efficient in calculation than dense surface models and do not require human input. L_1-regularization also provides feature selection and in the sex classification setting, the pseudo-landmarks around the eyebrows were selected as the most important features.

4 Content-Based Retrieval for 3D Human Face Meshes

The availability of large amounts of medical data made content based image retrieval systems useful for managing and accessing medical databases. Such retrieval systems help a clinician through the decision-making process by providing images of previous patients with similar conditions. In addition to clinical decision support, retrieval systems have been developed for teaching and research purposes [20].

Retrieval of 3D objects in a dataset is performed by calculating the distances between the feature vector of a query object and the feature vectors of all objects in the dataset. These distances give the dissimilarity between the query and every object in the dataset; thus, the objects are retrieved in the order of increasing distance. The retrieval performance depends on the features and the distance measure selected for the system. In order to evaluate the features introduced in Sect. 3.2, a synthetic database was created using the dense surface correspondence method [12]. 3D surface meshes of 907 healthy Caucasian individuals were used to create a synthetic database. The principle components of the data were calculated

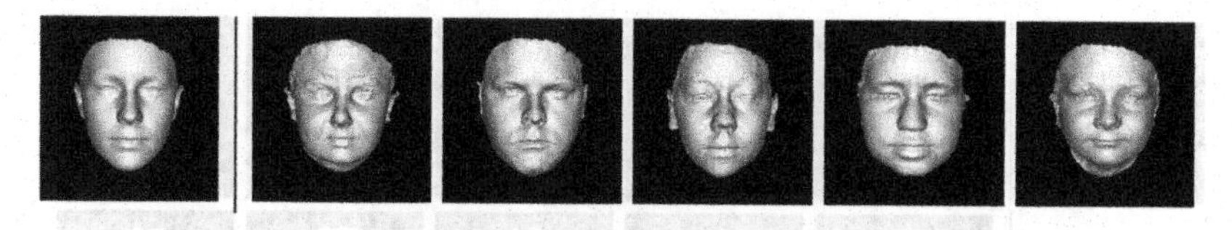

Fig. 15 Average face (*left*) and some examples (*right*) from the synthetic database

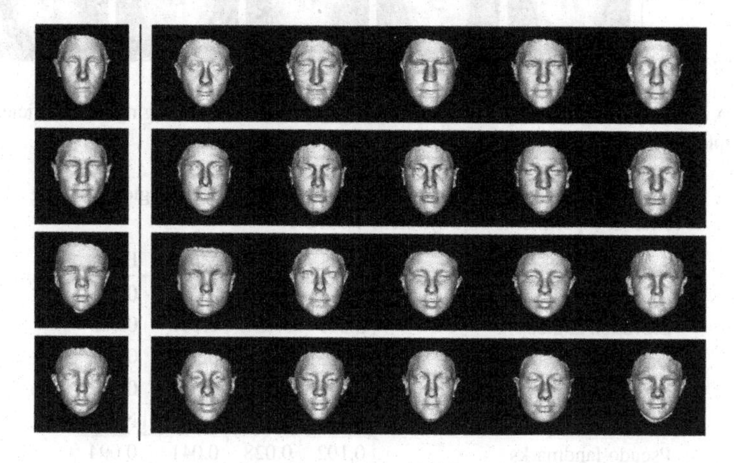

Fig. 16 Some queries made on the randomly produced synthetic dataset with the pseudo-landmark feature calculated from the whole face. The query is an adult female in the *first row*, an adult male in the *second row*, a young female in the *third row* and a young male in the *fourth row*

and 100 random synthetic faces were created by combining principle components with coefficients randomly chosen from a multivariate normal distribution modeling the population. Figure 15 shows the average face of the population and some example synthetic faces. Figure 16 shows four example queries made on the random dataset with adult female, adult male, young female and young male samples as queries. Although the retrieval results are similar to the query in terms of age and sex, it is not possible to evaluate the retrieval results quantitatively using randomly produced faces, since there is no "ground truth" for the similarity.

In a controlled experiment, the performance of the retrieval system can be measured by using a dataset that contains a subset of similar objects. Then, using the rank of the similar object in a query, a score based on the average normalized rank of relevant images [19] is calculated for each query:

$$score(q) = \frac{1}{N \times N_{rel}} \times \left(\sum_{i=1}^{N_{rel}} R_i - \frac{N_{rel} \times (N_{rel} + 1)}{2} \right) \tag{12}$$

where N is the number of objects in the database, N_{rel} is the number of objects relevant to the query object q, and R_i is the rank assigned to i-th relevant object. The evaluation scores range from 0 to 1, where 0 is the best and indicates that all relevant

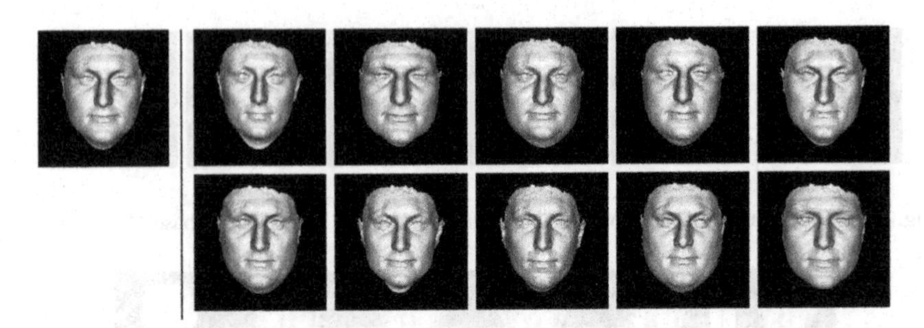

Fig. 17 A query face (*left*) and ten similar faces (*right*) produced by changing the coefficients of the principle components of the query face

Table 1 Evaluation results for different features and face regions in retrieval experiments on a synthetic database

	Face	Nose	Mouth	Eyes
Azimuth angles	0.077	0.150	0.226	0.168
Elevation angles	0.099	0.159	0.107	0.185
Gaussian curvature	0.223	0.386	0.333	0.385
2D azimuth-elevation histogram	0.046	0.109	0.087	0.142
Landmarks	0.164	NA	NA	NA
Pseudo-landmarks	0.102	0.028	0.041	0.094
Pseudo-landmarks (sized)	0.054	0.041	0.042	0.118

objects are retrieved before any other objects. To create similar faces in a controlled fashion, the coefficients of the principle components were selected carefully, and ten similar faces were produced for each query. For the synthesis of similar faces, we changed the coefficients of ten randomly chosen principle components of the base face by adding or subtracting 20 % of the original coefficient. These values are chosen experimentally by taking the limits of the population coefficients into consideration. Figure 17 shows a group of similar faces. Adding 10 new face sets with 1 query and 10 similar faces in each, a new dataset of 210 faces was obtained. The new larger dataset was used to evaluate the performance of shape features by running 10 queries for each feature-region pair. The features were calculated in four regions: the whole face, nose, mouth and eyes. Low-level features azimuth angles, elevation angles and curvature values were used to create histograms with 50 bins. 2D azimuth-elevation histograms were calculated at 8×8 resolution. Landmarks were calculated with our auto-landmarking technique. Pseudo-landmarks were calculated with 35 planes and 35 points. Both landmarks and pseudo-landmarks were aligned with Procrustes superimposition, and pseudo-landmarks were size normalized to remove the effect of shape size. Table 1 shows the average of the evaluation scores for each feature-region pair. Figure 18 shows a sample retrieval.

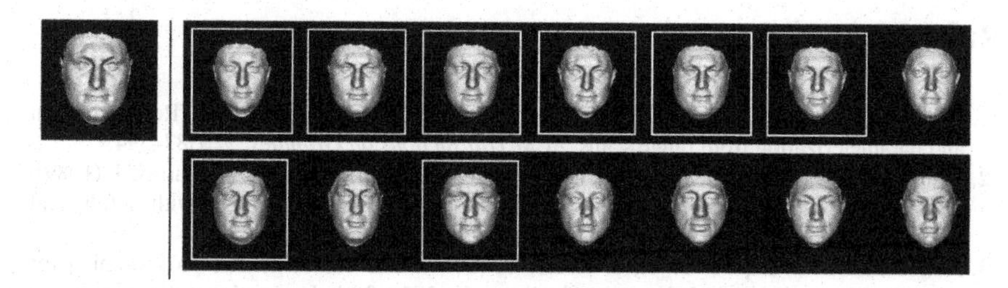

Fig. 18 Top 30 results of a query with the pseudo-landmark feature calculated from the whole face. The *top left* face is the query and the manually produced similar faces are marked with *white rectangles*. The faces without *white rectangle* are random faces in the database that happen to be similar to the query

The pseudo-landmarks obtained the best (lowest) retrieval scores for the nose, mouth and eyes, while the 2D azimuth-elevation angle histogram obtained the best score for the whole face. However, the sized pseudo-landmarks were a close second for whole faces.

5 Conclusions

This paper presents several techniques that automate the craniofacial image analysis pipeline and introduces methods to diagnose and quantify several different craniofacial syndromes. The pipeline starts with the preprocessing of raw 3D face meshes obtained by a stereo-photography system. Wu et al. [25] provided an automatic preprocessing method that normalizes the pose of the 3D mesh and extracts the face. After preprocessing, features can be calculated from the 3D face meshes, including azimuth and elevation angles, several curvature measures, symmetry scores [24], anthropometric landmarks [15] and pseudo-landmarks [18]. The extracted features have been used in the quantification of craniofacial syndromes and in classification tasks. Our new work, a content-based retrieval system built on multiple different features, was introduced, and the retrievals of similar faces from a synthetic database were evaluated.

Medical imaging has revolutionized medicine by enabling scientists to obtain lifesaving information about the human body—non-invasively. Digital images obtained through CT, MR, PET and other modalities have become standards for diagnosis and surgical planning. Computer vision and image analysis techniques are being used for enhancing images, detecting anomalies, visualizing data in different dimensions and guiding medical experts. New computational techniques for craniofacial analyses provide a fully automatic methodology that is powerful and efficient. The techniques covered in this paper do not require human supervision, provide objective and more accurate results, and make batch processing of large amounts of data possible.

References

1. Allen, B., Curless, B., & Popović, Z. (2003). The space of human body shapes: Reconstruction and parameterization from range scans. *ACM Transactions on Graphics, 22*, 587–594.
2. Ardinger, H. H., Buetow, K. H., Bell, G. I., Bardach, J., VanDemark, D., & Murray, J. (1989). Association of genetic variation of the transforming growth factor-alpha gene with cleft lip and palate. *American Journal of Human Genetics, 45*(3), 348.
3. Atmosukarto, I., Shapiro, L. G., & Heike, C. (2010). The use of genetic programming for learning 3d craniofacial shape quantifications. In *2010 20th International Conference on Pattern Recognition (ICPR)* (pp. 2444–2447). Los Alamitos, CA: IEEE Press.
4. Atmosukarto, I., Shapiro, L. G., Starr, J. R., Heike, C. L., Collett, B., Cunningham, M. L., et al. (2010). Three-dimensional head shape quantification for infants with and without deformational plagiocephaly. *The Cleft Palate-Craniofacial Journal, 47*(4), 368–377.
5. Benz, M., Laboureux, X., Maier, T., Nkenke, E., Seeger, S., Neukam, F. W., et al. (2002). The symmetry of faces. In *Vision Modeling and Visualization (VMV)* (pp. 43–50).
6. Besl, P. J., & Jain, R. C. (1985). Three-dimensional object recognition. *ACM Computing Surveys (CSUR), 17*(1), 75–145.
7. Boehringer, S., Vollmar, T., Tasse, C., Wurtz, R. P., Gillessen-Kaesbach, G., Horsthemke, B., et al. (2006). Syndrome identification based on 2d analysis software. *European Journal of Human Genetics, 14*(10), 1082–1089.
8. Claes, P., Walters, M., Vandermeulen, D., & Clement, J. G. (2011). Spatially-dense 3d facial asymmetry assessment in both typical and disordered growth. *Journal of Anatomy, 219*(4), 444–455.
9. Freund, Y., & Schapire, R. E. (1997). A decision-theoretic generalization of on-line learning and an application to boosting. *Journal of Computer and System Sciences, 55*(1), 119–139.
10. Glasgow, T. S., Siddiqi, F., Hoff, C., & Young, P. C. (2007). Deformational plagiocephaly: Development of an objective measure and determination of its prevalence in primary care. *Journal of Craniofacial Surgery, 18*(1), 85–92.
11. Gower, J. C. (1975). Generalized procrustes analysis. *Psychometrika, 40*(1), 33–51.
12. Hammond, P., et al. (2007). The use of 3d face shape modelling in dysmorphology. *Archives of Disease in Childhood, 92*(12), 1120.
13. Hutchison, B. L., Hutchison, L. A., Thompson, J. M., & Mitchell, E. A. (2005). Quantification of plagiocephaly and brachycephaly in infants using a digital photographic technique. *The Cleft Palate-Craniofacial Journal, 42*(5), 539–547.
14. Lanche, S., Darvann, T. A., Ólafsdóttir, H., Hermann, N. V., Van Pelt, A. E., Govier, D., et al. (2007). A statistical model of head asymmetry in infants with deformational plagiocephaly. In *Image analysis* (pp. 898–907). Berlin: Springer.
15. Liang, S., Wu, J., Weinberg, S. M., & Shapiro, L. G. (2013). Improved detection of landmarks on 3d human face data. In *2013 35th Annual International Conference of the IEEE Engineering in Medicine and Biology Society (EMBC)* (pp. 6482–6485). Los Alamitos, CA: IEEE Press.
16. Lin, H., Ruiz-Correa, S., Shapiro, L., Hing, A., Cunningham, M., Speltz, M., et al. (2006). Symbolic shape descriptors for classifying craniosynostosis deformations from skull imaging. In *27th Annual International Conference of the Engineering in Medicine and Biology Society, IEEE-EMBS 2005* (pp. 6325–6331). Los Alamitos, CA: IEEE Press.
17. McKinney, C. M., Cunningham, M. L., Holt, V. L., Leroux, B., & Starr, J. R. (2008). Characteristics of 2733 cases diagnosed with deformational plagiocephaly and changes in risk factors over time. *The Cleft Palate-Craniofacial Journal, 45*(2), 208–216.
18. Mercan, E., Shapiro, L. G., Weinberg, S. M., & Lee, S. I. (2013). The use of pseudo-landmarks for craniofacial analysis: A comparative study with l 1-regularized logistic regression. In *2013 35th Annual International Conference of the IEEE Engineering in Medicine and Biology Society (EMBC)* (pp. 6083–6086). Los Alamitos, CA: IEEE Press.

19. Müller, H., Marchand-Maillet, S., & Pun, T. (2002). The truth about corel—evaluation in image retrieval. In *Proceedings of the Challenge of Image and Video Retrieval (CIVR2002)* (pp. 38–49).
20. Müller, H., Michoux, N., Bandon, D., & Geissbuhler, A. (2004). A review of content-based image retrieval systems in medical applications clinical benefits and future directions. *International Journal of Medical Informatics, 73*(1), 1–23.
21. Perez, E., & Sullivan, K. E. (2002). Chromosome 22q11. 2 deletion syndrome (digeorge and velocardiofacial syndromes). *Current Opinion in Pediatrics, 14*(6), 678–683.
22. Plank, L. H., Giavedoni, B., Lombardo, J. R., Geil, M. D., & Reisner, A. (2006). Comparison of infant head shape changes in deformational plagiocephaly following treatment with a cranial remolding orthosis using a noninvasive laser shape digitizer. *Journal of Craniofacial Surgery, 17*(6), 1084–1091.
23. Ruiz-Correa, S., Sze, R. W., Lin, H. J., Shapiro, L. G., Speltz, M. L., & Cunningham, M. L. (2005). Classifying craniosynostosis deformations by skull shape imaging. In *Proceedings of the 18th IEEE Symposium on Computer-Based Medical Systems, 2005* (pp. 335–340). Los Alamitos, CA: IEEE Press.
24. Wu, J., Tse, R., Heike, C. L., & Shapiro, L. G. (2011). Learning to compute the symmetry plane for human faces. In *Proceedings of the 2nd ACM Conference on Bioinformatics, Computational Biology and Biomedicine* (pp. 471–474). New York, NY: ACM Press.
25. Wu, J., Tse, R., & Shapiro, L. G. (2014). Automated face extraction and normalization of 3d mesh data. In *2014 36th Annual International Conference of the IEEE Engineering in Medicine and Biology Society (EMBC)*. Los Alamitos, CA: IEEE Press.
26. Wu, J., Tse, R., & Shapiro, L. G. (2014). Learning to rank the severity of unrepaired cleft lip nasal deformity on 3d mesh data. In *2014 24th International Conference on Pattern Recognition (ICPR)*. Los Alamitos, CA: IEEE Press.
27. Yang, S., Shapiro, L. G., Cunningham, M. L., Speltz, M., & Le, S. I. (2011). Classification and feature selection for craniosynostosis. In *Proceedings of the 2nd ACM Conference on Bioinformatics, Computational Biology and Biomedicine* (pp. 340–344). New York, NY: ACM Press.
28. Zhang, M., Bhowan, U., & Ny, B. (2007). Genetic programming for object detection: A two-phase approach with an improved fitness function. *Electronic Letters on Computer Vision and Image Analysis, 6*(1), 2007. URL http://elcvia.cvc.uab.es/article/view/135
29. Zhu, X., & Ramanan, D. (2012) Face detection, pose estimation, and landmark localization in the wild. In *2012 IEEE Conference on Computer Vision and Pattern Recognition (CVPR)* (pp. 2879–2886). Los Alamitos, CA: IEEE Press.

Mammographic Mass Description for Breast Cancer Recognition

Khalifa Djemal, Imene Cheikhrouhou, and Hichem Maaref

1 Introduction

Breast cancer is one of the paramount issues in the field of public health. In fact, about one in ten women is affected by this disease during her lifetime. To ensure early detection of such tumors, radiologists were asked to increase the frequency of mammograms especially for the age of the most concerned group. For example, many countries such as France have implemented campaigns for routine screening every 2 years. It has been shown that this approach is very effective and can reduce the death rate to 35 %. According to the campaigns, two or four mammograms were achieved per patient at a rate of one or two mammograms per breast. This has resulted in an exponential increase in the number of mammograms performed. Thus, the task of interpretation has become difficult to manage by radiologists. Indeed, interpretation is a difficult task and depends on the expertise of the radiologist. Moreover, the detection rate of breast cancer is improved by about 15 % using a second reading. With the increasing number of mammograms in recent decades, various research were developed that make the effort to automatically detect breast lesions through Computer Aided Detection systems or to automatically interpret mammograms through Computer Aided Diagnosis systems. In this context, various treatment methods have been developed [9, 10, 13]. Most of these methods use the BIRADS standard [14] to classify mammographic images into two classes: malignant or benign and into six classes according the ACR classification (American College of Radiology).

In this chapter, we are interested in Computer Aided Diagnosis system which essentially consist of three stages namely features extraction, description and

K. Djemal (✉) • I. Cheikhrouhou • H. Maaref
IBISC Laboratory, University of Evry Val d'Essonne, 40 rue du Pelvoux, 91020 Evry, France
e-mail: khalifa.djemal@ibisc.univ-evry.fr; Imene.Cheikhrouhou@ibisc.univ-evry.fr; hichem.maaref@ibisc.univ-evry.fr

A. Briassouli et al. (eds.), *Health Monitoring and Personalized Feedback using Multimedia Data*, DOI 10.1007/978-3-319-17963-6_3

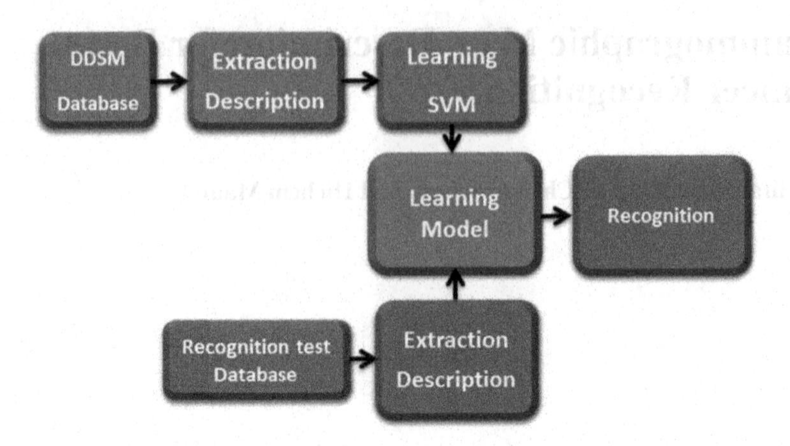

Fig. 1 Computer aided diagnosis system with its different steps: extraction, description and recognition

classification (Fig. 1). The classification result of breast masses is strongly linked to an appropriate choice of methods which carry out these three steps. The description step constitutes the major part of the presented work. Such an approach requires some knowledge about breast pathologies. According to the standard BIRADS [14], benign masses often have a round or oval shape and a circumscribed or micro-lobulated contour. Malignant masses often have a lobulated or irregular shape and hidden contour parts, indistinct or spiculated (Fig. 2). However, it is usually difficult to distinguish benign masses with lobulated contour, from malignant ones with spiculated contour. A detailed description of the contour is then necessary to avoid false positives (in order to reduce unnecessary biopsies) and avoid false negatives (passing a patient with breast cancer for non-sick person). In this context, various shape and texture descriptors have been proposed.

These methods are specifically concerned with shape descriptors as they are able to characterize the contour of masses and meet the selection criteria provided by the BIRADS. The most used descriptors in the field of pattern recognition are the area, perimeter, circularity and compactness [1]. However, they characterize the contour in the global way without taking into account its peculiarities. Then, they are often insensitive to slight differences between the ambiguous cases, such as benign masses of lobulated contour, and malignant ones with spiculated contour. Several methods for characterizing the morphology of the masses have been proposed using an appropriate description. Each description is focused on a very specific detail of the contour such that the direction of spiculations [16] or information regarding the concavities [8, 18]. However, these descriptors do not simultaneously take into account all the particularities of the contour. Furthermore, they are not always invariant to geometric transformations. Such a criterion is highly recommended, since it allows identifying the contours on a reliable and unique way without considering their geometrical positions. Therefore, developed a robust, accurate and invariant descriptor remains a challenge in the field of pattern recognition and

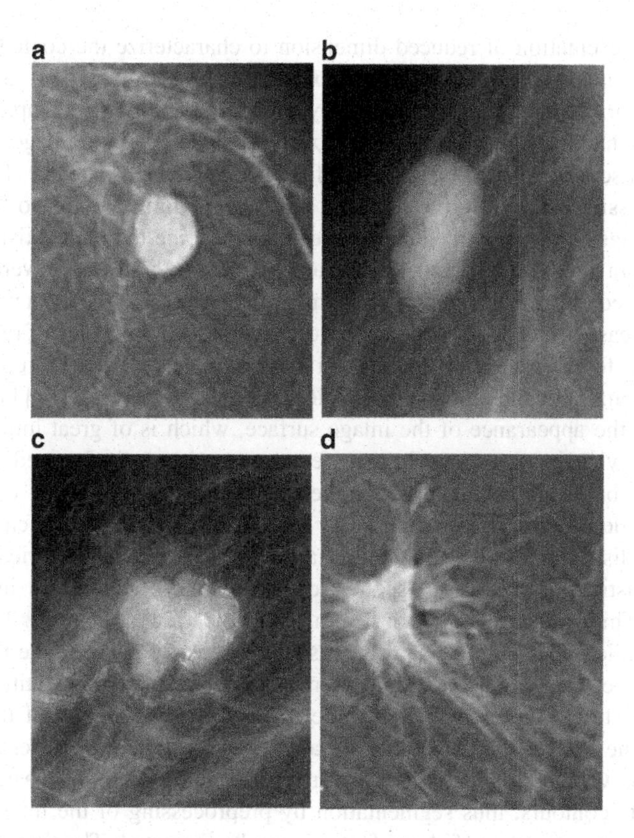

Fig. 2 Samples of DDSM database used in evaluation. The *first line* shows circular and oval masses. Lobulated and spiculated masses are shown in the *second line*

in mammographic masses description. In this work different descriptors of breast masses are presented and discussed. The assessment was carried performances through Computer Aided Diagnosis system (CAD). The implementation required the study and comparison of various methods that deal with different steps in this system i.e. extraction, description and classification.

In Sect. 2, the role and the need of the features extraction step in Computer Aided Diagnosis system are presented. Different kinds of mass descriptions are presented in Sect. 3. Indeed, geometric, morphologic and spiculated descriptors are discussed and compared. All recognition results of begin and malignant masses obtained by each descriptors through an SVM classifier are presented in Sect. 4.

2 Feature Extraction

The images are rich digital objects in terms of information. In addition to the required huge memory space, the direct manipulation of these images in an image recognition system does not get realistic response time. It is therefore necessary to use a representation of reduced dimension to characterize the content of these images. The main objective of feature extraction is determined for each image, a representation (signature) that is quickly accessible and easily comparable, and on the other hand sufficiently complete to fully characterize the image. However, robust and discriminant characterization of the images remains a major challenge in image processing. Generally, the so-called low-level features are most often used to describe images by their content. These features describe the main existing visual features in an image, i.e. color, texture and shape. Color features were the first descriptor used in image recognition systems, and they are still the most widely used due to their ease of extraction, richness of description and efficiency of recognition. These color features depend directly on the color space used for color image representation. In the literature, several color spaces have been studied [11]. Texture is linked to the appearance of the image surface, which is of great importance in any area of visual perception. Texture features are increasingly used in images description, because they decrease some problems with the color description. Indeed, the description of texture is very effective especially in the case of very close color distributions. Texture features are divided into two categories: The first is deterministic and refers to a spatial repetition of a basic pattern in different directions. This structural approach is a macroscopic view of textures. The second approach, called microscopic, is probabilistic and seeks to characterize the chaotic aspect that does not include localized pattern or frequency of main repetition. Unlike color and texture features, which are interested in the description of the general content of the image, shape features are able to characterize the different objects in the image. Generally, this kind of feature indicates the general appearance of an object, as its contours, thus segmentation by preprocessing of the image is often necessary. Two categories of shape features can be extracted: The first category is based on the geometry of the image regions. The second is based on statistics of pixel intensities of different regions in the image. Geometric transformations such as rotation, translation and scaling, may operate on an image. To ensure a robust and efficient description, shape features cover generally all levels of representation that includes an object. Moreover, they are often insensitive to variations caused by different geometric transformations. To extract the shapes of masses, segmentation is the most often used technique to achieved this task. Indeed, the mammographic segmentation consists of separating the abnormalities from the background of the mammographic image. The involvement of experts in drawing accurate boundaries around masses is extremely time consuming. Also, such manual drawing prevents the fully automation of CAD systems. For this reason, many segmentation tools are used to automatically detect masses. Global thresholding has been widely used in [2]. It is based on global informations such as the histogram. However, such technique is not adequate. In fact, mammograms are the 2D projections of the 3D overlapping tissues which may be brighter than the masses. Local thresholding can refine the results of global thresholding [17], since it is determined locally for each pixel based on the intensity values of the surrounding pixels. However, it is a pixel based operation and cannot accurately separate pixels into suitable sets. Also, an adaptive clustering process is needed to refine the result attained from the localized adaptive thresholding. Several research efforts are oriented to the region growing technique for segmenting masses [19]. The basic idea of the algorithm is to find a set of seed pixels which will be aggregated with the pixels that have similar properties. Nevertheless, the segmentation result depends on finding suitable seeds and it is generally sensitive to noise.

Other investigations in the breast cancer segmentation field are based on classical active contour models [3]. They can be easily formulated under a principled energy minimization framework and allow incorporation of various prior knowledge such as shape and intensity distribution. Also, they can provide smoothed and closed contours, which are necessary in our application. Existing active contour models can be categorized into two major classes: edge-based and region-based models. For the first approach, the active contour evolves to the object boundaries having the strongest gradient of intensity. Nevertheless, these methods require good initialization. Region-based approaches aim to identify each region of interest by using a certain region descriptor to guide the evolution equation of an active contour [12]. However, popular region-based active contour models tend to rely on intensity homogeneity in each of the regions to be segmented, while intensity inhomogeneity often occurs in medical images such as mammographies. In this study, we adopt the region-based active contour model as proposed by Li et al. [20]. The proposed model is able to segment images with intensity inhomogeneity. Also, it achieves good performance for images with weak object boundaries such the case of ill-defined and obscured margins. With the level set regularization term in the proposed formulation, the regularity of the level set function is intrinsically preserved to ensure accurate computation and avoid expensive reinitialization procedures. Figure 3 presents the segmentation results achieved by Li et al. in [20]. It shows the convergence to the final contour, where a,b,c and d show the different levels of spiculated masses.

3 Mass Description

In the pattern recognition domain, many specific shape descriptors have been proposed. These proposal, aim to obtain robust descriptors more accurate and informative about the shape details. Information extracted from an image as shape features needs to be described in order to use them in different applications. The description is often conditioned by the application domain. Indeed, to recognize for example, a malignant or benign breast mass, the description is based especially on information provided by experts in the breast cancer domain.

3.1 Geometric Mass Descriptors

According to the shape differences between benign and malignant masses, geometric descriptors such as area (A), perimeter (P), circularity (C) and com-

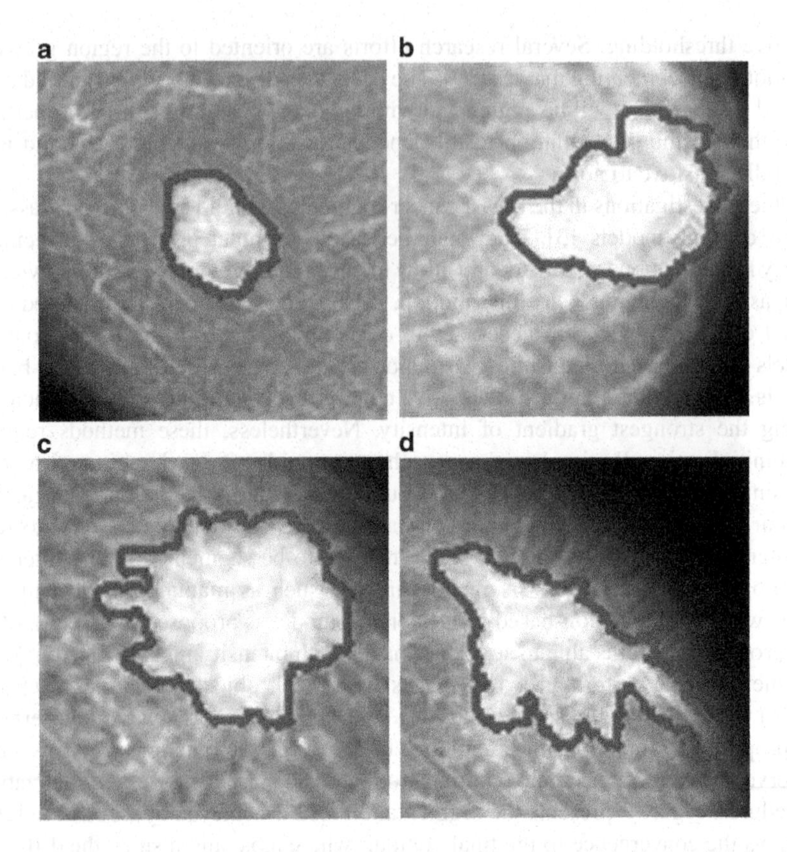

Fig. 3 Mammographic image segmentation. The figure show the final contours obtained on four different masses

pactness (Com) have been developed [1]. Nevertheless, these descriptors could provide satisfying classification results especially when they are associated to other features [6]. The squareness is used as a geometric descriptor, however, the standard squareness equation also called rectangularity (Rect) is not invariant under rotation. In order to overcome this sensitivity, we consider the minimum bounding box in the direction of the object. For this purpose, we define first the equivalent ellipse with the same central moments. These moments μ_{02}, μ_{20} and μ_{11} are obtained from gravity center of the lesion which is defined by (x_g, y_g):

$$\mu_{pq} = \sum_{i=0}^{n} \sum_{j=0}^{m} (i - x_g)^p (j - y_g)^q \tag{1}$$

with $(p, q) = \{0, 1 \text{ or } 2\}$, and (n, m) are the matrix dimensions.

The considered ellipse, having the same moments as the studied lesion, is defined by: the major axis a_1, minor axis a_2 and the rotation angle α of the lesion relative to the horizontal. These different parameters are obtained as follow:

$$a_1^2 = \frac{2(\mu_{02} + \mu_{20} + \sqrt{(\mu_{20} - \mu_{02})^2 + 4\mu_{11}^2})}{m_{00}} \tag{2}$$

$$a_2^2 = \frac{2(\mu_{02} + \mu_{20} - \sqrt{(\mu_{20} - \mu_{02})^2 + 4\mu_{11}^2})}{m_{00}} \tag{3}$$

with m_{00} the moment of order zero, which represents the area of considered lesion.

$$t = \frac{\mu_{02} - \mu_{20} + \sqrt{(\mu_{20} - \mu_{02})^2 + 4\mu_{11}^2)}}{2\mu_{11}} \tag{4}$$

$$\alpha = arctan(t) \tag{5}$$

From the obtained equivalent ellipse of the lesion (mass), we can easily obtain the improved squareness expression, which has named modified squareness *MRect*. The modified squareness is always obtained by the ratio between the lesion area and the its bounding box area.

3.2 *Morphologic Mass Descriptors*

In this section we present different kinds of mass descriptors, which are the most used in breast cancer recognition. These different description methods are compared with our spiculated mass descriptors in the results section.

3.2.1 Normalized Radial Length

Kilday et al. [18] developed a set of six shape features based on the Normalized Radial Length (NRL) from the objects centroid to the points on the boundary. The NRL features have had a good success in CAD applications and provide satisfying results with a generally round boundary as demonstrated in [7, 10].

$$d(i) = \frac{\sqrt{(x(i) - x_g)^2 + (y(i) - y_g)^2}}{max(d(i))}, \ i = \{1, 2, \ldots, N\} \tag{6}$$

with $(x(i), y(i))$ and (x_g, y_g) the coordinates of ith pixel and respectively the gravity center. N the mass perimeter.

a. ***NRL average*** (d_{avg}): The average of the NRL is defined as follow:

$$d_{avg} = \frac{1}{N}\sum_{i=1}^{N} d(i) \tag{7}$$

b. ***The standard deviation of the*** *NRL* (σ): This measure describes the irregularity:

$$\sigma = \sqrt{\frac{1}{N}\sum_{i=1}^{N}(d(i) - d_{avg})^2} \tag{8}$$

c. ***The entropy*** (E): Entropy is obtained from the histogram of the radial length. The perimeter p_k is the probability that LRN to be between $d(i)$ and $d(i) + 1/N_{bins}$, with N_{bins} the bins number of the normalized histogram, varying in the interval [0,1] was divided into $N_{bins} = 100$. Entropy measurement incorporates simultaneously the notion of circularity and irregularity:

$$E = \sum_{k=1}^{100} p_k log(p_k) \tag{9}$$

d. ***The area ratio*** (A_1): The area ratio is a measure of the percentage of the lesion part of the circular region defined by the average of NRL:

$$A_1 = \frac{1}{d_{avg}.N}\sum_{i=1}^{N}(d(i) - d_{avg}) \tag{10}$$

e. ***The roughness*** (R): The roughness aims to isolate the macroscopic shape of the lesion from the small structures of the contour. It provide information about the average between neighboring pixels:

$$R = \frac{1}{N}\sum_{i=1}^{N}(d(i) - d(i+1)) \tag{11}$$

f. ***The zero crossing count*** (ZC_1): It calculates the number of times that the line defined by the *NRL* average intercepts the contour lesion. It provides the contour spiculation degree.

3.2.2 The Modified Normalized Radial Length (MNRL)

From the NRL properties Chen et al. [7] proposed improved descriptors which had shown higher performance than basic NRL features. The normalized radial length $d(i)$ is filtered using a moving average filter and the filtered curve is noted $d_{ma}(i)$. The processed descriptors are the difference of standard deviation (σ_{diff})), the entropy of the difference between $d(i)$ and $d_{ma}(i)$ named E_{diff}, the area ratio A_2 and the Zero Crossing Count (ZC_2).

a. ***Difference of the standard deviations*** (σ_{diff}): The σ_{diff} can estimate the irregularity degree of the contour, so that, if the contour becomes more irregular, σ_{diff} reaches higher values.

$$\sigma_{diff} = |\sigma - \sigma_{ma}| \tag{12}$$

b. ***Modified entropy*** (E_{diff}): This descriptor provides the distribution of the difference between $d(i)$ and $d_{ma}(i)$. p_k is the probability that $|d(i) - d_{ma}(i)|$ is between $|d(i) - d_{ma}(i)|$ and $|d(i) - d_{ma}(i)| + 1/N_{bins}$.

$$E_{diff} = \sum_{k=1}^{100} p_k log(p_k) \tag{13}$$

c. ***Modified area ratio*** (A_2): The description is obtained by the following equation:

$$A_2 = \frac{1}{d_{avg}.N} \sum_{i=1}^{N} (d(i) - d_{ma}(i)) \tag{14}$$

d. ***The modified zero crossing count*** (ZC_2): This is the measure of the number of times that the curve $d(i)$ intercepts $d_{ma}(i)$.

3.2.3 The Curvature

The curvature noted $Curv$ was commonly used in the context of the shapes analysis in several domains. It has been recognized for its ability to characterize the object shapes. Many proposed approaches suggest minimizing the curvature while respecting the geometric constraints from directions tangents calculated on the contour. This method is robust and used in CAD system. In general, the curvature at a given point A, a curve is defined as the inverse of the radius of the osculating circle in A. The osculating circle can be obtained as follows: given two points B and C near A, we calculate the single circle through A, B and C. If these points are colinear, the circle has an infinite radius and the curvature is then zero.

$$Curv = \frac{1}{R} \tag{15}$$

The osculating circle is defined as following:

$$R = \frac{a.b.c}{\sqrt{(a+b+c)(a-b+c)(a+b-c)(b-a+c)}} \tag{16}$$

with $a = |AB|$, $b = |BC|$ and $c = |AC|$.

3.3 Spiculated Mass Descriptors

In the breast cancer domain and as explained in the previous sections, geometrical invariance in shape analysis is recommended to preserve the same feature value and then the same classification output for similar shapes. For this reason, we try to conceive an efficient and an invariant shape descriptor able to quantify the degree of mass spiculation and to further improve the classification performance.

3.3.1 Number of Substantial Protuberances and Depressions

Morphologic features to diagnose sonographic images was proposed in [8] using five nearly setting-independent. They provide reliable performance for different sample sizes. However, the proposed Number of Substantial Protuberances and Depressions (NSPD) descriptor could not consider all protuberances and depressions in the contour. For their detection, authors apply an approximation which affects the NSPD robustness. An improved equivalent protuberance selection descriptor was proposed in [4].

3.3.2 Elliptic Normalized Skeleton

The Elliptic Normalized Skeleton (ENS) was proposed by Chen et al. [8] to describing masses in echographic images. This descriptor consider the number of skeleton points, which are, the End Points (EP), the Simple Points (SP) and the Multiple Points (MP). The sum of the skeleton points is normalised by the perimeter of the equivalent ellipse noted (EqEP) for better considering the scaling invariance. the ENS descriptor is defined as follow:

$$ENS = \frac{EP + SP + MP}{EqEP} \tag{17}$$

3.3.3 Skeleton End Points and Protuberance Selection Descriptors

To describe spiculated masses, the Skeleton End Points (SEP) and the Protuberance Selection descriptor (PS) are proposed [4]. These two descriptors provided good results for spiculated masses recognition. The Skeleton End Points (SEP) is based on the number of end points of the skeleton. The choice of the number of end points to quantify the skeleton is based on the fact that this number is independent of the skeleton size whilst the number of single points increases with the size of the shape. The Protuberance Selection descriptor (PS) is based on the calculation of protuberance (or spiculations) contour. In this context, a preliminary step of deriving the contour is necessary to extract the stationary points and check their changes signs. After extraction of the protuberance and depressions, a test on neighboring pixels allows to preserve the protuberance so that we can extract all spiculations. We can differentiate between simple benign and irregular malignant masses. This descriptor managed not only to differentiate between simple and complex shapes, but also the invariance to known geometric transformations such as translation, rotation and scaling.

3.3.4 Spiculated Mass Descriptor

The spiculated mass descriptor (SMD) proposed in [5], improves the description rate of highly spiculated lesions. Indeed, the SMD descriptor is based on simple geometric procedures to detect lobulations whose length and width are implicitly considered. The SMD satisfy invariance to the geometric transformations (scaling, rotation and translation). The spiculated mass descriptor is defined as:

$$SMD = \frac{1}{K+1}\sum_{k=0}^{K} T^{k.\beta} \quad (18)$$

with $T^{k.\beta}$ template variations and β the step-size angle.

4 Classification and Recognition Results

In this section, we present first the protocol used to evaluate each descriptor on DDSM database. The classification and comparison of recognition results are presented and discussed in Sects. 4.2 and 4.3.

4.1 Digital Database for Screening Mammography (DDSM)

In order to validate extracted features and descriptors, we chose Digital Database for Screening Mammography (DDSM). It was assembled by a group of researchers from the University South Florida and was completed in 1991, [15]. DDSM contains 2,620 cases collected from Massachusetts General Hospital, Wake Forest University

Table 1 Distribution of the used database extracted from DDSM

Total number of masses	Number of training examples	Number of test examples
242	130	112
128 Benign	70 Benign	58 Benign
114 Malignant	60 Malignant	54 Malignant

School of Medicine, Sacred Heart Hospital and Washington University of St. Louis School of Medicine. DDSM was widely used by the scientific community in the field of breast cancer; it has the advantage of using the same lexicon standardized by the American College of Radiology in BI-RADS. The different patient records were made in the context of screening and were classified into three cases: normal case (no lesions), benign case and malignant case, each file is composed of four views containing the oblique incidence external and the incidence Cranio Caudal of each breast. These files are also provided with annotations provided by expert radiologists. Figure 2 shows samples DDSM used in the evaluation. The DDSM is composed by two kinds which describing breast cancer, the microcalcifications and masses.

In this chapter, we have selected images from the DDSM database containing only the masses to be used in the experimental step [5]. Subset from DDSM was created consisting of 242 masses: 128 benign and 114 malignant. These examples are partitioned into 130 images for training and 112 for testing. Table 1 summarizes the distribution of used images.

4.2 *Classification*

Several classification approaches have been treated in the literature. Some researchers are focused on linear discriminant analysis [21]. The main idea of this technique is to construct the decision boundaries directly by optimizing the error criterion. However, this method is not adapted to nonlinear separable data. The artificial neural network method is extensively used in classification [22]. Indeed, it is able to model a complex nonlinear system using hidden units in a compact range. But the number of hidden layers and the number of neurons on each layer have to be empirically determined on the basis of the training data set. Other researchers are oriented to logistic regression for its ability to perform probability estimation using a logistic formula. Therefore, since such classifier fits the data to a formulated function, it requires much more data than discriminant analysis for example to achieve stable regression coefficients [22]. The support vector machine SVM learning developed by Vapnik [23], has found a wide range of real world applications including object recognition, face identification and breast cancer diagnosis. Indeed, SVM is based on the principle of structural risk minimization. It aims at minimizing the bound on the generalization error and maximizing the margin i.e., the distance between it and the nearest data point in each class. Thus, the SVM classifier is adopted in this work.

Table 2 The used mass descriptors

Mass descriptors	
The area A	The number of large protuberances and depressions $NSPD$
Perimeter P	The average of the normalized radial length NRL d_{avg}
Circularity C	The standard deviation of the NRL σ
Squareness $Rect$	The entropy E
The modified squareness $MRect$	The ratio of surface A_1
Compactness Com	Roughness R
Curvature $Curv$	The rate of zero crossing ZC_1
The standard elliptical skeleton ENS	The difference of the standard deviations σ_{diff}
The report of surface modified A_2	Modified entropy E_{diff}
The rate of modified crossing in zero ZC_2	Skeleton end points SEP
Protuberance selection PS	Spiculated masses descriptor SMD

4.3 Recognition Results

The 22 descriptors presented in Sect. 3 and summarized in Table 2 are tested through the CAD system. Table 3 presents the obtained results considering each descriptor and its category which are geometric, morphologic and spiculated masses. Indeed, considering each feature individually and compared to each other, we notice that the perimeter P ($A_z = 0.67$), the zero crossing ZC_2 ($A_z = 0.74$), and also the difference of standard deviation σ_{diff} ($A_z = 0.78$) relatively fail to well classify circumscribed/spiculated lesions and provide areas under ROC less than $A_z = 0.8$. While, the area A ($A_z = 0.81$), the area ratio A_2 ($A_z = 0.82$), the compactness Com ($A_z = 0.84$) and the entropy of the difference between $d(i)$ and $d_{ma}(i)$ E_{diff} ($A_z = 0.87$) provide satisfying results with A_z ranging from 0.8 to 0.9. The best results are obtained using the circularity C ($A_z = 0.92$) and the spiculated mass descriptor SMD ($A_z = 0.97$) with areas under ROC curves upper than 0.9. Although the circularity achieves a considerable classification performance, the SMD descriptor in the present study has clearly outperformed all the other shape descriptors and seems to be the most effective in the benign and malignant recognition of breast masses. Figure 4 presents all recognition results obtained with each descriptors through ROC curves.

5 Conclusion

In this chapter, we have presented the helpful of mammographic images in the breast cancer recognition domain. The main steps of the CAD system are presented which the features extraction, description and classification which lead to the

Table 3 Recognition results obtained for each mass descriptor

Geometric descriptors (%)	Morphologic descriptors (%)	Spiculated descriptors (%)
$A : 81$	$d_{avg} : 93$	$ENS : 90$
$P : 67$	$\sigma : 86$	$NSPD : 92$
$C : 92$	$E : 88$	$SEP : 92$
$Rect : 71$	$A_1 : 86$	$PS : 93$
$MRect : 73$	$R : 92$	$SMD : 97$
$Com : 84$	$Curv : 76$	
	$ZC_1 : 72$	
	$\sigma_{diff} : 78$	
	$E_{diff} : 87$	
	$A_2 : 82$	
	$ZC_2 : 74$	

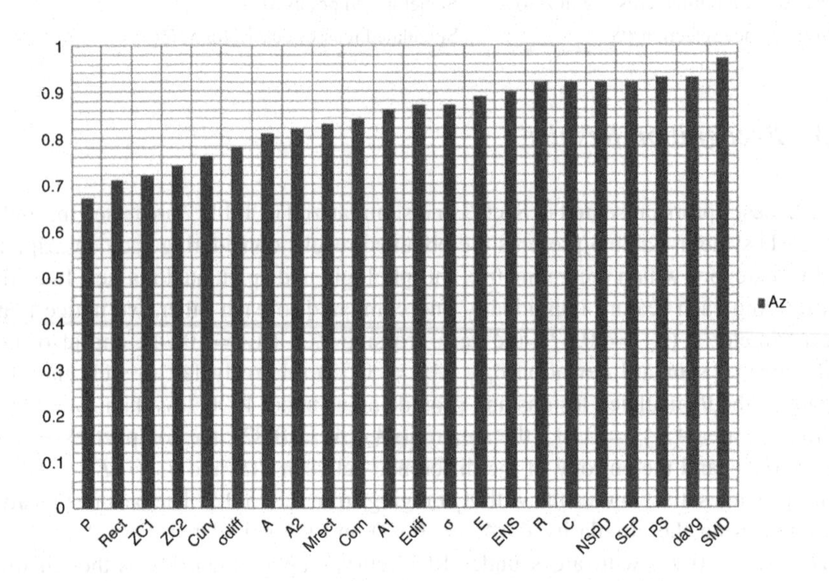

Fig. 4 Classification rate obtained by each descriptor using the SVM classifier

lesions recognition. We have presented three description categories dedicated to masses description. Through the obtained recognition results, the spiculated mass descriptors provide the best recognition rates. Through the results are good, the masses description remains challenge problem in the breast cancer diagnosis and recognition especially in the ACR classification system.

References

1. Bottigli, U., Cascio, D., Fauci, F., Golosio, B., Magro, R., Masala, G. L., et al. (2006). Massive lesions classification using features based on morphological lesion differences. In *Proceedings of World Academy of Science, Engineering and Technology* (Vol. 12, pp. 20–24).

2. Brzakovic, D., Luo, X. M., & Brzakovic, P. (1990). An approach to automated detection of tumors in mammograms. *IEEE Transactions on Medical Imaging, 9*(3), 233–241.
3. Chan, T., & Vese, L. (2001). Active contours without edges. *IEEE Transactions on Image Processing, 10*(2), 266–277.
4. Cheikhrouhou I., Djemal, K., & Maaref, H. (2011). Protuberance selection descriptor for breast cancer diagnosis. In *3rd European Workshop on Visual Information Processing (EUVIP)*, Paris (pp. 280–285).
5. Cheikhrouhou I., Djemal, K., & Maaref, H. (2012). Characterization of mammographic masses using a new spiculated mass descriptor in computer aided diagnosis systems. *International Journal of Signal and Imaging Systems Engineering, Inderscience, 5*(2), 132–142.
6. Cheikhrouhou, I., Djemal, K., Sellami, D., Maaref, H., & Derbel, N. (2009). Empirical descriptors evaluation for mass malignity recognition. In *The First International Workshop on Medical Image Analysis and Description for Diagnosis Systems - MIAD 2009* (pp. 91–100).
7. Chen, C. Y., Chiou, H. J., Chou, Y. H., Chiou, S. Y., Wang, H. K., Chou, S. Y., et al. (2009). Computer-aided diagnosis of soft tissue tumors on high-resolution ultrasonography with geometrical and morphological features. *Academic Radiology, 16*(5), 618–626.
8. Chen, C. M., Chou, Y. H., Han, K. C., Hung, G. S., Tiu, C. M., Chiou, H. J., et al. (2003). Breast lesions on sonograms. Computer-aided diagnosis with nearly setting-independent features and artificial neural networks. *Radiology, 226*, 504–514.
9. Ciatto, S., Turco, M. R. D., Risso, G., Catarzi, S., et al. (2003). Comparison of standard reading and computer aided detection (CAD) on a national proficiency test of screening mammography. *European Journal of Radiology, 37*(2), 135–138.
10. Delogu, P., Fantaccia, M. E., Kasae, P., & Retico, A. (2007). Characterization of mammographic masses using a gradient-based segmentation algorithm and a neural classifier. *Computers in Biology and Medicine, 37*(10), 1479–1491.
11. Djemal, K., Cocquerez, J. P., & Precioso, F. (2012). Visual feature extraction and description. In *Visual indexing and retrieval book* (pp. 5–20). New York: Springer. ISBN 978-1-4614- 3587-7.
12. Djemal, K., Puech, W., & Rossetto, B. (2006). Automatic active contours propagation in a sequence of medical images. *International Journal of Images and Graphics, 6*(2), 267–292.
13. Djemal, K., & Maaref, H. (2011). Intelligent information description and recognition in biomedical image databases. In B. Igelnik (Ed.), *Computational modeling and simulation of intellect: Current state and future perspectives*. Hershey, PA: IGI Global. ISBN: 978-1-60960-551-3.
14. D'Orsi, C. J., Bassett, L. W., Berg, W. A., Feig, S. A., Jackson, V. P., Kopans, D. B., et al. (2003). American college of radiology (Breast imaging reporting and data system). Troisième édition française réalisée par SFR (Société Française de Radiologie).
15. Heath, M., Bowyer, K., Kopans, D., Moore, R., & Kegelmeyer, P. (2000). The digital database for screening mammography. In *5th International Workshop on Digital Mammography*, Toronto, Canada.
16. Jiang, H., Tiu, W., Yamamoto, S., & Iisaku, S. I. (1997). Automatic recognition of spicules in mammograms. In *International Conference on Image Processing* (pp. 520–523).
17. Kallergi, M., Woods, K., Clarke, L. P., Qian, W., & Clark, R. A. (1992). Image segmentation in digital mammography: Comparison of local thresholding and region growing algorithms. *IEEE Transactions on Computerized Medical Imaging and Graphics, 16*, 231–323.
18. Kilday, J., Palmieri, F., & Fox, M. D. (1993). Classifying mammographic lesions using computer-aided image analysis. *IEEE Transactions on Medical Imaging, 12*(4), 664–669.
19. Lee, Y. J., Park, J. M., & Park, H. W. (2000). Mammographic mass detection by adaptive thresholding and region growing. *IEEE Transactions on International Journal of Imaging Systems and Technology, 11*(5), 340–346.
20. Li, C., Kao, C. Y., Gore, J. C., & Ding, Z. (2008). Minimization of region-scalable fitting energy for image segmentation. *IEEE Transaction of Image Processing, 17*(10), 1940–1949.

21. Petrick, N., Chan, H. P., Sahiner, B., Wei, D., Helvie, M. A., Goodsitt, M. M., et al. (1995). Automated detection of breast masses on digital mammograms using adaptive density-weighted contrast-enhancement filtering. *Proceedings of SPIE, Medical Imaging, Image Processing, 2434*, 590–597.
22. Song, J. H., Venkatesh, S. S., Conant, E. A., Arger, P. H., & Sehgal, C. M. (2005). Comparative analysis of logistic regression and artificial neural network for computer-aided diagnosis of breast masses. *Academic Radiology, 12*, 487–495.
23. Vapnik, V. (1998). *Statistical learnig theory*. New York: Wiley.

Development, Debugging, and Assessment of ParkinsonCheck Attributes Through Visualisation

Vida Groznik, Martin Možina, Jure Žabkar, Dejan Georgiev, Ivan Bratko, and Aleksander Sadikov

1 Introduction

Parkinson's disease (PD) is a chronic, progressive neurological disease. It is the second most common neurodegenerative disorder after Alzheimer's disease. PD is estimated to affect between four and six million individuals over the age of 50 worldwide, and that number is expected to double by the year 2030 [32]. The costs associated with 1.2 million PD patients in European Union were estimated at 13,934 million EUR in the year 2010 [13]. Newer research suggests that early detection of PD is beneficial in terms of treatment [1, 24]. The disease often presents itself with a Parkinsonian tremor (PT), a motor symptom that is frequently overlooked or misdiagnosed in the early stages of the disease. Some motor symptoms are similar to those of essential tremor (ET)—the most prevalent movement disorder. The tremor and other movement disorders considerably affect the patient's quality of life. This is especially so with the action tremor, which affects basic tasks such as writing, drinking, eating, etc. About three quarters of the patients with tremor

V. Groznik (✉) • M. Možina • J. Žabkar • I. Bratko • A. Sadikov
Faculty of Computer and Information Science, University of Ljubljana,
Večna pot 113, Ljubljana, Slovenia
e-mail: vida.groznik@fri.uni-lj.si; martin.mozina@fri.uni-lj.si; jure.zabkar@fri.uni-lj.si; ivan.bratko@fri.uni-lj.si; aleksander.sadikov@fri.uni-lj.si

D. Georgiev
Department of Neurology, University Medical Centre Ljubljana,
Zaloška cesta 2, Ljubljana, Slovenia
e-mail: dejan.georgiev@kclj.si

A. Briassouli et al. (eds.), *Health Monitoring and Personalized Feedback using Multimedia Data*, DOI 10.1007/978-3-319-17963-6_4

report impairment of their daily life and disability due to the tremor. Digitalised spirography is a relatively new method for detection and evaluation of different types of tremors; however, only a DaTSCAN test can reliably differentiate between PT and ET. The downside of DaTSCAN is its cost, invasiveness and it can usually be performed only in a larger clinical centre.

ParkinsonCheck mobile application takes the digitalised spirography to a new level and brings it to patients' homes. It exploits the powerful multimedia interaction of modern mobile devices—like smartphones and tablets—to detect the signs of tremors. The application asks the patient to draw an Archimedean spiral from which it calculates different quantitative parameters and estimates the risk of PT for the patient.

This chapter focuses primarily on knowledge extraction from user's spiral drawings, development, and validation of attributes used in the decision support system of ParkinsonCheck application. However, in separate sections, we describe three interconnected bodies of work: (1) a neurological decision support system (DSS) for differentiating between PT and ET implemented for ordinary computers [11], (2) the ParkinsonCheck mobile application, and (3) the use of visualisation for developing, debugging, and appraising the ParkinsonCheck's attributes.

The DSS takes as input both clinical and spirography data. Besides the practical value of a system differentiating between healthy, PD, and ET affected people, part of the research motivation was to investigate whether spirography can offer additional help (information) with such a task. When the DSS was completed, we realised that over a half of the decision rules in its final model included spirographic attributes. Moreover, some other clinical attributes could be approximated with either spirography or other simple tests that an individual could perform him or herself. This realisation prompted the idea of researching the possibility of a self-test for signs of PD and ET on a mobile platform based on spirography, eventually culminating in the ParkinsonCheck application.

ParkinsonCheck analyses the spiral drawings and textual user data automatically using a built-in expert system. We relied on experts' background knowledge while developing the DSS. Manual attribute extraction, however, can be greatly facilitated by having a good visualisation of extracted attributes. While the physicians explained important characteristics of the spirals, they did not provide us with exact implementable knowledge, but merely with some vague descriptions that still needed to be operationalised. Here, the ability to visually assess the attribute contributions and to be able to debug and modify them became critically important.

Written informed consent of all the participants was acquired before enrolling them in the study.

2 Preliminaries

In this section we briefly relate some points that will enable the reader to more easily follow the remainder of this chapter.

2.1 Parkinsonian and Essential Tremor

Parkinson's disease (PD) is rather difficult to accurately diagnose and effectively treat. While clinical manifestations of PD are diverse, in this paper we focus on the main motor symptoms, which can potentially be detected by spirography. The main motor symptoms are: tremor, bradykinesia or slowness of movements, increase in muscular tone or rigidity, and impaired postural reflexes. In the continuation we will rather broadly refer to these symptoms as Parkinsonian tremor (PT).

Essential tremor (ET), the most prevalent movement disorder [6], is characterised by postural and kinetic tremor with a frequency between 6 and 12 Hz. Although it is regarded as a symmetrical tremor, ET usually starts in one upper limb and then spreads to the other side affecting the contralateral upper limb, consequently spreading to the neck and vocal cords, giving rise to the characteristic clinical picture of the disorder. However, there are many deviations from this classical presentation of ET, e.g. bilateral tremor onset, limb tremor only, head tremor only, isolated voice tremor. Although distinct clinical entities, ET is very often misdiagnosed as PT [31]. Results from clinical studies show that ET is correctly diagnosed in 50–63 %, whereas PT in 76 % of the cases. Co-existence of both disorders is also possible [27]. In addition, PT can be very often observed when the upper limbs are stretched (postural tremor) and even during limb movement (kinetic tremor), which further complicates the differential diagnosis of the tremors. It is, however, important to differentiate between PD and ET as the treatment of the disorders is different.

2.2 DaTSCAN

DaTSCAN is the current golden standard test for differentiating between PD and ET. DaTSCAN is a single photon emission computed tomography of the dopamine transporter (DAT) in the striatum. During the procedure, a radioactive agent (ioflupane (123)I-FP-CIT) is injected in the blood. In PD there is a remarkable loss of DAT activity (as labelled by ioflupane (123)I-FP-CIT) while DAT activity in ET is normal. Although it has been proven as a useful tool for the differentiation between ET and PD with high sensitivity (93.7 %) and specificity (97.3 %), the main disadvantages of the method are its high cost, invasiveness, and limited access to the method, as it is usually available in bigger hospitals only [33]. Due to aforementioned disadvantages, DaTSCAN is not suitable for use as a screening test for general population.

2.3 *Spirography*

In this subsection we give some background on spirography as it presents the basis of all research described in this chapter. Spirography is a drawing method; usually a spiral is the preferred shape drawn as it tests the subject's motor ability in all directions.

Spiral drawing has been recommended by the Movement Disorder Society [5] as a method for the assessment of tremors. It has been demonstrated that the results of spirography correlate with the UPDRS (Unified Parkinson's Disease Rating Scale) and may be useful in the assessment of severity of Parkinsonian signs [29].

The application of spirography is quite widespread. It is used in assessment of akinesia in Parkinson's disease [10, 30, 35], drug-induced dyskinesias [18], tremor in multiple sclerosis [2, 9, 19], and essential tremor [7, 20]. It can also be useful in evaluating the therapeutic effect of deep brain stimulation [14], a neurosurgical procedure involving the implantation of a medical device called a brain pacemaker to specific parts of the brain for the treatment of movement and affective disorders.

Traditionally, spirography is done on paper, by simply drawing the spirals with a pencil. These drawings are assessed by visual evaluation. However, there are no clear guidelines regarding the judgement and interpretation, resulting in the basic method being somewhat subjective. A published protocol with the highest recognition for evaluation of tremor through spiral drawing is that by Bain and Findley [3]. This paper and pencil version of spirography does not provide any data beyond the drawing itself, the timestamps of various points of the spiral cannot be recorded as well as the data on pencil pressure, etc.

There were attempts to objectively measure the tremor amplitude of the pencil drawn spirals via scanning them and analysing the scanned images [17, 21]. The main disadvantage still remains, though: as there are no times recorded, several important aspects (e.g. speed of drawing, frequency analysis) cannot be analysed/observed.

Of particular note in relation to the visual debugging and assessment described in this chapter is the work of Wang et al. [34]. They observed that when the centre of the spiral is incorrectly identified, the polar expression becomes irregular with superimposed low frequency oscillations, which may result in less accurately measured spiral indices. Poorly executed spirals and abnormal movements such as tremors, superimposed on the spiral shape, cause artefacts that distort the spiral data, alter the polar coordinates, and result in improperly defined spiral centers. They attempted to improve the detection of the centre of the Archimedean spiral based on regression analysis. In Sect. 5 we describe how this same problem was detected using visual debugging.

Digitalised spirography [8, 26] uses a digital capture system to record the drawing. The system is usually composed of a computer with a dedicated software, a graphic tablet and a stylus. Digitalised spirography enables physicians to store the exact timestamp of each point in a two-dimensional area. This is useful for further analyses of the time series (e.g. frequency analysis). Figure 1 shows some textbook spiral examples recorded with digitalised spirography.

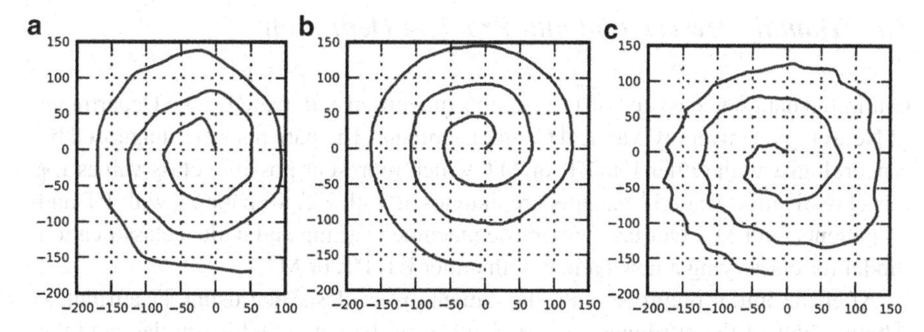

Fig. 1 Typical examples of spirals drawn by PT affected, ET affected and healthy subject. (**a**) Parkinsonian tremor. (**b**) Healthy subject. (**c**) Essential tremor

In the first application, described in Sect. 3, standard digitalised spirography test data were available, i.e. drawn using a stylus on a graphical pad. However, due to the nature of PARKINSONCHECK, described in Sect. 4, the spirography data for this application were obtained by drawing the spirals using fingers and not a stylus.

3 Decision Support System to Distinguish Between Parkinsonian and Essential Tremor from Clinical and Spirographic Attributes

The section describes the process of building a decision support system (DSS) for diagnosing and differentiating between three types of tremors, namely ET, PT, and mixed tremor (MT; both ET and PT at the same time). A successful DSS, acting as a second opinion, mostly for difficult cases, is expected to increase the combined diagnostic accuracy, reducing the need for patients to undergo an invasive, and very expensive further examination (DaTSCAN). This will also save both patients' and doctors' time.

Although several sets of guidelines for diagnosing both ET and PT do exist [15, 25], none of them enjoys general consensus in the community. Furthermore, none of these guidelines takes into account additional information from spirography. Our DSS combines all sources of knowledge, experts' background knowledge, machine-generated knowledge, and spirography data in an attempt to improve prediction accuracy. Indeed, as shown later in the text, spirography adds valuable information and the results of the described experiment served as a decisive motivation to build the PARKINSONCHECK mobile application.

The section mainly focuses on the task of knowledge acquisition as this is usually the most challenging part of building a DSS. Our knowledge acquisition process was based on argument-based machine learning (ABML) [23]. Here we give only a brief summary of the work on the project, with an emphasis on parts that influenced work described in the next sections. A detailed description of our work can be found in [11].

3.1 Domain Description and Problem Definition

Our initial data set consisted of 122 patients diagnosed and treated at the Department of Neurology, University Medical Centre Ljubljana. The patients were diagnosed by a neurologist with either ET, PT, or MT which represent possible class values for our classification task: 52 patients are diagnosed with ET, 46 patients with PT and 24 patients with MT. Our task was to use machine learning and build a classification model for classifying a new patient with either ET, PT, or MT.

After the initial preprocessing, the patients were described using 47 attributes. About a half of the attributes were derived from the patient's history data and the neurological examination, the other half included data from spirography.

Digitalised spirography is a technique used to evaluate different types of tremor through acquisition and analysis of spiral drawings. For this task, the doctors at the Department of Neurology are using a computer, a tablet for digital acquisition of the signal and a special pencil. A patient is asked to draw an Archimedean spiral on the tablet, and the resulting spiral is then included in the process of diagnosis.

Spiral drawing of patients with ET, PT, and MT can be seen in Fig. 2, respectively. Besides the raw signal received from the tablet, a spectral analysis is performed, which provides information about the tremor frequency. Commonly used are also speed-time and radius-angle transforms, where the latter depicts changes of the radius as a function of changes of the angle during spiral drawing. However, although the result of spirography comprises several time-series, in machine learning experiment in [11] we only used the qualitative assessments of the spirals provided by the physicians. We used the "raw" data in later experiments, as it is described in the following sections.

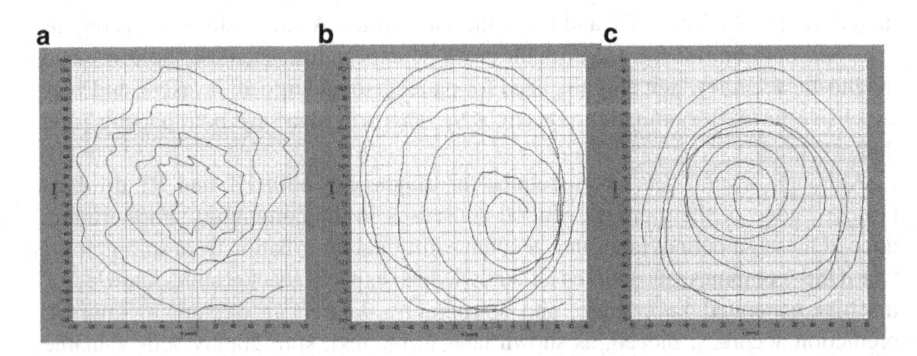

Fig. 2 Three spirals drawn by three patients with different types of tremor using their left hand. The spirals were drawn on a specialized tablet using a pen. (**a**) Essential tremor. (**b**) Parkinsonian tremor. (**c**) Mixed tremor

3.2 Methodology for Acquisition of Neurological Knowledge

Knowledge acquisition is a task that often proves to be a very difficult one, especially so if the goal is to acquire the knowledge in a comprehensible form. Our knowledge acquisition process was based on argument-based machine learning (ABML) [23]. ABML seamlessly combines the domain expert's knowledge with machine-induced knowledge, and is very suitable for the task of knowledge elicitation as it involves the expert in a very natural dialogue-like way [22]. The expert is not required to give general knowledge of the domain (which can be hard), but is only asked to explain concrete examples which the machine cannot correctly classify on its own. The process usually results in improved accuracy and comprehensibility [23]. Such focused knowledge elicitation also saves a lot of expert's time.

The reader can find more about ABML in [23] and at its website www.ailab.si/martin/abml. In [11] we used the ABCN2 method, an argument based extension of the well-known CN2 method [4], that learns a set of unordered probabilistic rules from examples with attached arguments, also called *argumented examples*. The attached arguments constrain rule learner to learn a set of rules that are consistent with given arguments. The following loop, called the ABML loop, defines the scheme of a dialogue between an expert and a computer in the ABML learning process. The loop contains four steps:

Step 1: Learn a hypothesis with ABCN2 using given data (in the initial step, data contains no arguments).

Step 2: Find the "most critical" example and present it to the expert. A critical example is an example that the current set of rules failed to explain correctly. If a critical example can not be found, stop the procedure.

Step 3: The expert explains the example; the explanation is encoded in arguments and attached to the learning example.

Step 4: Return to step 1.

The third step can turn into a mini-dialogue between the expert and the computer, if the provided arguments are in conflict with the data. Here we describe in details step 3 of the above loop:

Step 3a: Explaining critical example. First, the expert articulates a set of reasons suggesting the example's class value.

Step 3b: Adding arguments to example. An argument is given in natural language and needs to be translated into domain description language (attributes). If the argument mentions concepts currently not present in the domain, these concepts need to be included in the domain (as new attributes) before the argument can be added to the example.

Step 3c: Discovering counter examples. Counter examples are used to spot if an argument is sufficient to successfully explain the critical example or not. If not, ABCN2 will select a counter example. A counter example has the opposite class of the critical example, however it is covered by the rule induced from the given arguments.

Step 3d: Improving arguments with counter examples. The expert has to revise his initial argument with respect to the counter example.
Step 3e: Return to step 3c if counter example found.

We will continue by describing a sample iteration from the actual dialogue between ABML and the neurologist. We believe that the idea of such team work between a computer and an expert is also interesting from the multi-media perspective, as information on critical examples is usually presented in different formats (media). The expert needs to construct an argument using all media and then translate it to the language understood by the ABML algorithm.

In one of the iterations, a critical example was found, which was diagnosed as a MT (suffers from both, PT and ET) by a physician. In this case, the neurologist was asked to describe which features of the example are in favor of ET *and* which features are in favor of PT. The neurologist explained that the presence of postural tremor speaks in favor of ET, while the presence of rigidity speaks in favor of PT. These two conditions (postural tremor and rigidity) are assessed with a clinical test and are typical symptoms of ET and PT, respectively.

Unfortunately, these two conditions were not directly present in our feature space. However, we could derive their values from some other related features. Therefore, features POSTURAL.TREMOR.UP and RIGIDITY.UP were introduced into the domain and their values where automatically computed from the original ones that describe features postural tremor and rigidity. The former was used as an argument for ET and the latter was used as an argument for PT—both of these arguments were added to the critical example. While no counter examples were found for the expert's argument in favor of ET, the method selected a counter example for his argument in favor of PT.

The method, afterwards, found a counter example classified as ET, however having rigidity, which is typical for a PT. The neurologist was now asked to compare the critical example with the counter example; he needed to explain what is the most important feature in favor of PT that appears in the critical example and does *not* appear in the counter example. The crucial difference was discovered while inspecting the spirals drawn by both patients. Figure 3 shows the spirals drawn by patients that correspond to the critical and counter examples, and Fig. 4 shows spectral graphs of these two spirals. The neurologist spotted a reoccurring pattern of frequency spikes that is called harmonic, and according to the expert's judgement, it was the presence of harmonics in the counter example (or their absence in critical example), which are typical of ET. Therefore, the attribute HARMONICS that was added into the domain earlier with possible values of *true* and *false* was added to the previous argument. However, the method then found another counter example having ET, however with rigidity and no harmonics. The expert explained that the tremor in the new counter example did not have symmetrical onset, as opposed to the one in the critical example. The argument was further extended using the attribute SIM.TREMOR.START and added to the critical example. No new counter examples were found and this particular iteration was therefore concluded.

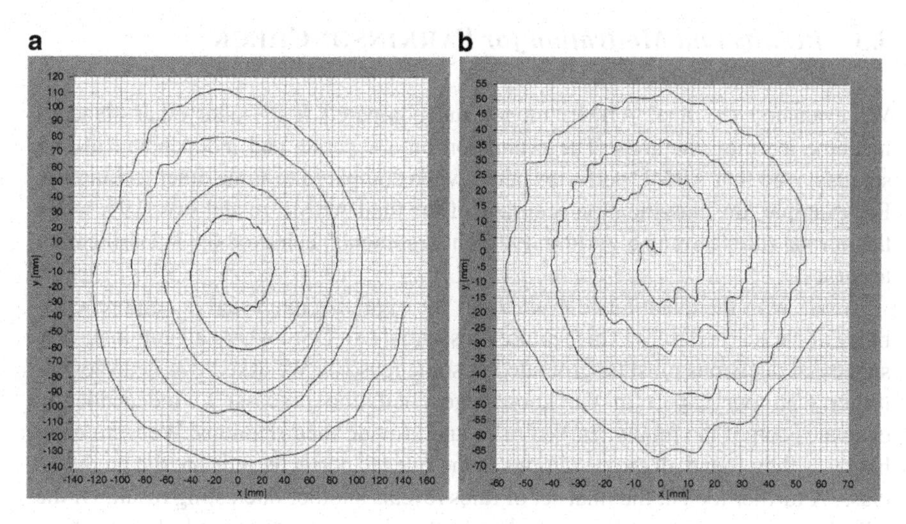

Fig. 3 Spirals drawn by a MT and by an ET patient. These two patients were the main subject in a dialogue between a neurologist and ABML. (**a**) Critical example. (**b**) Counter example

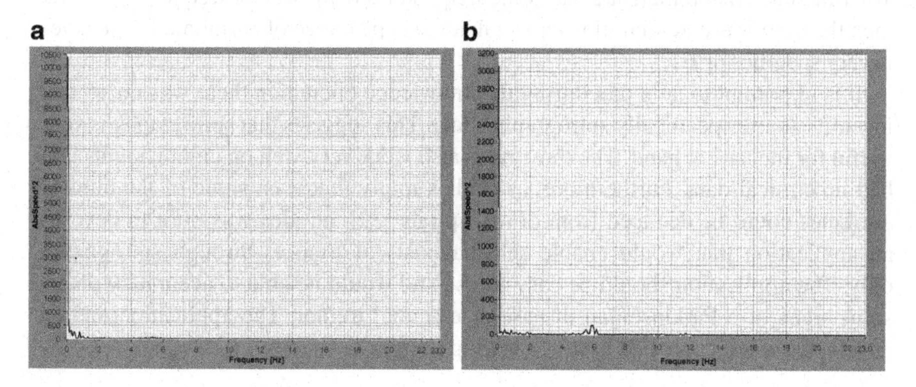

Fig. 4 Spectral analysis of spirals from Fig. 3. The spikes at around 6 and 12 Hz in the second picture suggest that this patient has a particular condition called harmonics, which is usually a symptom for the ET disease. (**a**) Critical example. (**b**) Counter example

The knowledge elicitation process consisted of 19 iterations. During the process, 17 new attributes were included into the domain. All new attributes were based on the explanations given by the expert and were derived either from the original attributes or from the qualitative evaluations of the drawn spirals.

3.3 *Results and Motivation for* ParkinsonCheck

We evaluated the final ABML model and compared it to some other standard machine learning methods. The evaluation on a separate and independent testing set displayed that ABML performs better when compared to some other techniques. For example, the classification accuracy of the final ABML model was 0.91, while the normal rule learning algorithm (without arguments) achieved 0.82 classification accuracy.

Along with the quantitative evaluation, we were also interested in comprehensibility of the learned rules. This feature is essential to a DSS, where all automatically suggested decisions must be grounded in some reasons. We asked two neurologists (other than our expert in the knowledge elicitation process) to independently evaluate each of the finally learned rules and provide their opinions. We were quite happy as they found all the conditions in the rules to correctly indicate the predicted class. A further look at the final set of rules reveals another interesting result. Of the 13 rules, nine contain attributes dealing with spirography tests, and of those nine, three are based exclusively on spirography. The following rule is one of them:

IF Harmonics = *true* THEN *class* = ET or MT;

The rule states that if there are harmonic frequencies in the tremor frequency spectra, then the tremor is essential. It is known that the appearance of harmonic frequencies is very specific for ET.

It is important to note that during the knowledge elicitation there was no special incentive to use specifically spirographic data. This suggests that spirography is very useful for the task at hand. The three rules also work very well on eight patients that had no clinical data. Furthermore, as studies imply that even some of the clinical attributes could be deduced from drawn spirals [29], an idea was born to develop an application that would enable self-diagnosis of tremors through spirography only. The application should be easy-to-use and would be able to accurately detect some signs of a Parkinsonian disease or essential tremor. The application and its development are described in the following section.

4 ParkinsonCheck Mobile Application

One of the results of the research described in the previous section, perhaps somewhat surprising, was how prominently the spirographic attributes featured in the constructed DSS for differentiating between PT and ET. The fact that these attributes were observed in more than a half of the decision rules and the fact that spirography is a relatively easy-to-do-yourself test gave rise to the idea of creating a mobile application for detecting signs of PT and ET. This was additionally made possible by the steady advancement in capabilities of modern multimedia mobile devices and by the observation that even some "non-spirographic" attributes of the DSS could be approximated with spirography or other self-tests. This latter observation was already discussed in Sect. 2.3.

This section briefly describes the application itself, and some selected attributes of its decision support system that are key to understanding the next section. An interested reader can learn more about the inner workings of PARKINSONCHECK's DSS in [28].

4.1 The Application

The PARKINSONCHECK application[1] is a decision support system used to detect indicators of either PT or ET. It was implemented for most of the operating systems used on mobile devices, which makes it generally accessible to the public. We hope that a preemptive use of this application will result in sooner identifications of the disease, making physicians able to observe and treat patients in the early stages of the disease. Figure 5 shows the graphical user interface of the version implemented for mobile devices.

PARKINSONCHECK is based on spirography along with some additional data: age, handedness, and medications taken being the most important. The spirography results in several time series, measuring X and Y coordinate of each drawn point and its exact timestamp. For our purposes, the spirography test was made using mobile devices with touch screens. Unlike the standard digitalised spirography, the subjects were asked to draw the spirals using their fingers and not a stylus. The reason

Fig. 5 The graphical user interface of the Parkinson*Check* application. The Figure shows the main menu, the window for drawing a spiral, and the visualisation of diagnostic results

[1] A web service version is available at http://www.parkinsoncheck.net/pc.

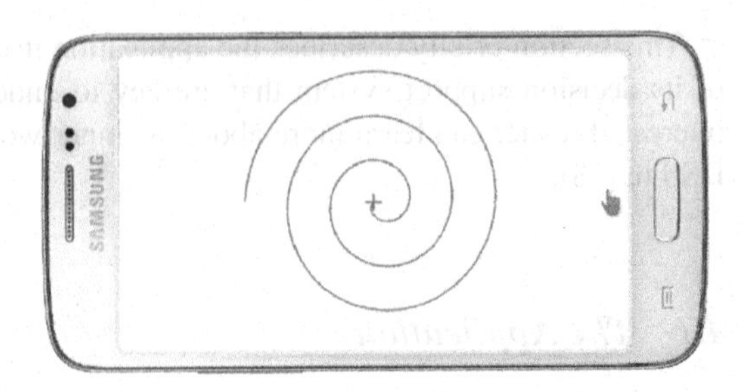

Fig. 6 Drawing area of the ParkinsonCheck app with the displayed template

was that only a minority of people own styluses for touch screen phones. Another significant change is the limited size of the drawing area. For this reason, our final spiral templates consisted of only three turns. Our comparison with larger templates (of five turns) revealed no difference in the accuracy of the results. A screenshot of the drawing area with the template displayed in it is shown in Fig. 6.

4.2 *How the Application is Used?*

While the application should be properly used to give valid results, it was designed with ease-of-use first mindset. It includes brief but comprehensive instructions at the beginning and at every step of the testing procedure, also available as spoken instructions. One very important component is also the practice mode where the user familiarises him or herself with the drawing task. The user is guided to first read the instructions, then practice drawing, and only then take the test by the layout of the main menu.

The application is equipped with automatic detection of some drawing mistakes by the user that would critically affect the test results. Two such mistakes that were also exposed by the visualisation methods are drawing in the wrong direction (clockwise) and starting the spiral too far from the centre of the screen marked with a cross.

Every user is asked to provide data on age, handedness, and medications taken, and to draw four spirals on a mobile phone and/or tablet touch screen. The first task is to follow the line of an already drawn Archimedean spiral (a template) with three turns using a right hand finger. The second task is to draw an Archimedean spiral without a template using the right hand finger. Tasks three and four are the same as one and two except that they are performed with the left hand finger.

After the user completes all four drawing steps, the results are immediately calculated and presented. If the results exhibit signs of any disorders, the user is encouraged to retake the test, and consult the physician. The results, including the given data and drawn spirals, are stored locally on the device and can be managed (also deleted) by the user at a later time.

4.3 The Prediction Model

The prediction model of the DSS was built using knowledge elicitation, feature construction and machine learning. The main learning data set consisted of 124 subjects which were diagnosed/checked at the University Medical Centre Ljubljana. These subjects included both PD and ET patients as well as healthy individuals. The latter were also checked by the neurologists to confirm that they do not exhibit any movement disorders (some mostly harmless tremors do occur in healthy people as well). The testing data also included the larger templates with five turns which are omitted from the final application as they do not improve the application's accuracy.

From the gathered data we first computed five transformations of the original spirals that are usually used in spirography: *absolute speed* is the actual speed of drawing, *polar coordinates* represents spirals in angle-radius graph (an optimal Archimedean spiral corresponds to a line in the polar coordinates), *radial speed* is the change in radius over time, *angular speed* measures the change in angle over time, and the *frequency analysis* plots the power spectrum density of the absolute speed signal.

From the six representations of a spiral (the original and five transformations), we implemented 207 descriptive attributes that can be categorised into six groups: symmetry group, extrema group, error group, frequency group, radial group and multi-test group. The implementation of these attributes was mostly motivated by medical background knowledge. Experts' background knowledge is based on their experience with digitalised spirography, which they have been using in clinical practice for several years. A large extent of this knowledge was successfully elicited in the study described in Sect. 3. We were aware of many characteristics of the spirals associated with certain diagnoses, e.g. saw-like patterns, sharp square-like edges, presence of peaks and harmonics in the spectral density graph, compressions and asymmetry of the spirals on one side, asymmetry between left/right hand spirals, etc.

To select the best classifying method, we tested several machine learning methods on the provided data and selected a logistic regression model as the one to be implemented in the system. The classification accuracy of the final method evaluated with cross-validation was 0.835, and the area under curve (AUC) equaled 0.930. We preferred the logistic regression not only because it performs best, but as it learns a linear model. This makes it easier to analyse which of the implemented attributes are deemed more important by the model, hence which of the attributes should be used in the visual explanation of classifications.

4.4 Description of Selected Attributes

In the following paragraphs, we give a short description of the attributes that are discussed in Sect. 5, where we detail some interesting cases of visualisation. We selected these attributes based on their importance in the prediction model (their corresponding linear coefficients where amongst the highest) and since they adhere to the constraints of our visualisation method (only attributes that have cumulative properties can be visualised with the proposed method). An interested reader can find full descriptions of all attributes and their corresponding groups in [28].

The attribute *Rad.sp.stdevP* (from the extrema group) is related to the direction changes (peaks) in the *radial speed* graph. A peak is a local extreme (minima and maxima). The attribute measures the standard deviation of the absolute differences between two neighbouring extremes. A large value *Rad.sp.stdevP* implies larger fluctuation in the radial speed.

The attributes `err` (from the error group) compute the root mean squared error (RMSE) between an "optimal spiral" and the actual spiral in polar coordinate space. As we actually do not know the optimal spiral, however we know that it should be a line in polar coordinates, we approximated it with a least squares fit to the data of the actual spiral represented in polar coordinates.

The attribute `percNeg` (from the radial group) measures the percentage of the spiral length when the user is drawing towards the centre. This is simply the percentage of the length when the radial speed is below zero. When drawing the ideal Archimedean spiral, the radius is constantly increasing, thus the drawing is never towards the centre. Obviously, even the healthy users occasionally have to correct the direction and draw towards the centre, however, this is much less frequent than with users with impaired movement.

5 Visual Debugging of ParkinsonCheck Attributes

In this section we introduce and discuss an attribute-based visualisation of spirals from the digitalised spirography as used for development, debugging, and assessment of ParkinsonCheck attributes, building on [12]. To design an expert system to automatically classify the spiral drawings and user data, knowledge extraction from the experts is necessary. However, such knowledge is initially in the form of relatively vague definitions and needs to be properly translated into precisely defined attributes. For this purpose, the possibility to visually appraise the attributes and their contributions to classification (decision making) is extremely helpful. This is a natural consequence of the power of visualisation in humans.

To visually explain the classification of a time-series (the spiral in our particular case), we need to identify parts of the spiral that significantly affect the values of attributes. For example, if we have an attribute measuring the distance between the drawn and the optimal spiral, the selected parts would be those that significantly differ from the optimal spiral. Hence, when these parts are identified, they are provided as an explanation for the classification, and a knowledge engineer or a physician needs to check the explanation. If the explanation is in accordance with the background knowledge, we can expect that attributes were properly defined. In the other case, we need to re-implement the attributes considering the problems in the explanation.

It turned out that visual debugging is not the only benefit of attribute visualisation of time-series. The visualization itself is also a valuable tool for physicians to understand the reasoning of the decision support system. Potentially, it allows them to discover new knowledge when a computer-generated explanation is reasonable, however different it may be to a traditional physician's explanation.

5.1 *Method for Visualisation of Contributing Factors of Attributes*

The type of visualisation depends on the type of an attribute one would like to show. In this section we will present some of the methods for visualisation that we find appropriate.

5.1.1 Discovery of the Most Contributing Sectors Using a Floating Window

One of the approaches is to visualise the sectors of the spiral, which contribute the most to the final value of the attribute in question. We used this kind of visualisation for attributes `radSpeed.stdevP`, `radSpeed.percNeg`, and `err` already presented in Sect. 4.4. However, the approach slightly differs for some attributes.

For the first two attributes, we used the data describing the change of radial speed in time. We calculated the attribute values using only the data from a sliding window interval. The results were stored together with the starting points of the sliding window.

This is a bit different for `err` and similar attributes, which represent a result of comparing a drawn time-series to an optimal time-series. For calculating `err` attribute we transform the drawn spiral from Cartesian (x vs. y) coordinate system to polar (radius vs. angle) coordinate system by calculating the values for radius and angle as shown below.

$$radius = \sqrt{x^2 + y^2} \tag{1}$$

$$angle = tan^{-1}\left(\frac{y}{x}\right) \tag{2}$$

The Archimedean spiral, which is in our case deemed as the optimal spiral, is represented as an ascending straight line in the polar coordinate system. If we would calculate the values of the `err` attribute on the intervals, we would not be able to compare the points to the desired optimal time-series. To compare the drawn spiral with the optimal one, we first had to calculate the values of the desired (optimal) spiral and then calculate the mean squared error (MSE) for each point in a time-series. We used this data to calculate the sum of MSEs of all the points on an interval using a sliding window.

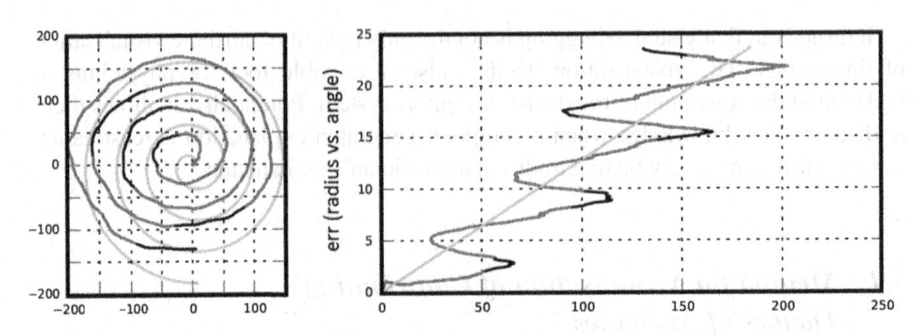

Fig. 7 An example of the drawn spiral in the cartesian and polar coordinate system (*medium grey*) plotted together with the optimal spiral (*light grey*). Highlighted with *black* is an `err` attribute

After calculating the attribute values on each interval, we sorted the intervals by their contribution to the total attribute value. We plotted the points of n intervals which had the largest contribution to the attribute value on the graph.

Plotting the intervals was done only for the attributes, which were relevant for the diagnosis of each case. We visualised the attribute only if its value for the considered case was over the threshold defined in the diagnostic model.

Since it is not easy to understand the `err` attribute and what it actually measures, we plotted the optimal spiral on the same picture as shown in Fig. 7. This made the `err` attribute clearer.

It is not always possible to calculate the attribute value on a small number of points (e.g. for drawing frequencies). In this case we would like to use every data we can get. We decided to try this approach on `radSpeed.stdevP` attribute. First we calculated the average value of an attribute on the whole time-series. We used this value for replacing the value of the points in a floating window. The value of an attribute was then calculated on the "corrected" time-series. When plotting the most contributing parts we now took the starting points of n lowest attribute values, as these actually represent the most contributing parts of the time-series.

5.1.2 Discovery of the Most Contributing Sectors Without Using a Floating Window

This approach is used for visualising all points of a time-series which increase the value of an attribute. In our case, those were the attributes `radSpeed.percNeg0` and `radSpeed.percNeg005`. Both attributes have also been visualised using a method presented in Sect. 5.1.1. The reason for two different visualisations of the same attribute lies in its informative value. Patients with ET usually draw saw-like spirals, which means they often draw towards the center of the spiral. With this approach we can mark all these irregular parts. An example is shown in Fig. 12. However, if the spiral has been drawn by a PD patient, using the method from Sect. 5.1.1 is more appropriate as the spirals are usually smoother.

For this type of visualisation we took all points and the corresponding times from the time-series whose values for radial speed were below zero (or below −0.05 for `radSpeed.percNeg005`). We plotted these points on a radial speed graph and on the spiral as shown later on Fig. 12.

5.2 *Benefits of Visualisation*

In this section we describe the benefits of visualisation through a series of real-life examples encountered while designing and later improving the PARKINSONCHECK application [28]. We see three distinct types of benefits: (a) for explanation of DSS's classification, (b) for "visual debugging", and (c) for discovering new knowledge in the domain.

5.2.1 Example 1: Explanation of Classification

The first example presents a classic use of visualisation for the purpose of explanation of DSS's classification of a spiral. Figure 8 shows a typical Parkinsonian spiral. One of the visual clues for a neurologist are the condensed lines on one side of the spiral. The medical explanation for this visual effect is rigidity (stiffness and resistance to limb movement), which in turn is one of the four principal motor symptoms of PD (and not a symptom of ET). A person affected with limb rigidity experiences trouble drawing in a given direction which results in spiral lines being close together in that direction. This is clearly visible in the upper right part of the spiral in Fig. 8. Note that the outermost turn of the spiral to the bottom also looks condensed, however, this is due to the smartphone's screen limitation; this is easily detected from the spiral's coordinates.

The sections of the spiral in Fig. 8 highlighted in black represent parts of the spiral contributing the most to the high value of the attribute `err`. They coincide remarkably well with the condensed lines in the upper right quadrant of the spiral.

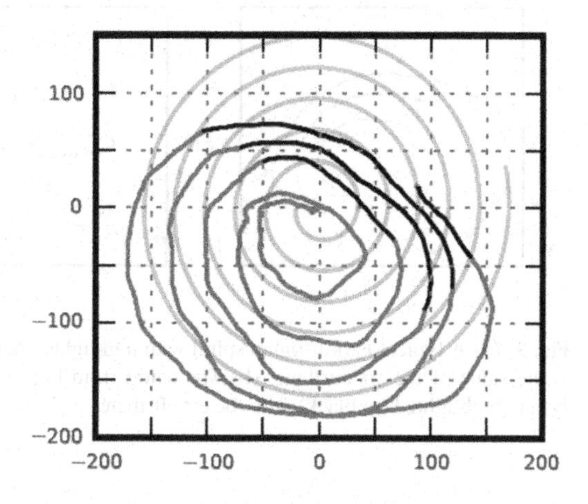

Fig. 8 A typical Parkinsonian spiral highlighted with the `err` attribute

Such a visualisation of the attribute `err` can immediately alert the physician to an important detail, which in some cases is not as obvious as in this typical example. It should be noted that this attribute does not target condensed parts alone, this is just one irregularity that it can detect (also see next paragraph).

This particular visualisation proved very interesting also to the developers of the DSS. The attribute `err` was not designed specifically to describe condensed parts of the spiral. Its intention was to describe the general misfit of the drawn spiral with the ideal (template) spiral. The logic behind this attribute is that people with motor disorders will have a harder time following a template: an unwanted move due to tremor would cause the finger to go astray of the ideal course. However, the visualisation showed that this attribute is also very useful for finding the condensed parts. This promptly explained why all attempts to design an attribute specifically targeted at detecting the condensed parts failed to further improve the classification accuracy of the DSS.

Figure 9a visually explains how (or why) the attribute `err` actually detects the condensed parts of the spiral. On the figure we see an optimal spiral along with the one drawn by the user. The optimal spiral is a reverse transformation into Cartesian coordinate system of the ideal fit (what the `err` attribute actually calculates) to the drawn spiral in polar coordinates. Part (b) of the figure shows the polar coordinate representation where the straight line is the best fit. Highlighted in black are the regions that most contribute to the high values of `err`. The reason why `err` detects condensed parts is now obvious: these are the parts that are not even close to their supposed turn. This was interesting to DSS developers.

The spiral in Fig. 9a was drawn without the template (freehand test). The same person's spiral with the template is shown in Fig. 10. If we compare the two spirals, we see that only the freehand drawing is condensed. The visualisation nicely confirms (and conforms to) the known fact that "Bradykinesia, a consequence of dysfunction of the basal ganglia, is more profound when no visual cues are available as the involvement of the cerebellar pathways are then largely redundant [16, 18]."

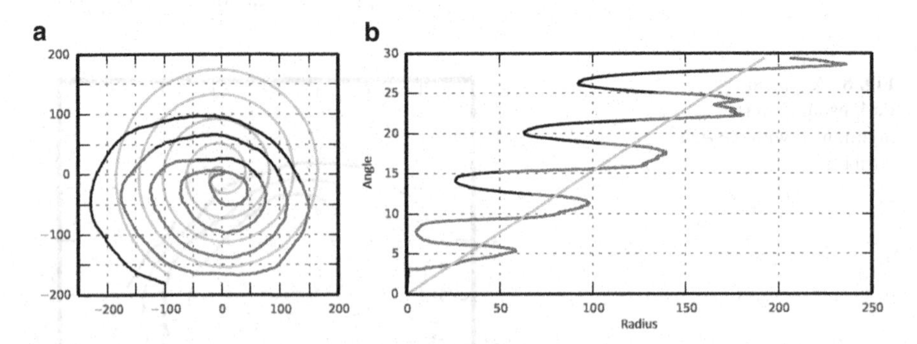

Fig. 9 (**a**) A typical Parkinsonian spiral with highlighted parts (in black) that contribute the most to the value of the err attribute. (**b**) The spiral from Fig. 9a represented in the polar coordinate system highlighted (in black) with the err attribute

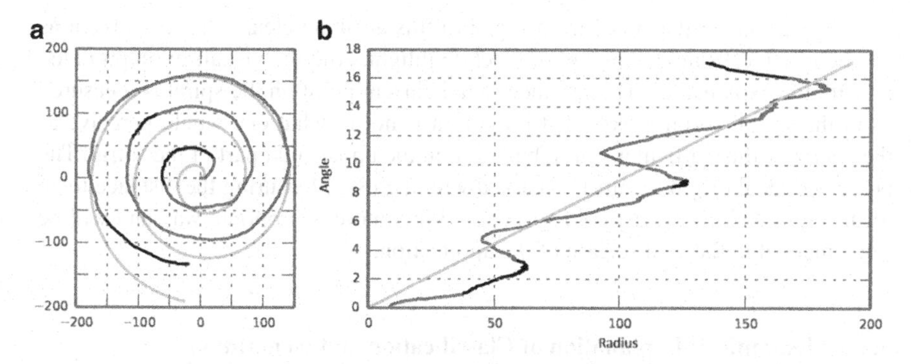

Fig. 10 (**a**) A spiral drawn by the same subject as the one in Fig. 9 with the template highlighted with the err attribute. (**b**) The spiral from Fig. 10a represented in the polar coordinate system highlighted with the err attribute

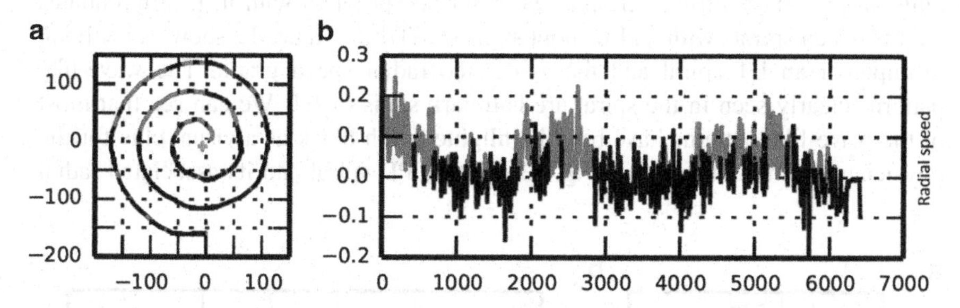

Fig. 11 (**a**) An almost perfect spiral and (**b**) the associated radial speed graph both highlighted with the `radSpeed.percNeg` attribute. The cross in the centre of the Fig. 11a represents an ideal origin of the spiral

5.2.2 Example 2: Visual Debugging

This example demonstrates the use of visualisation for the purpose of visual debugging. In Fig. 11a, we can see an example of an almost perfect spiral. So, if the spiral is almost perfect, why is half of it highlighted in black, signifying something is wrong with it? The answer lies in the graph in Fig. 11b. The graph plots the radial speed of drawing the spiral over time in milliseconds. If the user is drawing *towards* the centre of the spiral, the radial speed is negative and vice versa. The black highlight on both the spiral and the radial speed graph in Fig. 11 are sections where the radial speed is negative.

The attribute `radSpeed.percNeg` measures the percentage of time the user draws with negative radial speed. It proved to be an important attribute, because normally the distance from the spiral's origin (radius) should be monotonically increasing. Any decrease is either due to unwanted movement or a healthy user trying to compensate for increasing the radius too much in the previous section of the spiral. Nonetheless, this latter case manifests itself in smaller percentage of negative speed than the case with unwanted movement due to illness.

However, given the spiral from Fig. 11a this attribute clearly has a problem as can be inferred from the amount of black highlight. And the visualisation provided us with the explanation. The distance to the current point on the spiral is measured from the spiral's origin. Note that the origin is not in what one would perceive as the centre of this particular spiral, but more in the upper right part of the spiral. The perceived ideal origin (centre) is marked with a cross. Measuring the distance to an ideal origin defines `radSpeed.percNeg` in a more appropriate way and will be implemented in the next release of the application.

5.2.3 Example 3: Explanation of Classification and Generation of New Knowledge

For reasons mentioned in the previous Example (5.2.2), `radSpeed.percNeg` is quite an important attribute despite the described problem with it. It differentiates well between spirals with and without signs of ET/PD. Figure 12 shows a textbook example of an ET spiral and the associated radial speed graph. The wave-like patterns clearly seen in the spiral are hallmark signs of ET. We can see that most of the wave-like patterns have been highlighted with black as a visual clue for the physicians. On the other hand, Fig. 13 shows a PD spiral and its associated radial

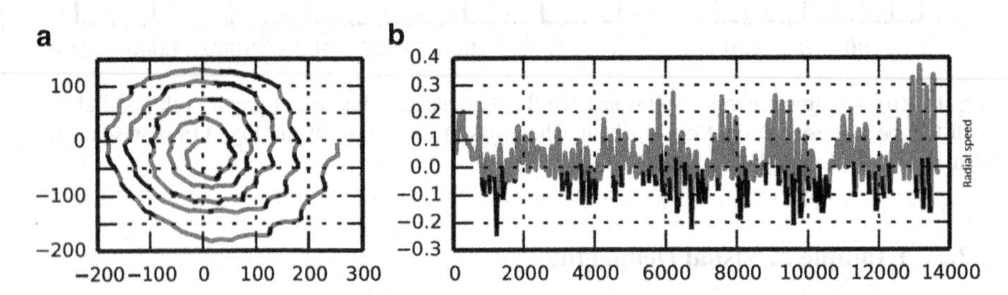

Fig. 12 (**a**) A typical essential spiral and (**b**) the associated radial speed graph both highlighted with the `radSpeed.percNeg` attribute

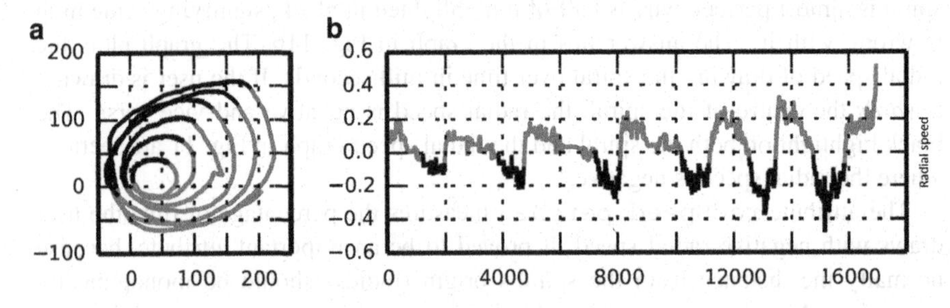

Fig. 13 (**a**) A typical Parkinsonian spiral and (**b**) the associated radial speed graph both highlighted with the `radSpeed.percNeg` attribute

speed graph. The black highlight in this case marks sections on the spiral just before the condensed parts (explained in Example 5.2.2) occur.

This example also demonstrates new knowledge generation. In this case, it is not new medical knowledge, but an idea how a new attribute based on `radSpeed.percNeg` can be constructed to differentiate between ET and PD. A high value of `radSpeed.percNeg` is indicative of the spiral containing signs of either PD or ET, but it does not differentiate between the two conditions. Yet looking at Figs. 12 and 13 we can see that the black areas marking when the user was drawing towards the centre are distributed quite differently. We can observe that for ET, the total value of `radSpeed.percNeg` is composed of many small regions on the spiral (waves), while for PD it is composed of a smaller number of longer regions. An attribute taking this difference into account could thus differentiate between ET and PD.

In fairness, the general idea about this new attribute surfaced before the visualisation of `radSpeed.percNeg`, yet the latter vividly confirmed our reasoning. This new attribute will be implemented in the next release of the application.

5.2.4 Example 4: Explanation of Classification

Figure 14 presents another typical ET spiral and the accompanying radial speed graph. This time the black highlights on the graph (and consequently the spiral) represent the parts most contributing to the high value of the attribute `radSpeed.stdevP`. This attribute measures how erratic are the local extrema on the radial speed graph; in essence how often and how much the user changes the direction (increase/decrease) of radius while drawing the spiral. It can be seen in the figure how well the highlighted areas correspond with wave-like distortions on the spiral. Note that only the areas most contributing to `radSpeed.stdevP` are highlighted, so not all the distortions are marked. We believe this is a better way of showing the visual clues to the physician, pointing out just the most important parts.

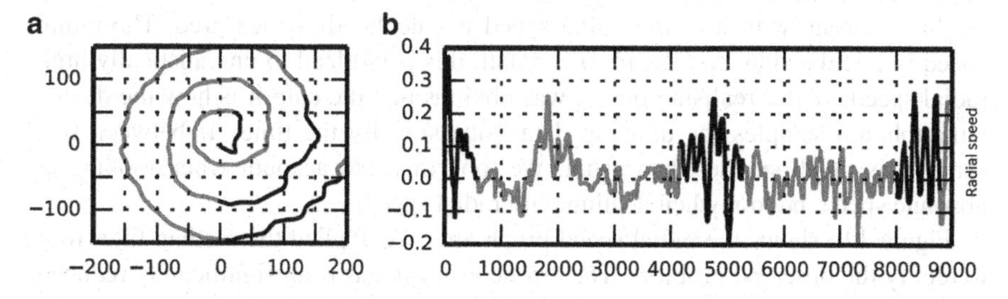

Fig. 14 (**a**) A typical essential spiral and (**b**) the associated radial speed graph both highlighted with the `radSpeed.stdevP` attribute (with the median filter applied)

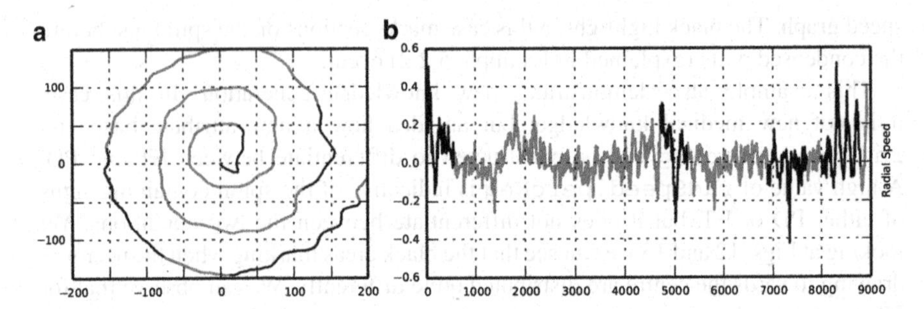

Fig. 15 (**a**) A typical essential spiral and (**b**) the associated radial speed graph both highlighted with the `radSpeed.stdevP` attribute (without the median filter applied)

In an interesting twist, this particular visualisation helped explain what is the contribution of `radSpeed.stdevP` to the developers of the DSS. The attribute `radSpeed.stdevP` was initially included only as an easy-to-program prototype and was not given much thought as it was believed that it will later be replaced with a more sophisticated (or targeted) one. To everyone's surprise it remained important in the final model. Our hypothesis is that such a primitive attribute is more robust than most targeted ones. While seemingly an odd case, such a situation with initially incomprehensible attributes can arise quite often: automatic feature construction is an obvious example of how this can happen.

5.2.5 Example 5: Visual Debugging

The previous example's highlighted areas (given in Fig. 14) initially looked as shown in Fig. 15. Comparing the two figures (spirals) one notices that initially a relatively unimpaired section in the lower left quadrant of the spiral was highlighted. This seemed strange and prompted us to investigate why this occurs.

The radial speed graph in Fig. 15b shows a very big negative peak at around 7,400 ms. As this large peak does not correspond to a significant decrease of radius in the spiral (in the lower left highlighted section) we concluded that something might be wrong with how the radial speed is calculated or measured. The radial speed is calculated as $\Delta r/\Delta t$; if Δt is small, this could lead to unrealistically high radial speeds. After realising this, it was obvious that the culprit is how the device (smartphone) samples the drawing data—occasionally the time Δt between two samples is much smaller than usual. We took this into account when looking at absolute speed, but not when dealing with radial speed.

Figure 14b shows the radial speed graph after we applied the median filter to it to rectify the observed problem. The erroneous peak has been significantly reduced and the highlighted areas show the improved functioning of `radSpeed.stdevP`.

6 Conclusions

In 2011 we wrote "Spirography was revealed to have real potential for tremor diagnosis, in conjunction with clinical data, but also as a standalone early screening method." This was one of the conclusions of the first study [11] briefly presented in this chapter. Unbeknownst at that time, this idea was later put into practice in no small part also due to advancement of mobile multimedia devices. The increased capabilities of such devices enabled spirography, an easy-to-do-yourself test for signs of PD and ET, become available to everyone in the comfort of their own home. Multimedia, sensors, graphics, and artificial intelligence (AI) all played an important part in achieving this.

There are several benefits of PARKINSONCHECK. As anyone concerned can use it at any time, the application can act as an early warning system. It can also be used for triage. The analysis is performed on the device itself, and there is no need to send any data anywhere or to anyone. This allows full privacy and reduces the cost as there is no need for a physician to evaluate the drawings and related data. As the analysis is performed without any human intervention, it is available immediately after taking the test. The application can serve as a spirographic test anywhere and anytime for free (in Slovenia for now), replacing other equipment. This latter point is the result of modern mobile multimedia devices and can play an even bigger role with the intended continuation of the project: an upgrade for monitoring the patients objectively. As PD is a chronic disease, monitoring is a very important issue. Even more so, because the treatment dosage is not easy to optimally determine as it changes with disease progression.

The monitoring objective, but also the original intention of PARKINSONCHECK is subject of further research and improvement. Sound is one of the possibilities to improve the accuracy of the application for both, detection and monitoring. PD also affects vocal cords and this manifests itself in the way the patient speaks, this can also be picked up by smartphones and tablets for analysis.

The main focus of the chapter, however, was the use of visualisation in PARKINSONCHECK's attribute construction and assessment. The relatively simple method proved very helpful for debugging the attributes and the model of a concrete decision support system by means of provided visual explanations. Note that not all spirals are so easily classified as the ones presented in this paper—these are more or less textbook examples.

As people are very visual beings, similar methods of visual debugging can be extremely powerful when dealing with complex systems where various subsystems/calculations are entangled or depend on one another. A good illustration might be Example 5.2.5 in Sect. 5.2. The attribute in the example depended on several calculations (and in itself all were correct) and assumptions about measurement (small interval between two samples was detected). While not very easy to debug in a conventional way, it was quite straightforward to analyse the situation once it was adequately visualised. Without the visualisation such an attribute could easily be discarded as not promising by feature subset selection for example. We believe this warrants further research unlocking the potential of visual debugging.

Another benefit of visualisation is generation of new domain knowledge—be it for domain experts or knowledge engineers coding such knowledge. Sometimes the attributes are unintelligible as they were not obtained using background knowledge; one can easily imagine the situation with automated feature construction. If such attributes prove powerful and can be nicely visualised, new knowledge for domain specialists is immediately gained.

Acknowledgements The research was partly funded by Slovenian Research Agency (ARRS). The development of ParkinsonCheck was partly funded by Slovenian Ministry of Education, Science and Sport and European Regional Development Fund.

References

1. Alberio, T., & Fasano, M. (2011). Proteomics in parkinson's disease: An unbiased approach towards peripheral biomarkers and new therapies. *Journal of Biotechnology, 156*(4), 325–337. Special issue: {IBS} 2010 - Industrial Biotechnology.
2. Alusi, S. H., Worthington, J., Glickman, S., Findley, L. J., & Bain, P. G. (2000). Evaluation of three different ways of assessing tremor in multiple sclerosis. *Journal of Neurology, Neurosurgery & Psychiatry, 68*(6), 756–760.
3. Bain, P. G., & Findley, L. J. (1993). *Assessing tremor severity: A clinical handbook*. Standards in Neurology. London: Smith-Gordon.
4. Clark, P., & Boswell, R. (1991). Rule induction with CN2: Some recent improvements. In *EWSL* (pp. 151–163).
5. Deuschl, G., Bain, P., & Brin, M. (1998). Consensus statement of the movement disorder society on tremor. *Movement Disorders, 13*(S3), 2–23.
6. Deuschl, G., Wenzelburger, R., Loffler, K., Raethjen, J., & Stolze, H. (2000). Essential tremor and cerebellar dysfunction clinical and kinematic analysis of intention tremor. *Brain, 123*(8), 1568–1580.
7. Elble, R. J., Brilliant, M., Leffler, K., & Higgins, C. (1996). Quantification of essential tremor in writing and drawing. *Movement Disorders, 11*(1), 70–78.
8. Elble, R. J., Sinha, R., & Higgins, C. (1990). Quantification of tremor with a digitizing tablet. *Journal of Neuroscience Methods, 32*(3),193–198.
9. Feys, P., Helsen, W., Prinsmel, A., Ilsbroukx, S., Wang, S., & Liu, X. (2007). Digitised spirography as an evaluation tool for intention tremor in multiple sclerosis. *Journal of Neuroscience Methods, 160*(2), 309–316.
10. Filipová, M., Filip, V., Macek, Z., Müllerová, S., Marková, J., Kás, S., et al. (1988.) Terguride in parkinsonism a multicenter trial. *European Archives of Psychiatry and Neurological Sciences, 237*(5), 298–303.
11. Groznik, V., Guid, M., Sadikov, A., Možina, M., Georgiev, D., Kragelj, V., et al. (2013). Elicitation of neurological knowledge with argument-based machine learning. *Artificial Intelligence in Medicine, 57*(2), 133–144.
12. Groznik, V., Sadikov, A., Možina, M., Žabkar, J., Georgiev, D., & Bratko, I. (2014). Attribute visualisation for computer-aided diagnosis: A case study. In *Proceedings of 2014 IEEE International Conference on Healthcare Informatics (ICHI 2014)* (pp. 294–299). IEEE Computer Society, US Conference Publishing Services.
13. Gustavsson, A., Svensson, M., Jacobi, F., Allgulander, C., Alonso, J., Beghi, E., et al. (2011). Cost of disorders of the brain in Europe 2010. *European Neuropsychopharmacology, 21*, 718–779.
14. Hubble, J. P., Busenbark, K. L., Wilkinson, S., Penn, R. D., Lyons, K., & Koller, W. C. (1996). Deep brain stimulation for essential tremor. *Neurology, 46*(4), 1150–1153.

15. Hughes, A. J., Daniel, S. E., Kilford, L., & Lees, A. J. (1992.) Accuracy of clinical diagnosis of idiopathic Parkinson's disease: A clinico-pathological study of 100 cases. *Journal of Neurology, Neurosurgery, and Psychiatry, 55*(3), 181–184.
16. Jueptner, M., & Weiller, C. (1998). A review of differences between basal ganglia and cerebellar control of movements as revealed by functional imaging studies. *Brain, 121*(8), 1437–1449.
17. Kraus, P. H., & Hoffmann, A. (2010). Spiralometry: Computerized assessment of tremor amplitude on the basis of spiral drawing. *Movement Disorders, 25*(13), 2164–2170.
18. Liu, X., Carroll, C. B., Wang, S.-Y., Zajicek, J., & Bain, P. G. (2005). Quantifying drug-induced dyskinesias in the arms using digitised spiral-drawing tasks. *Journal of Neuroscience Methods, 144*(1), 47–52.
19. Longstaff, M. G., & Heath, R. A. (2006). Spiral drawing performance as an indicator of fine motor function in people with multiple sclerosis. *Human Movement Science, 25*(45), 474–491; Advances in graphonomics: Studies on fine motor control, its development and disorders.
20. Louis, E. D., Yu, Q., Floyd, A. G., Moskowitz, C, & Pullman, S. L. (2006). Axis is a feature of handwritten spirals in essential tremor. *Movement Disorders, 21*(8), 1294–1295.
21. Miralles, F., Tarong, S., & Espino, A. (2006). Quantification of the drawing of an archimedes spiral through the analysis of its digitized picture. *Journal of Neuroscience Methods, 152*(12), 18–31.
22. Možina, M., Guid, M., Krivec, J., Sadikov, A., & Bratko, I. (2008). Fighting knowledge acquisition bottleneck with argument based machine learning. In M. Ghallab, C. D. Spyropoulos, N. Fakotakis & N. M. Avouris (Eds.), *Proceedings of the 2008 Conference on ECAI 2008: 18th European Conference on Artificial Intelligence. Frontiers in Artificial Intelligence and Applications* (Vol. 178, pp. 234–238). Amsterdam, The Netherlands: IOS Press.
23. Možina, M., Žabkar, J., & Bratko, I. (2007). Argument based machine learning. *Artificial Intelligence, 171*(10/15), 922–937.
24. Murman, D. L. (2012). Early treatment of parkinson's disease: Opportunities for managed care. *American Journal of Managed Care, 18*(7), 183–188.
25. Pahwa, R., & Lyons, K. E. (2003). Essential tremor: Differential diagnosis and current therapy. *American Journal of Medicine, 115*, 134–142.
26. Pullman, S. L. (1998). Spiral analysis: A new technique for measuring tremor with a digitizing tablet. *Movement Disorders, 13*(S3), 85–89.
27. Quinn, N. P., Schneider, S. A., Schwingenschuh, P., & Bhatia, K. P. (2011). Tremor-some controversial aspects. *Movement Disorders, 26*(1), 18–23.
28. Sadikov, A., Žabkar, J., Možina, M., Groznik, V., Georgiev, D., & Bratko, I. (2014). *Parkinsoncheck: A decision support system for spirographic testing*. Technical report, University of Ljubljana, Faculty of Computer and Information Science.
29. Saunders-Pullman, R., Derby, C., Stanley, K., Floyd, A., Bressman, S., Lipton, R. B., et al. (2008). Validity of spiral analysis in early Parkinson's disease. *Movement Disorders, 23*(4), 531–537.
30. Stanley, K., Hagenah, J., Brggemann, N., Reetz, K., Severt, L., Klein, C., et al. Digitized spiral analysis is a promising early motor marker for parkinson disease. *Parkinsonism and Related Disorders, 16*(3), 233–234.
31. Thanvi, B., Lo, N., & Robinson, T. (2006). Essential tremor—the most common movement disorder in older people. *Age and Ageing, 35*(4), 344–349.
32. World Health Organization (2008). The global burden of disease: 2004 update. WHO Press, World Health Organization, Geneva, Switzerland.
33. Towey, D. J., Bain, P. G., & Nijran, K. S. (2011). Automatic classification of 123I-FP-CIT (DaTSCAN) SPECT images. *Nuclear Medicine Communications, 32*(8), 699–707.
34. Wang, H., Yu, Q., Kurtis, M. M., Floyd, A. G., Smith, W. A., & Pullman, S. L.(2008). Spiral analysis improved clinical utility with center detection. *Journal of Neuroscience Methods, 171*(2), 264–270.
35. Westin, J., Ghiamati, S., Memedi, M., Nyholm, D., Johansson, A., Dougherty, M., et al. (2010). A new computer method for assessing drawing impairment in parkinson's disease. *Journal of Neuroscience Methods, 190*(1), 143–148.

Meaningful Bags of Words for Medical Image Classification and Retrieval

Antonio Foncubierta Rodríguez, Alba García Seco de Herrera, and Henning Müller

1 Introduction

Image retrieval and image classification have been extremely active research domains with hundreds of publications in the past 20 years [1–3]. Content-based image retrieval has been proposed for diagnosis aid, decision support and enabling similarity-based easy access to medical information [4, 5] ranging from similar case, to similar images and similar regions of interest.

One of the main domains of image retrieval has been the medical literature with millions of images being available [6, 7]. ImageCLEFmed[1] (the medical image retrieval task of the Cross Language Evaluation Forum) is an annual evaluation campaign on retrieval of images from the biomedical open access literature [8]. In the ImageCLEF medical task, usually 12–17 teams compare their approaches each year from 2004–2013 based on a variety of search tasks [9].

The Bag-of-Visual-Words (BoVW) is a visual description technique that aims at shortening the semantic gap by partitioning a low-level feature space into regions of the features space that potentially correspond to visual topics. These regions are called visual words in an analogy to text-based retrieval and the bag of words approach. An image can be described by assigning a visual word to each of the feature vectors that describe local regions or patches of the images (either via a dense

[1] http://www.imageclef.org/.

A.F. Rodríguez • A.G.S. de Herrera • H. Müller (✉)
University of Applied Sciences and Arts Western Switzerland, Technoark 3, 3960 Sierre, Switzerland
e-mail: antonio.foncubierta@hevs.ch; alba.garcia@hevs.ch; henning.mueller@hevs.ch

A. Briassouli et al. (eds.), *Health Monitoring and Personalized Feedback using Multimedia Data*, DOI 10.1007/978-3-319-17963-6_5

grid sampling or interest points often based on saliency), and then representing the set of feature vectors by a histogram of the visual words. One of the most interesting characteristics of the BoVWs is that the set of visual words is created based on the actual data and therefore only topics present in the data are part of the visual vocabulary [10].

The creation of the vocabulary is normally based on a clustering method (e.g. k-means, DENCLUE) to identify local clusters in the feature space and then assigning a visual word to each of the cluster centers. This has been investigated previously, either by searching for the optimal number of visual words [11], by using various clustering algorithms [12] instead of the k-means or by selecting interest points to obtain the features [13].

Although the BoVW is widely used in the literature [14, 15] there is a strong performance variation within similar experiments when considering different vocabulary sizes [11]., making the choice of vocabulary size a crucial aspect of visual vocabularies that can vary very strongly. We hypothesize that this variance of the BoVW method is strongly related to the quality of the vocabulary used, understanding quality as the ability of the vocabulary to accurately describe useful concepts for the task. Therefore, we try to reduce the size of the vocabulary without reducing the performance of the method. The use of supervised clustering [16, 17] to force the clusters to a known number of classes was also considered as an option but it is against the notion of learning a variety of topics present in the data. Instead, we compute the latent semantic topics in the dataset in an unsupervised way by analyzing the probability of each word to occur. This allows to extract concepts or topics from a combination of various visual word types, since the topics are discovered based on the probability of co-occurrence of a set of visual words regardless of their origin. The resulting reduced vocabularies present two benefits over the full ones. First, a reduction of the descriptors leads to reduction of the computational cost of the online phase of retrieval but also in the offline indexing phase. This reduction becomes important in the context of large-scale databases or *Big Data*. The second benefit of the approach is that by removing non-meaningful visual words, the dataset description becomes more compact. A compact representation makes it easier to use neighbourhood-based classifiers, which tend to fail in high dimensional feature spaces due to the curse of dimensionality. Finally, a transformation of the descriptor is proposed combining the pruning of meaningless visual words and weighting meaningful words accordingly to their importance for the visual topics.

The rest of this chapter is organized as follows: Sect. 2 explains in details the materials and methods used with focus on the data set, the probabilistic latent semantic analysis and how it is used to remove meaningless visual words from the vocabulary. Section 3 contains factual details of results of the experiments run on the dataset, while Sect. 4 discusses them. Conclusions and future work are explained in Sect. 5.

2 Materials and Methods

In this section, further details on the data set and the techniques employed are given.

2.1 Data Set

Image modality filters are one of the characteristics of medical image retrieval that practitioners would like to see included in existing image search systems [18]. Medical image search engines such as GoldMiner[2] and Yottalook[3] contain modality filters to allow users to focus retrieval results. Whereas DICOM headers often contain metadata that can be used to filter modalities, this information is lost when exporting images for publication in journals or conferences where images are stored as JPG, GIF or PNG files and are usually further processed (fewer grey levels cropping, ...). In this case visual appearance is key to identify modalities or the caption text can be analyzed for respective keywords to identify modalities. The ImageCLEFmed evaluation campaign contains a modality classification task that is regarded as an essential part for image retrieval systems. In 2012, the modality classification data set contained 2,000 images from the medical literature organized in a hierarchy of 31 categories [19]. Figure 1 shows the hierarchical structure of modalities. All images in the dataset belong to a single leaf node in the hierarchy.

The modality classification dataset is divided into two subsets of 1,000 images each, one for training and one for testing. The training set and its corresponding ground truth are made public for the groups to train and optimize their methods but the comparison is performed on a test set of which the ground truth is not known by the groups. Figure 2 shows the distribution of images across modalities in the training and test sets.

Besides modality classification, an image retrieval task is also performed during the benchmarking event where independent assessors judge the relevance of each document in the pool of results submitted by the groups. The retrieval task is performed on a dataset containing the full ImageCLEFmed data set, which in 2012 consisted of more than 306,000 images form the open access biomedical literature.

Both data sets were used in the experiments described in this article. Methods were first tested on the modality classification data set (training and testing) to investigate the effect of parameters on the system. Then, fewer parameter combinations were tested on the retrieval task with a larger data base.

---

[2]http://goldminer.arrs.org/.

[3]http://www.yottalook.com/.

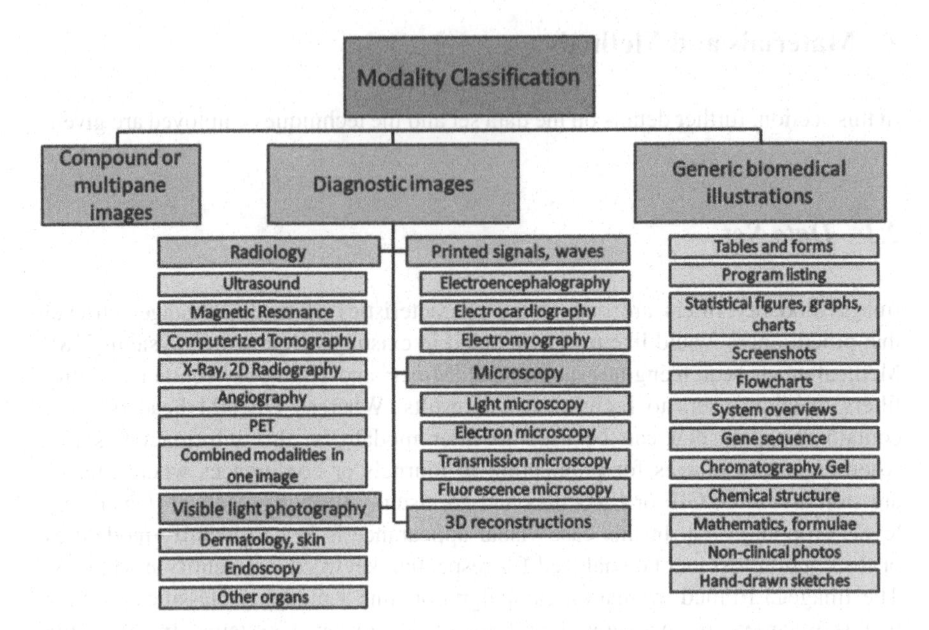

Fig. 1 Hierarchy of modalities or image types considered in the modality classification task

2.2 Descriptors

In this section, the descriptors used in our experimental evaluation are presented. Scale Invariant Feature Transform (SIFT) and Bag-of-Colors (BoC) were chosen as images descriptors.

2.2.1 SIFT

In this work, images are described with a BoVW based on their SIFT [20] descriptors. This representation has been commonly used for image retrieval because it can be computed efficiently [15, 21, 22]. The SIFT descriptor is invariant to translations, rotations and scaling transformations and robust to moderate perspective transformations and illumination variations. SIFT encodes the salient aspects of the greylevel-images gradient in a local neighbourhood around each interest point.

2.2.2 Bag of Colors

BoC is used to extract a color signature from the images [23]. The method is based on BoVW image representation, which facilitates the fusion with the SIFT-BoVW

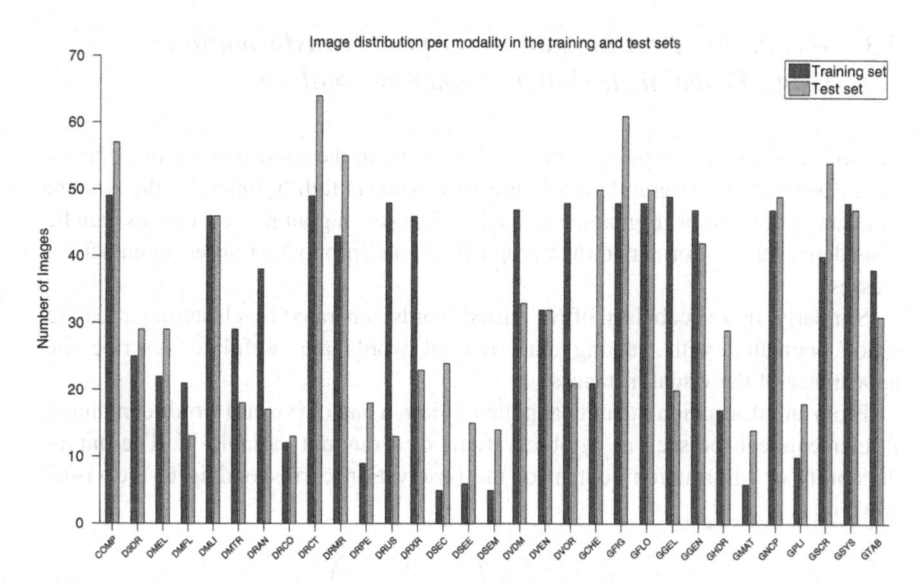

Fig. 2 Distribution of images across modalities for the modality classification training and test sets

descriptor. The CIELab[4] color space was used since it is a perceptually uniform color space [24]. A color vocabulary $\mathscr{C} = \{c_1, \ldots, c_{100}\}$, with $c_i = (L_i, a_i, b_i) \in$ *CIELab*, is defined by automatically clustering the most frequently occurring colors in the images of a subset of the collection containing an equal number of images from the various classes.

The BoC of an image I is defined as a vector $BoC = \{\bar{c}_1, \ldots, \bar{c}_{100}\}$ such that, for each pixel $p_k \in I$:

$$\bar{c}_i = \sum_{k=1}^{P} \sum_{j=1}^{P} g_j(p_k)$$

with P the number of pixels in the image I, where

$$g_j(p) = \begin{cases} 1 \; if \; d(p, c_j) \leq d(p, c_l) \\ 0 \; otherwise \end{cases} \tag{1}$$

and $d(x, y)$ is the Euclidean distance between x and y.

[4]CIELab is a color space defined by the International Commission on Illumination (Commission Internationale de l'Éclairage) describing all colors visible for humans while trying to mimic the nonlinear response of the eye.

2.3 *Vocabulary Pruning and Descriptor Transformation Using Probabilistic Latent Semantic Analysis*

In spoken or written language, not all words contain the same amount of information. Specifically, the grammatical class of a word is tightly linked to the amount of meaning it conveys. E.g. nouns and adjectives (open grammatical classes) can be considered more informative than prepositions and pronouns (closed grammatical classes).

Similarly, in a vocabulary of N_W visual words generated by clustering a feature space populated with training data, not all words are useful to describe the appearance of the visual instances.

From an information theoretical point of view, a bag of (visual) words containing L_i elements can be seen as L_i observations of a random variable W. The unpredictability or information content of the observation corresponding to the visual word w_n is

$$I(w_n) = log\left(\frac{1}{P(W = w_n)}\right) \tag{2}$$

This explains why nouns or adjectives contain, in general, more information than prepositions or pronouns. Words belonging to a closed class are more probable than those belonging to a much richer class. According to Eq. 2, information is related to unlikelihood of a word.

In a bag of visual words scheme for visual understanding it is important to use very specific words with high discriminative power. On the other hand, using very specific words alone does not always allow to establish and recognize similarities. This can be done by establishing a concept that generalizes very specific words that share similar meanings into a less specific *visual topic*. E.g. in order to recognize the similarities between the (specific) words *bird* and *fish* we need a less specific *topic* such as *animal*.

A visual topic z is the representation of a generalized version of the visual appearance modeled by various visual words. It corresponds to an intermediate level between visual words and the complete understanding of visual information. A set of visual topics $\mathscr{Z} = \{z_1, \ldots, z_{N_Z}\}$ can be defined in a way that every visual word can belong to none, one or several visual topics, therefore establishing and possibly quantifying the relationships among words (see Fig. 3).

2.3.1 Probabilistic Latent Semantic Analysis

Visual words are often referred to as an extension of the bag of words technique used in information retrieval from textual to visual data. Similarly, language modelling techniques have also been extended from text to visual words-based techniques [25, 26].

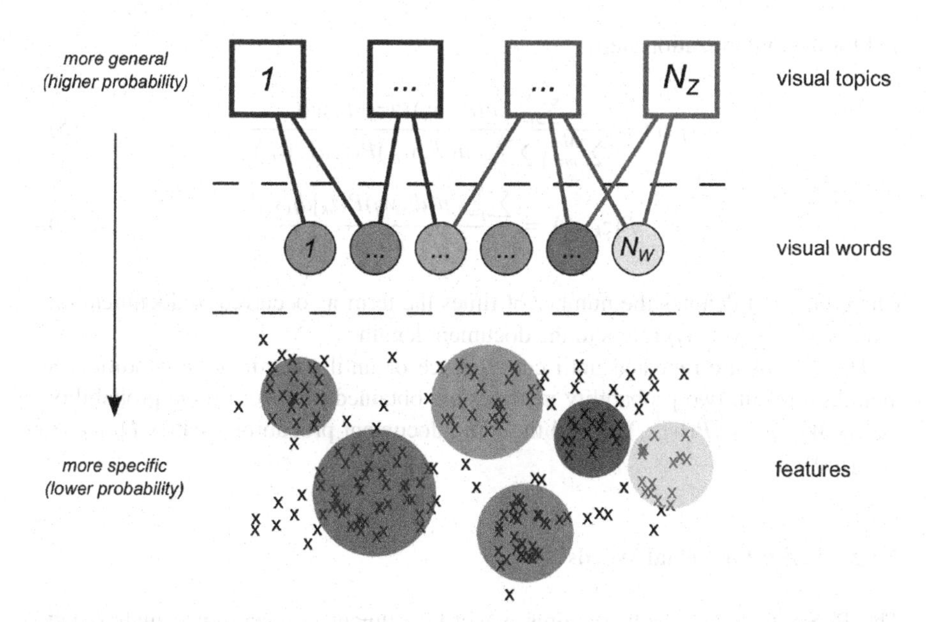

Fig. 3 Conceptual model of visual topics, words and features. Whereas continuous features are the most informative descriptors from an information theoretical point of view, visual words generalize feature points that are close in the feature space. We propose visual topics as a higher generalization level, modelling partially shared meanings among words

Latent Semantic Analysis (LSA) [27] is a language modelling technique that maps documents to a vector space of reduced dimensionality, called *latent semantic space*, based on a Singular Value Decomposition (SVD) of the terms-documents co-occurrence matrix. This technique was later extended to statistical models, called *Probabilistic* Latent Semantic Analysis (PLSA), by Hofmann [28]. PLSA removes restrictions of the purely algebraic former approach (namely, the linearity of the mapping).

Hofmann defines a generative model that states that the observed probability of a word or term $w_j, j \in 1, \ldots, M$ occurring in a given document $d_i, i \in 1, \ldots, N$, is linked to a latent or unobserved set of concepts or topics $\mathscr{Z} = \{z_1, \ldots, z_K\}$ that happen in the text:

$$P(w_j|d_i) = \sum_{k=1}^{K} P(w_j|z_k)P(z_k|d_i). \tag{3}$$

The model is fit via the EM (Expectation-Maximization) algorithm. For the expectation step:

$$P(z_k|d_i, w_j) = \frac{P(w_j|z_k)P(z_k|d_i)}{\sum_{l=1}^{K} P(w_j|z_l)P(z_l|d_i)}. \tag{4}$$

and for the maximization step:

$$P(w_j|z_k) = \frac{\sum_{i=1}^{N} n(d_i, w_j)P(z_k|d_i, w_j)}{\sum_{m=1}^{M}\sum_{i=1}^{N} n(d_i, w_m)P(z_k|d_i, w_m)}, \tag{5}$$

$$P(z_k, d_i) = \frac{\sum_{j=1}^{M} n(d_i, w_j)P(z_k|d_i, w_j)}{n(d_i)}. \tag{6}$$

where $n(d_i, w_j)$ denotes the number of times the term w_j occurred in document d_i; and $n(d_i) = \sum_j (d_i, w_j)$ refers to the document length.

These steps are repeated until convergence or until a termination condition is met. As a result, two probability matrices are obtained: the word-topic probability matrix $W_{M\times K} = (P(w_j|z_k))_{j,k}$ and the topic-document probability matrix $D_{K\times N} = (P(z_k|d_i))_{k,i}$.

2.3.2 PLSA for Visual Words

The PLSA technique only requires a word-document co-occurrence matrix and therefore the technique can be referred to as feature-agnostic. Since it does not set any requirements on the nature of the low level features that yield these co-occurrence matrices (other than being discrete), the extension to visual words is simple. PLSA in combination with visual words for classification purposes was also applied in [29, 30].

In our approach, images are described in terms of a BoC in the CIELab color space and a BoVW based on SIFT descriptors. Therefore, the dataset can be described using the following co-occurrence matrices:

$$C_{N\times N_C} = (n(d_i, c_j))_{i,j}, \tag{7}$$

$$S_{N\times N_S} = (n(d_i, s_l))_{i,l}, \tag{8}$$

where N is the number of images in the dataset, N_C the length of the color vocabulary, N_S the length of the SIFT-based vocabulary and $n(d_i, c_j)$ or $n(d_i, s_l)$ is the number of occurrences of the color word c_j or SIFT word s_l occurring in the image d_i.

2.3.3 Vocabulary Pruning

The key idea in our approach is that not only the color and SIFT vocabularies are over-complete and redundant individually for the dataset, but they may as well contain visual words that model the same latent topics. Therefore, a full color-SIFT representation of the dataset is obtained by concatenating the two matrices C and S into a single $N \times (N_C + N_S)$ visual features matrix V.

The matrix V is then analysed using the PLSA technique with a varying number of topics K and the resulting visual word-topic conditional probability matrices $W_{(N_C+N_S)\times K}$ are used to find the meaningless visual words that need to be removed from the vocabulary.

A visual word is considered meaningless if its conditional probability is below the *significance threshold* T_k for every latent topic. Since each topic can be linked to a different number of visual words, the significance threshold is not an absolute value, but relative to each topic. In our approach, T_k takes the value of the p_T-th percentile of each topic. This allows to keep only the $(100-p_T)\%$ most significative visual words for each topic while removing the remaining visual words. A visual word can signify several topics (*polysemic words*) and several visual words can be equally significative for a given topic (*synonyms*). These factors, which are common in language modelling, have as a result that the vocabulary reduction cannot be estimated directly using the value of p_T, since it depends on the distribution of synonyms and polysemic words in the experimental data model.

The number of latent topics as well as the value of the significant percentile are parameters of the technique presented in this paper. Section 3 explains the results of the experimental evaluation of the technique for various values of K and p_T.

2.3.4 Meaningfulness-Based Descriptor Transformation

Instead of using a hard decision based on a meaningfulness threshold, a transformation can be defined to weight visual words according to their meaningfulness. The visual meaningfulness of a visual word w_n is its maximum topic-based significance level:

$$m_n = \begin{cases} \max_j \left\{t_{n,j}\right\} & \text{if } \max_j \left\{t_{n,j}\right\} \geq T_{meaning} \\ 0 & \text{otherwise} \end{cases}$$

Let $\mathbf{h}$ be a histogram vector where each component represents the multiplicity of a visual word, and $\mathbf{M}$ a meaningfulness transformation matrix:

$$\mathbf{h} = (n(w_1), n(w_2), \ldots, n(w_{N_W}))^T \tag{9}$$

$$\mathbf{M} = \begin{pmatrix} m_1 & 0 & \cdots & 0 \\ 0 & m_2 & \cdots & 0 \\ \vdots & \vdots & \ddots & \vdots \\ 0 & 0 & \cdots & m_{N_W} \end{pmatrix} \tag{10}$$

Then, the vector $\mathbf{h^M} = (n(w_1^M), n(w_2^M), \ldots, n(w_{N_W}^M))^T$ is the histogram vector of visual words in the meaningfulness-transformed space.

$$\mathbf{h^M} = \mathbf{Mh} \tag{11}$$

$$n(w_i^M) = m_i \cdot n(w_i) \tag{12}$$

2.4 Experiments

Several experiments were run to evaluate the performance of the vocabulary pruning technique. In this section, the experiments are described.

2.4.1 Classification with a Truncated Descriptor

Preliminary experiments on the vocabulary pruning technique over the training set were based on removing meaningless visual words from the descriptors but not from the vocabulary (i.e. the histogram values for meaningful visual words remain the same and therefore histograms are no longer normalized).

By running a twofold cross validation on the modality classification training set, the effect of the parameters K (number of latent topics) and p_T (significant percentile threshold) was investigated. All descriptors were computed using the full vocabulary and visual words below the significance threshold were later removed from the descriptors. No fusion rules were applied to the SIFT-BoVW and BoC descriptors.

2.4.2 Classification with a Reduced Vocabulary

In this experiment, meaningless visual words were removed from the vocabulary, histograms were recomputed and therefore stayed normalized. Due to the presence of very unbalanced classes in the dataset, experiments included twofold cross-validation on the training set and cross-validation based on separate training and test set. The same experiments were run with the full vocabularies.

Classification using the SIFT-BoVW and BoC can benefit from a fusion technique to include color and texture information. The similarity scores were calculated using both descriptors separately and the CombMNZ fusion rule [31] was used to obtain final scores. Images were classified using a weighted k-NN (k-Nearest Neighbors) voting [32]. Experiments were run with various k values for the voting.

2.4.3 Retrieval with a Reduced Vocabulary Over the Complete Data Set

In this experiment, the complete ImageCLEF dataset for medical images was indexed for retrieval. The number of images in the dataset (306,000) is sufficiently large to allow measures on speed gain when reducing the vocabulary. Retrieval was performed using the fusion rule described in Sect. 2.4.2. The retrieval experiment consisted of 22 topics (each consisting of 1–7 query images), corresponding to the ImageCLEF 2012 medical track.

2.4.4 Classification Using Descriptor Transformation

In order to assess the impact of vocabulary size and meaningfulness-based weighting of visual words, an experimental evaluation based on the SIFT description of the images was performed. Images were described with a BoVW based on SIFT [20] descriptors. This representation has been commonly used for image retrieval because it can be computed efficiently [15, 21, 22]. The SIFT descriptor is invariant to translations, rotations and scaling transformations and robust to moderate perspective transformations and illumination variations. SIFT encodes the salient aspects of the grey-level images gradient in a local neighborhood around each interest point.

Evaluation with separate training and test sets was performed using all combinations of the following parameters:

1. Two SIFT-based visual vocabularies with 100 and 500 visual words.
2. A varying number of visual topics from 25 to 350 in steps of 25.
3. A varying meaningfulness threshold from 50 to 100 %.

3 Results

In this section a summary of the results for each experiment is given.

3.1 Truncated Descriptor

This section explains the results of the experiment described in Sect. 2.4.1. Since the descriptor requires the full vocabulary before performing the truncation of meaningless words no speed gain in the offline phase was obtained.

Figure 4a shows the results of the accuracy obtained using a 1-NN classifier compared to the effect of truncating descriptors on vocabulary size in Fig. 4b. The number of latent topics K varies from 10 to 100 in steps of 10 and the significant percentile threshold for each topic p_T from 1 to 99.

The effect of increasing the significant percentile is much stronger on the number of visual words used than on the classification accuracy. Similarly, the number of latent topics has a limited impact on accuracy while having a strong impact on the vocabulary size. Rather unsurprisingly, the fewer latent topics considered, the easier it becomes to find meaningless visual words. Also, vocabulary sizes tend to be more similar for various K values when p_T is high.

Statistical significance tests were run to compare the results distributions using the truncated descriptors. These tests failed to show a statistically significant difference between classification using the full descriptor or any of the reduced descriptors over the training set.

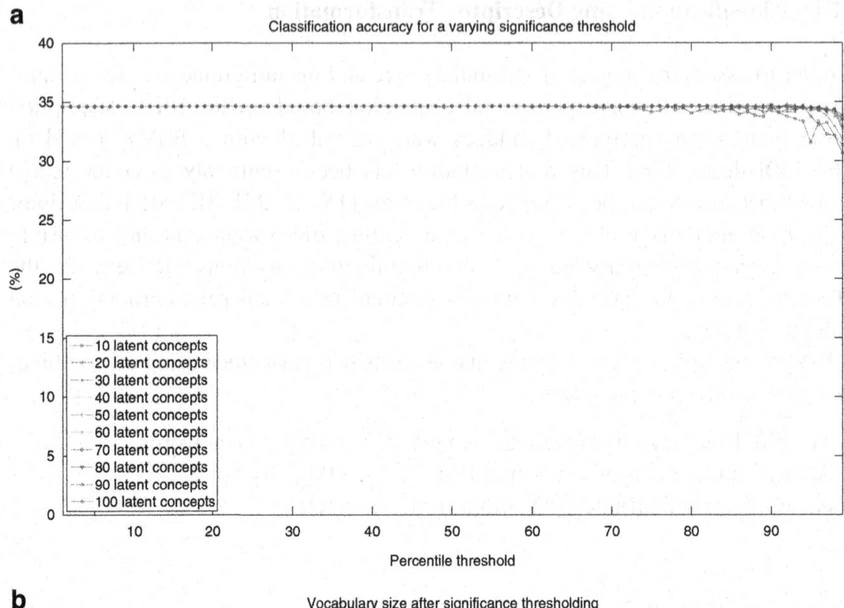

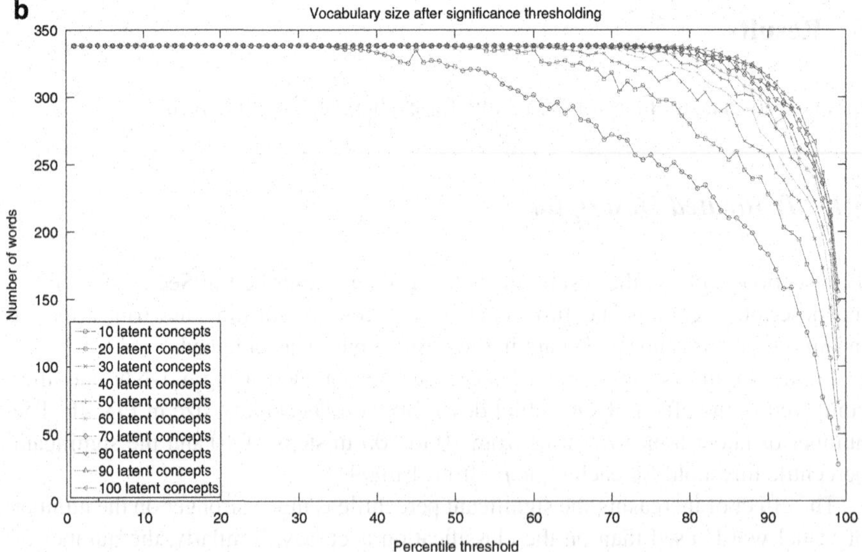

Fig. 4 Evaluation of descriptor truncation over the modality classification training set using cross-validation. 1-NN classification was performed for a varying number of latent topics K and significant percentile p_T. (**a**) Effect on classification accuracy. (**b**) Effect on effective vocabulary size

Table 1 Best classification results (varying the k-NN voting) over the training set for varying number of latent topics and a fixed significant percentile $p_T = 80$

Latent topics	Removed words (%)	Accuracy (reduced vocabulary) (%)	Accuracy (complete vocabulary) (%)
10	**27.22**	**44.20**	43.79
20	**17.16**	**44.20**	43.79
30	**6.8**	**43.99**	43.79
40	**3.25**	**43.79**	43.79
50	**2.96**	**43.99**	43.79
60	**2.07**	**43.99**	43.79
70	**1.18**	**43.79**	43.79
80	**0.59**	**43.79**	43.79
90	**0.59**	**43.79**	43.79
100	**0.3**	**43.79**	43.79

The last column contains the accuracy when using the complete vocabulary with the same classifier. Results are shown in bold when a reduced vocabulary produces better or equal classification than the complete vocabulary

3.2 *Reduced Vocabulary Over Modality Classification Training and Test Sets*

This section contains a summary of the results of the experiments described in Sect. 2.4.2.

Table 1 contains a summary of the best results for a significant percentile $p_T = 80$ and a varying number of topics. It also includes the results obtained with the full vocabulary using the same classifier. Although it is not shown in the table, all of the removed words for $p_T = 80$ belonged to the SIFT-BoVW vocabulary.

Table 2 contains the corresponding results for a 99-percentile as significance threshold. In this experiment meaningless words were found in both the BoC and the SIFT-BoVW vocabularies.

Tables 3 and 4 contain the corresponding results over the test set when performing cross-validation with separate test and training sets. The vocabularies used are the same as those from Tables 1 and 2.

3.3 *Reduced Vocabulary for the Retrieval Task*

Based on the results in Sect. 3.2, two vocabularies were selected for obtaining results in the ImageCLEFmed retrieval task. The smallest vocabulary corresponds to the $p_T = 99$ and 10 latent topics vocabulary, whereas the most accurate vocabulary was the $p_T = 80$ and 10 latent topics.

Table 2 Best classification results (varying the k-NN voting) over the training set for varying number of latent topics and a fixed significant percentile $p_T = 99$

Latent topics	Removed words (%)	Accuracy (reduced vocabulary) (%)	Accuracy (complete vocabulary) (%)
10	**91.72**	**41.55**	41.34
20	**84.32**	**44.20**	43.18
30	**78.99**	**43.79**	42.16
40	**72.78**	**45.01**	41.34
50	**67.75**	**44.81**	42.16
60	**61.83**	**44.60**	42.97
70	**59.47**	**43.81**	42.97
80	**54.73**	**45.62**	42.97
90	**53.85**	**43.99**	42.97
100	**50**	**43.79**	42.97

The last column contains the accuracy when using the complete vocabulary with the same classifier. Results are shown in bold when a reduced vocabulary produces better or equal classification than the complete vocabulary

Table 3 Best classification results (varying the k-NN voting) over the test set for varying number of latent topics and a fixed significant percentile $p_T = 80$

Latent topics	Accuracy (reduced vocabulary) (%)	Accuracy (complete vocabulary) (%)
10	**40.14**	38.94
20	**39.24**	38.94
30	**39.54**	38.64
40	**39.24**	38.24
50	**39.34**	38.94
60	**39.24**	38.94
70	**39.24**	38.94
80	**39.24**	38.94
90	**39.24**	38.94
100	**39.24**	38.94

The last column contains the accuracy when using the complete vocabulary with the same classifier. Results are shown in bold when a reduced vocabulary produces better or equal classification than the complete vocabulary

Table 5 contains a summary of the results in terms of time required for indexing the complete dataset for the most accurate configuration ($p_T = 80$ and 10 latent topics), the smallest vocabulary ($p_T = 99$ and 10 latent topics) and the complete vocabulary.

Table 4 Best classification results (varying the k-NN voting) over the test set for a varying number of latent topics and a fixed significant percentile $p_T = 99$

Latent topics	Accuracy (reduced vocabulary) (%)	Accuracy (complete vocabulary) (%)
10	36.44	37.94
20	36.24	37.94
30	36.84	38.64
40	38.44	38.94
50	37.24	38.64
60	37.34	38.94
70	**38.94**	38.94
80	37.94	38.94
90	**38.94**	38.94
100	**39.44**	38.94

The last column contains the accuracy when using the complete vocabulary with the same classifier. Results are shown in bold when a reduced vocabulary produces better or equal classification than the complete vocabulary

Table 5 Average indexing time per image for the smallest vocabulary, the most accurate and the complete vocabulary

(a) Average time per image for the reduced vocabulary with parameters p_T=99 and K=10

Feature type	Index time	Size
BoC	2.14 s	19 words
SIFT-BoVW	0.74 s	9 words

(b) Average time per image for the reduced vocabulary with parameters $p_T = 80$ and $K = 10$

Feature type	Index time	Size
BoC	4.86 s	100 words
SIFT-BoVW	1.15 s	146 words

(c) Average time per image for the complete vocabulary

Feature type	Index time	Size
BoC	4.86 s	100 words
SIFT-BoVW	1.67 s	238 words

Table 6 shows the results when performing the retrieval task on the complete ImageCLEFmed 2012 dataset with the selected vocabularies for each of the 22 topics or queries.

Table 6 Results of retrieval experiments for each vocabulary

(a) Retrieval results for each vocabulary and various queries. Results with higher recall are shown in bold

	Relevant items	Items retrieved (complete vocabulary)	Items retrieved ($p_T = 80$, $K = 10$)	Items retrieved ($p_T = 99$, $K = 10$)
Topic 1	21	7	**8**	**8**
Topic 2	33	**21**	20	16
Topic 3	47	**35**	**35**	29
Topic 4	22	15	**16**	15
Topic 5	58	**7**	**7**	4
Topic 6	13	7	7	**8**
Topic 7	11	2	2	**3**
Topic 8	6	**3**	**3**	2
Topic 9	2	0	0	0
Topic 10	17	**6**	**6**	**6**
Topic 11	72	17	**19**	8
Topic 12	27	5	6	**9**
Topic 13	147	**50**	48	38
Topic 14	521	**57**	56	48
Topic 15	0	0	0	0
Topic 16	3	**1**	**1**	**1**
Topic 17	7	0	0	**2**
Topic 18	4	0	0	0
Topic 19	6	**3**	**3**	2
Topic 20	5	0	0	0
Topic 21	49	5	5	**7**
Topic 22	19	**7**	**7**	5
Total	1090	248	**249**	211

(b) Mean Average Precision (MAP) across all topics

Vocabulary used	MAP (%)
Complete vocabulary	6.51
$p_T = 80, K = 10$	6.52
$p_T = 99, K = 10$	1.51

(c) Average execution times of the online phase for a single query image

Vocabulary used	Online retrieval time
Complete vocabulary	125 s
$p_T = 80, K = 10$	107 s
$p_T = 99, K = 10$	45 s

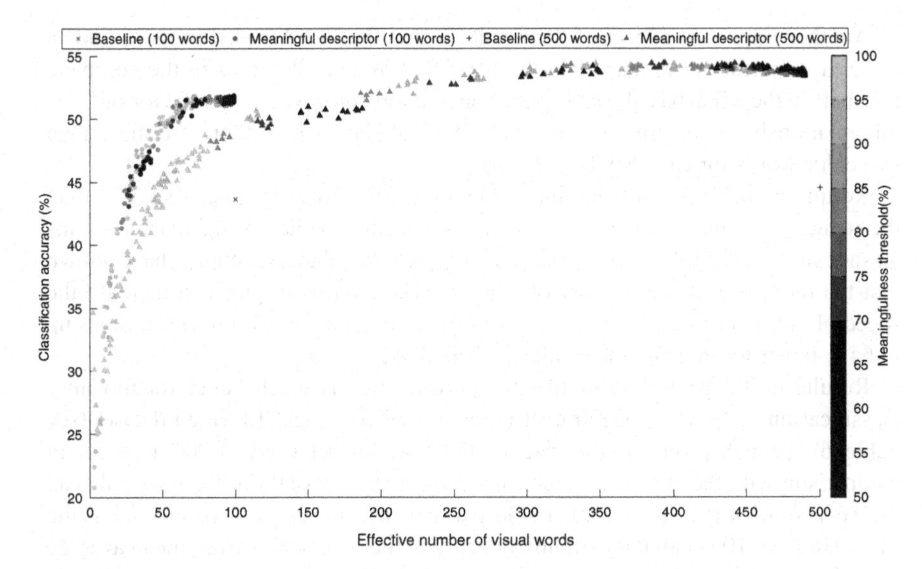

Fig. 5 Evaluation of descriptor transformation using the proposed meaningfulness transform over the modality classification task set using training and test sets. 1-NN classification was performed for a varying number of latent topics and meaningfulness threshold

3.4 Descriptor Transformation and Effect on Vocabulary Size

Using the parameters explained in Sect. 2.4.4 and applying the transformation proposed in Sect. 2.3.4, the effect of the initial vocabulary size and the meaningfulness threshold can be studied.

Figure 5 shows the effect of the transformation when using various meaningfulness thresholds on two vocabularies.

4 Discussion

As shown in Fig. 4 the impact of PLSA-based pruning has a stronger effect on the size of the vocabulary than on the performance of the classifiers. Table 2 shows that a vocabulary reduction of up to 91.72 % can be obtained with a comparable accuracy for the same classifier. For the 99-percentile, the best classification method with the reduced vocabulary always obtains higher accuracy than the same classification method on the full vocabulary.

However, significance tests have failed to show a statistically significant difference between the various accuracies obtained. Therefore, the main contribution of this work is a method that can enormously reduces visual word vocabularies while obtaining a comparable (and often slightly higher) accuracy.

Another important aspect of the results is that the PLSA-based pruning finds a more meaningful vocabulary than the SIFT-BoVW one. Whereas in the complete vocabulary the SIFT-based words outnumbered the color words by a factor of 2.38, this relationship is inverted in the smallest vocabulary where there are more than two color words for each SIFT-based word.

Results in Table 5 show that the reduction of the indexing time is smaller than the reduction in the number of words. However, the smallest vocabulary presents an indexing time 55.9 % lower than the complete vocabulary. Studies have shown that the reduction of the number of features used as a descriptor can increase the speed of online retrieval [33]. This is confirmed in Table 5c, with retrieval times up to 64 % lower when using the smallest vocabulary.

Results in Tables 1–4 show that the performance is much better for modality classification tasks than for retrieval in the complete ImageCLEFmed dataset (see Table 6), probably due to the size of the training set used (1,000 images) in comparison with the 306,000 images in the complete dataset. For the retrieval task, the vocabularies present a comparable performance in terms of recall, being the $p_T = 80, K = 10$ vocabulary slightly better than the others. However, mean average precision strongly varies between large vocabularies and the smallest vocabulary ($p_T = 99, K = 10$).

Evaluation of the proposed meaningfulness transformation shows an improvement in accuracy as well as the impact on the vocabulary size already found in the PLSA-based pruning. The increase of accuracy is non-negligible, and passes statistical significance tests. The accuracy is increased for both original vocabularies tested, and there is a slight *saturation effect* where the size of the descriptor can be safely reduced without impact on accuracy. Massive reductions of the descriptors, strongly reduce performance as well.

It can be discussed that de benefits of the PLSA-based pruning presented are not the ability to discover new and meaningful visual words for retrieval but the ability to recognize those visual words that convey most of the meaning among the ones present in the vocabulary. However, the meaningfulness transform is able to improve the accuracy by increasing the relative weight of the most meaningful visual words.

5 Conclusions and Future Work

In this work a vocabulary pruning and description transformation method based on probabilistic latent semantic analysis of visual words for medical image retrieval and classification is presented. The selection of optimal visual words is performed by removing visual words with a conditional probability over all learnt latent topics that is below a given threshold, the remaining (meaningful) words are weighted according to the largest conditional probability. The process is completely unsupervised, since the learning of the topics is performed without taking into consideration the number of classes or what is the actual class assigned to each image. Therefore, it can be used to reduce massive fine-grained vocabularies to smaller vocabularies that contain only the most meaningful visual words even before training the classifier. To obtain these fine-grained vocabularies, simple clustering algorithms can be used to produce a large number of small clusters that later will be pruned using the methods explained in this paper. Smaller clusters are supposed to encode subtle visual differences among images, which will be preserved by the PLSA-based pruning if they are meaningful for some latent topic. Future applications of the technique also include the use of multiple vocabularies that can be merged and pruned as a single set of discrete features.

We are currently extending the techniques to images obtained for clinical use, where the use of low-dimensional descriptors can achieve fast and accurate characterization of large-scale datasets of high-dimensional (3D, 4D, multimodal) images. This is expected to lead to different results as for the modality classification tasks and retrieval tasks from the literature color plays a more important roles than for most clinical images. Still, the possibility to reduce visual vocabularies strongly can lead to larger base vocabularies that can potentially capture the image content much better but can then be reduced for efficient retrieval.

Acknowledgements This work was partially supported by the Swiss National Science Foundation (FNS) in the MANY2 project (205320-141300), the EU 7th Framework Program under grant agreements 257528 (KHRESMOI) and 258191 (PROMISE).

References

1. Müller, H., Michoux, N., Bandon, D., & Geissbuhler, A. (2004). A review of content-based image retrieval systems in medicine-clinical benefits and future directions. *International Journal of Medical Informatics, 73*(1), 1–23.
2. Akgül, C., Rubin, D., Napel, S., Beaulieu, C., Greenspan, H., & Acar, B. (2011). Content-based image retrieval in radiology: Current status and future directions. *Journal of Digital Imaging, 24*(2), 208–222.
3. Tang, L. H. Y., Hanka, R., & Ip, H. H. S. (1999). A review of intelligent content-based indexing and browsing of medical images. *Health Informatics Journal, 5*, 40–49.
4. Demner-Fushman, D., Antani, S., Siadat, M.-R., Soltanian-Zadeh, H., Fotouhi, F., & Elisevich, K. (2007). Automatically finding images for clinical decision support. In *Proceedings of the Seventh IEEE International Conference on Data Mining Workshops, ICDMW '07* (pp. 139–144). Washington, DC: IEEE Computer Society.
5. Caputo, B., Müller, H., Mahmood, T. S., Kalpathy-Cramer, J., Wang, F., & Duncan, J. (2009). Editorial of miccai workshop proceedings on medical content-based retrieval for clinical decision support. In *Lecture Notes in Computer Science: Vol. 5853. Proceedings on MICCAI Workshop on Medical Content-Based Retrieval for Clinical Decision Support*. Heidelberg: Springer.
6. Müller, H., Kalpathy-Cramer, J., Kahn, Jr. C. E., & Hersh, W. (2009). Comparing the quality of accessing the medical literature using content-based visual and textual information retrieval. In *SPIE Medical Imaging*, Orlando, FL (Vol. 7264, pp. 1–11).
7. Deserno, T. M., Antani, S., & Long, L. R. (2009). Content-based image retrieval for scientific literature access. *Methods of Information in Medicine, 48*(4), 371–380.
8. Müller, H., de Herrera, A. G. S., Kalpathy-Cramer, J., Fushman, D. D., Antani, S., & Eggel, I. (2012). Overview of the ImageCLEF 2012 medical image retrieval and classification tasks. In *Working Notes of CLEF 2012 (Cross Language Evaluation Forum)*.

9. Müller, H., Clough, P., Deselaers, T., & Caputo, B., (Eds.). (2010). *ImageCLEF: Experimental evaluation in visual information retrieval*. The Springer International Series on Information Retrieval (Vol. 32). Berlin/Heidelberg: Springer.
10. Leibe, B., & Grauman, K. (2011). *Visual object recognition*. San Rafael, CA: Morgan & Claypool Publishers.
11. Foncubierta-Rodríguez, A., Depeursinge, A., & Müller, H. (2012). Using multiscale visual words for lung texture classification and retrieval. In H. Greenspan, H. Müller, & T. S. Mahmood, (Eds.), *Lecture Notes in Computer Sciences: Vol. 7075. Medical content-based retrieval for clinical decision support* (pp. 69–79) *MCBR-CDS 2011*.
12. Hinneburg, A., & Gabriel, H.-H. (2007). DENCLUE 2.0: Fast clustering based on kernel density estimation. *Advances in Intelligent Data Analysis VII, 4723/2007*, 70–80.
13. Haas, S., Donner, R., Burner, A., Holzer, M., & Langs, G. (2011). Superpixel-based interest points for effective bags of visual words medical image retrieval. In H. Greenspan, H. Müller & T. Syeda-Mahmood (Eds.), *Lecture Notes in Computer Sciences: Vol. 7075. Medical content-based retrieval for clinical decision support, MCBR-CDS 2011*.
14. Avni, U., Greenspan, H., Konen, E., Sharon, M., & Goldberger, J. (2011). X-ray categorization and retrieval on the organ and pathology level, using patch-based visual words. *IEEE Transactions on Medical Imaging, 30*(3), 733–746.
15. Markonis, D., de Herrera, A. G. S., Eggel, I., & Müller, H. (2012). Multi-scale visual words for hierarchical medical image categorization. In *SPIE Medical Imaging 2012: Advanced PACS-Based Imaging Informatics and Therapeutic Applications* (Vol. 8319, pp. 83190F–11).
16. Basu, S., Banerjee, A., & Mooney, R. (2002). Semi-supervised clustering by seeding. In *19th Internaional Conference on Machine Learning (ICML-2002)* (pp. 19–26).
17. Bilenko, M., Basu, S., & Mooney, R. (2004). Integrating constraints and metric larning in semi-supervised clustering. In *21st Internaional Conference on Machine Learning (ICML-2004)*.
18. Markonis, D., Holzer, M., Dungs, S., Vargas, A., Langs, G., Kriewel, S., et al. (2012). A survey on visual information search behavior and requirements of radiologists. *Methods of Information in Medicine, 51*(6), 539–548.
19. Müller, H., Kalpathy-Cramer, J., Demner-Fushman, D., & Antani, S. (2012). Creating a classification of image types in the medical literature for visual categorization. In *SPIE Medical Imaging*.
20. Lowe, D. G. (2004). Distinctive image features from scale-invariant keypoints. *International Journal of Computer Vision, 60*(2), 91–110.
21. Yang, Y., & Newsam, S. (2010). Bag-of-visual-words and spatial extensions for land-use classification. In *Proceedings of the 18th SIGSPATIAL International Conference on Advances in Geographic Information Systems, GIS '10* (pp. 270–279). New York, NY: ACM.
22. Ke, Y., & Sukthankar, R. (2004). Pca-sift: A more distinctive representation for local image descriptors. In *IEEE Computer Society Conference on Computer Vision and Pattern Recognition (CVPR 2004)*, Washington, DC. (Vol. 2, pp. 506–513).
23. Wengert, C., Douze, M., & Jégou, H. (2011). Bag-of-colors for improved image search. In *Proceedings of the 19th ACM International Conference on Multimedia, MM '11* (pp. 1437–1440). New York, NY: ACM.
24. Banu, M. S., & Nallaperumal, K. (2010). Analysis of color feature extraction techniques for pathology image retrieval system. IEEE.
25. Tirilly, P., Claveau, V., & Gros, P. (2008). Language modeling for bag-of-visual words image categorization. In *Proceedings of the 2008 International Conference on Content-Based Image and Video Retrieval* (pp. 249–258). New York: ACM.
26. Tian, Q., Zhang, S., Zhou, W., Ji, R., Ni, B., & Sebe, N. (2011). Building descriptive and discriminative visual codebook for large-scale image applications. *Multimedia Tools and Applications, 51*(2), 441–477.
27. Deerwester, S., Dumais, S. T., Furnas, G. W., Landauer, T. K., & Harshman, R. (1990). Indexing by latent semantic analysis. *Journal of the American Society for Information Science, 41*(6), 391–407.

28. Hofmann, T. (2001). Unsupervised learning by probabilistic latent semantic analysis. *Machine Learning, 42*(1–2), 177–196.
29. Bosch, A., Zisserman, A., & Munoz, X. (2006). Scene classification via plsa. In *Computer Vision-ECCV 2006* (pp 517–530). Heidelberg: Springer.
30. Elsayad, I., Martinet, J., Urruty, T, & Djeraba, C. (2012). Toward a higher-level visual representation for content-based image retrieval. *Multimedia Tools and Applications, 60*(2), 455–482.
31. Fox, E. A., & Shaw, J. A. (1993). Combination of multiple searches. In *Text Retrieval Conference* (pp. 243–252).
32. Hand, D. J., Mannila, H., & Smyth, P. (2001). *Principles of data mining (adaptive computation and machine learning)*. Cambridge: The MIT Press.
33. McG, D., Squire, Müller, H., & Müller, W. (1999). Improving response time by search pruning in a content-based image retrieval system, using inverted file techniques. In *IEEE Workshop on Content-Based Access of Image and Video Libraries (CBAIVL '99)* (pp. 45–49).

Multimedia Information Retrieval from Ophthalmic Digital Archives

Gwenolé Quellec, Mathieu Lamard, Béatrice Cochener, and Guy Cazuguel

1 Introduction

With the generalization of computer-based processing, storage of medical data and cloud solutions, there is now a universal consensus that stored medical data can (and should) be used to improve health care [21, 54]. Since a single hospital can process tens of thousands of medical examinations per year by itself, and that collected data are almost systematically stored, huge archives are becoming available. More and more, collected data (multidimensional images and videos, demographic data, biological test results, clinician reports, etc.) are structured as specialized health records, within digital archives. Eventually, these data should be structured as multispecialty patient records [37]. These archives of health records are a major asset to develop new decision support tools. This second use of medical data is already common practice: many epidemiological studies are performed to extract new knowledge (in the form of diagnosis rules) from medical archives. Case-based

G. Quellec (✉)
Inserm, UMR 1101 LaTIM, SFR ScInBioS, F-29200 Brest, France
e-mail: gwenole.quellec@inserm.fr

M. Lamard
Université de Bretagne Occidentale, F-29200 Brest, France
e-mail: mathieu.lamard@univ-brest.fr

B. Cochener
CHRU Brest, Service d'Ophtalmologie, F-29200 Brest, France
e-mail: beatrice.cochener@ophtalmologie-chu29.fr

G. Cazuguel
Institut Mines-Telecom, Telecom Bretagne, UEB, Dpt ITI, F-29200 Brest, France
e-mail: guy.cazuguel@telecom-bretagne.eu

A. Briassouli et al. (eds.), *Health Monitoring and Personalized Feedback using Multimedia Data*, DOI 10.1007/978-3-319-17963-6_6

reasoning (CBR) goes one step further: medical archives are used to develop patient-specific decision support tools (as opposed to the population-specific tools in epidemiology) [9]. CBR was introduced as a general way to reuse stored data for decision support. Content-based image, video or health record retrieval were introduced as general ways to find relevant images, videos or health records in large datasets, a critical step for reusing the data.

Ophthalmology is a branch of medicine that produces a lot of images, for diagnosis purposes, and a lot of videos, to document surgical procedures. In the past few years, we have developed various solutions to reuse ophthalmic digital archives for computer-aided diagnosis of retinal pathologies and for computer-aided decision during video-monitored eye surgeries. After a short review of content-based image, video or health record retrieval techniques, this chapter presents the solutions we have developed for these two applications.

2 Multimedia Information Retrieval in Medical Decision Support Systems: A Short Review

The process of solving a new problem using the solutions of similar past problems is called Case-Based Reasoning (CBR). In a medical application, the new problem is a medical case about which a decision has to be taken (a diagnosis, an operating procedure, etc.) [9]. The past problems are medical cases stored in a digital archive together with medical decisions (the solution to these past problems). CBR consists of four steps: (1) retrieval, (2) reuse, (3) revision, (4) retaining. Retrieval, the process of finding relevant previous cases in memory, is the most important part. Reuse involves adapting these previous cases to the new problem. Revision is the process of testing the proposed solution and revising it if necessary. Finally, retaining consists in storing the resulting experience as a new case in memory if the solution was successful. Content-Based Image Retrieval (CBIR) and Content-Based Video Retrieval (CBVR) were proposed to address the retrieval step when cases consist of images (Sect. 2.1) or videos (Sect. 2.2), respectively. When medical cases are full health records, the retrieval step is referred to as Content-Based Health Record Retrieval (CBHRR) in this chapter (Sect. 2.3).

2.1 Content-Based Image Retrieval (CBIR)

To retrieve images with CBIR, image contents (as opposed to metadata) are characterized and compared. Several features are extracted from images (e.g. textural, color or shape features). Image comparisons rely on similarity measurements between feature vectors. CBIR is a very active research area, particularly in medical applications where images often play a major role in decision making [29]. Popular medical CBIR systems include medGIFT[1] and IRMA.[2] The major challenge of CBIR systems is to bridge the so-called semantic gap, i.e. relate low-level features (texture, shape or color features) to the high-level concept of clinical similarity [29]. The most popular solution seems to be relevance feedback: clinicians indicate which retrieved cases are relevant and this feedback is used to refine the search. Our approach relies on machine learning (Sect. 3.2).

Nowadays, the most common approach to characterize images in a CBIR system is to extract *visual words* from images [13, 24, 27, 52, 65, 68]: each image is then described by a set of extracted visual words, referred to as a *bag of visual words* (BoW). Following the example of text retrieval, where images play the role of texts and visual words play the role of regular words, standard information retrieval techniques (such as tf-idf) can be used to retrieve similar images. One of the advantages of this trick is that fast image search in large image archives is now possible. Visual words are quantized visual feature vectors. The latter are usually extracted in the neighborhood of salient points, using SIFT, SURF or ORB detectors [53], for instance. This sparse approach is well suited to images where local shapes play a predominant role. When texture plays a major role, which is often the case in medical images, this approach is not optimal, although successful experiments have been reported [36]. In the competing approach, the dense approach, images are divided into patches [6] or regions [63] and a visual feature vector is extracted from each patch or region; this approach is better suited to textured images [6]. There are several solutions to associate visual feature vectors to visual words. These solutions often rely on unsupervised classification techniques such as K-means: each cluster is associated with a visual word [55]. We have proposed a supervised case-based solution [39].

In order to transform a variable size bag of words to a fixed size feature vector, and therefore facilitate image comparisons, a histogram of visual words is generally built. In the particular case of medical images, where only parts of the image are relevant (e.g. lesions), it can be a good idea to identify relevant visual words and discard the others. Multiple Instance Learning is an efficient solution to identify the (local) visual word that explains a diagnosis assigned to an image as a whole [2]. Boosting techniques exist to extend multiple-instance learning to the search of multiple relevant visual words [51]. We have proposed a case-based multiple instance learning solution able to define an infinite number of relevant visual words [39].

[1] http://medgift.hevs.ch/.

[2] http://ganymed.imib.rwth-aachen.de/irma/index_en.php.

2.2 Content-Based Video Retrieval (CBVR)

Initially popularized in broadcasting [31] and video surveillance [23], the use of CBVR is now emerging in medical applications as well [3, 4, 58]. Medical CBVR systems are used as a diagnosis aid, by analogy reasoning [4, 58], or as a training tool [3]. We propose to use it as surgical aid, also by analogy reasoning (Sect. 4).

CBVR systems differ by the nature of the objects placed as queries. First, queries can be images [34]. In that case, the goal is to select videos containing the query image in a video archive; these systems are very similar to image retrieval systems. Second, queries can be video shots [18, 31]. In that case, the goal is to find other occurrences of the query shot [31], or similar shots [18], in the video archive. Third, queries can be entire videos [3, 58]. In that case, the goal is to select the most similar videos, overall, in the archive. CBVR systems also differ by the way videos or video subsequences are characterized. Several systems rely mainly on the detection and characterization of key frames [25, 34]. Others characterize videos or video subsequences directly [18, 19]. The combination of multimodal (visual, audio and textual) information in a retrieval engine has also been proposed [11, 22]. Finally, CBVR systems differ by how flexible the distance metrics should be. First, several systems have been proposed to find objects that are almost identical to the query. For instance, a copy detection system was proposed to protect copyrighted videos [15]. Another system has been proposed to detect repeating shots in a video stream, in order to automatically structure television video content [31]. However, in most CBVR systems, we are interested in finding videos or video subsequences that are semantically similar but whose visual content can significantly vary from one sequence to another [3, 25, 64]. Distance metrics able to bridge the semantic gap are needed.

Following the example of CBIR systems, the BoW model is also popular in CBVR systems [5, 28, 66]. In order to characterize motion efficiently, some CBVR systems rely on spatiotemporal visual words [1, 7, 56].

2.3 Content-Based Health Record Retrieval (CBHRR)

A few systems have been proposed for the retrieval of multimodal documents containing images and texts [61]. One task, evaluated in the ImageClef evaluation forum, is the retrieval of medical articles from Pubmed [30]. The proposed solutions usually rely on the separate processing of images and textual information. A late fusion step is then performed to provide a unique ordering of the results. Medical cases and health records share similarities, but unlike medical articles, the content of health records in specialized archives is usually structured.

CBR systems for health records usually rely on ontologies [9]: well-known ontology-based CBR systems include ALEXIA, MNAOMIA and CARE-PATNER. Mémoire, a unified representation language for case-based ontologies in biomedicine, can be applied to any application provided that an ontology is available [8]. In particular, images and signals can be processed by such CBR systems thanks to image- or signal-specific ontologies [12, 35]. The image- and signal-specific ontologies depend either on descriptive keywords and textual metadata [12] or on automatic image analysis [62].

We proposed multimodal CBHRR systems for medical cases consisting of a varying-length sequence of images together with a sparse vector of demographic and biological data (Sect. 3.3). Research on multimodal health record retrieval has just been initiated: it could benefit to a much larger range of applications.

3 Multimedia Information Retrieval for Computer-Aided Retinal Diagnosis

The retina, a tissue lining the inner surface of the eye, is an essential part of the human visual system. It can be affected by several pathologies, including Diabetic Retinopathy (DR), age-related macular degeneration, retinal detachment, macular edema, retinitis pigmentosa and glaucoma. These pathologies can lead to blindness in the most advanced cases. Diabetic retinopathy is the leading cause of blindness in the working population of the European Union and the United States. Because early screening of DR insures efficient treatment of the disease, several screening programs have been set up throughout the world: in the United Kingdom (1.7 million diabetics have been screened in 2007–2008), in the Netherlands (in 2010, 30 % of all patients with diabetes in the country had been screened by the EyeCheck program), in the United States (120,000 diabetics have been screened in 2008 by the U.S. Department of Veterans Affairs) and in France (38,000 patients have been screened by the Ophdiat network in Paris between 2004 and 2008). These pathologies affect a large and growing part of the population, which would require more and more specialists. But, in the United States for instance, the number of ophthalmologists is expected to become inadequate within the next decade to meet the medical need. In order to increase the population that can benefit from eye fundus screening, ophthalmologists tend to be replaced by non-ophthalmologist readers. In that context, computer-aided diagnosis solutions are welcome. Fundus photograph is a cheap and convenient solution to image the retina: therefore, it is massively used in DR screening networks, resulting in large archives of fundus photograph examinations. Therefore, content-based image or health record retrieval solutions seem particularly relevant for the design of decision support systems.

3.1 Existing Computer-Aided Retinal Diagnosis Systems

Automated DR screening systems in eye examinations have been proposed by several research groups. The proposed methods rely on the detection of the first appearing lesions of DR: microaneurysms (MAs) and sometimes hemorrhages and exudates [32]. Some automated DR screening systems are already commercialized: iGradingM (Medalytix, United Kingdom),[3] CARA (Diagnos, Canada),[4] IDx-DR (ID, USA),[5] Hubble (Hubble Telemedical, USA)[6] and EYESTAR (VisionQuest Biomedical, USA),[7] The acceptance of such automated systems by ophthalmologists and health care organizations is not widespread yet. So limited success has been achieved so far. One limitation of those commercialized systems is that they usually focus on DR lesions and ignore other eye pathologies that ophthalmologists would detect. Another limitation is that they only take the eye fundus photographs into account: they ignore the contextual information that is required by ophthalmologists to make a reliable diagnosis (the demographic and clinical data stored in health records). This is what motivated our works: the first limitation is addressed through image mining (Sect. 3.2), the second one is addressed through information fusion (Sect. 3.3).

3.2 Our CBIR Approach: Adaptive Wavelet-Based Image Retrieval

In order to use medical archives for decision support, we proposed a set of solutions to characterize medical images contained in health records using their digital content. All these solutions rely on image compression techniques, which preselect relevant information in images. In particular, we focused on the wavelet transform, on which the JPEG-2000 compression standard is built. Instead of using general-purpose wavelet bases, the idea was to find the optimal wavelet basis for a given image dataset, which typically contains images of the same anatomical part, acquired with a similar device.

[3] www.medalytix.com.

[4] www.diagnos.ca/cara.

[5] www.eyediagnosis.net.

[6] www.hubbletelemedical.com.

[7] visionquest-bio.com/eyestar-tm.html.

3.2.1 Image Retrieval using a Dataset-Specific Wavelet Transform

A novel image characterization based on the wavelet transform was proposed: the idea was to characterize the overall distribution of wavelet transform coefficients at different scales and along different directions [44]. The wavelet basis used to decompose images is tuned, through a machine learning procedure, so that two images semantically similar are close in the proposed feature space. In other words, they are tuned so that the semantic gap is maximally narrowed.

The proposed approach relies on the lifting scheme, which was designed to construct second generation wavelets. Those wavelets are not necessarily translates and dilates of one fixed function, as opposed to first generation wavelets. After relaxing this constraint, the Fourier transform can no longer be used as a construction tool, hence the need for a new one. In the lifting scheme, one starts with a trivial multiresolution analysis and gradually improves it (through lifting steps) to obtain a multiresolution analysis with desired properties [57]. Besides the ability to construct second generation wavelets, the lifting scheme allows fast in-place calculation of the wavelet transform, hence its use in JPEG-2000.

In our wavelet adaptation approach, the wavelet bases are designed to be efficient for both image retrieval and compression, as illustrated in Fig. 1. Therefore, image characterizations can be used to easily retrieve images once they are archived. Note that the same wavelet transform could also be used for securing images once they are archived [33]. The proposed methodology was implemented in the case of separable wavelets (1-D wavelets applied to each image dimension separately) [44] and in the more complex case of nonseparable wavelets (n-D wavelets) [43].

3.2.2 Image Retrieval using a Query-Specific Wavelet Transform

Adapting the wavelet basis to a dataset is certainly better than using a general-purpose basis, but it may not be enough. Let's say this dataset contains images from patients with various pathologies. One wavelet basis may be suitable to differentiate pathology A from pathology B, but not to differentiate pathology B from pathology C. Of course, we could try to adapt multiple wavelet bases through a boosting strategy, but then it increases the complexity of the system. So we decided to go further and adapt the wavelet basis to each query image itself, which is more challenging because the query image is not known during the learning phase. The solution we adopted was to adapt the wavelet basis to each training image, but with a continuity constraint: two similar images must have a similar optimal wavelet. To define how similar two training images are, we compared their characterizations extracted using the overall best wavelet basis, obtained as described above. Thanks to that continuity constraint, it is possible to estimate the best wavelet basis for an unknown image, through interpolations in the initial feature space (the one defined by the overall best wavelet basis) [45]. To allow fast retrieval with a query-specific wavelet basis, a solution was proposed to estimate the characterization of an image instantly, using an arbitrary wavelet basis [50]. The proposed solution is suitable for both separable and non-separable wavelets.

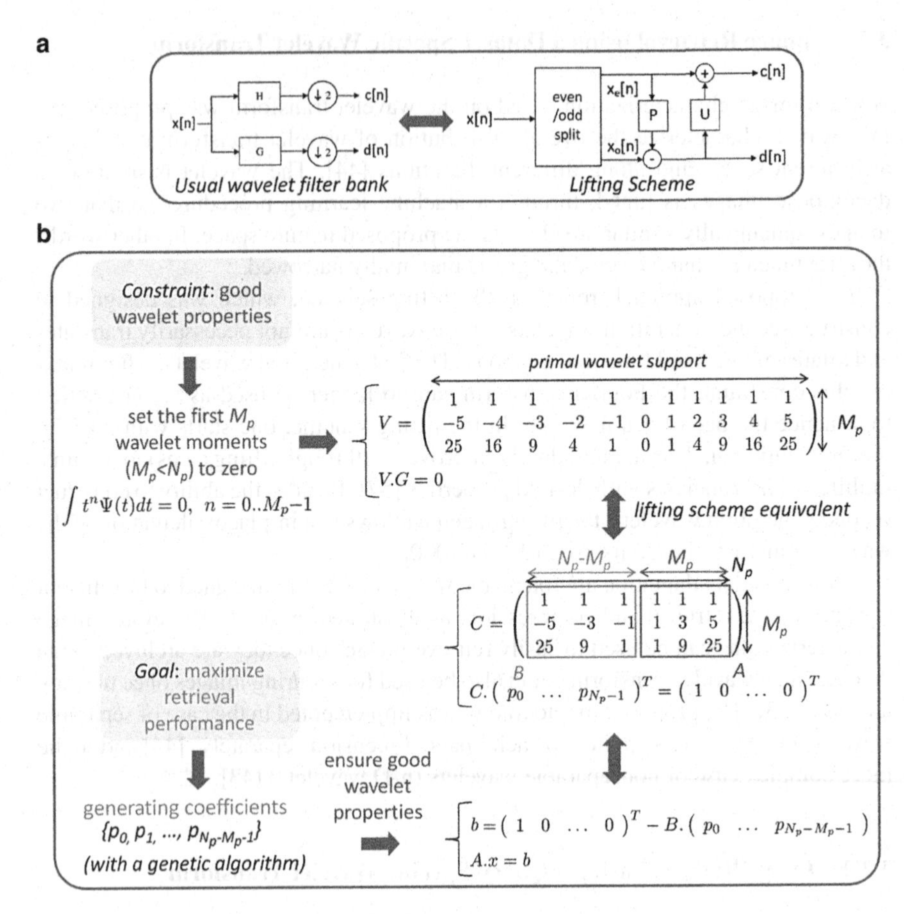

Fig. 1 Proposed wavelet adaptation strategy, in the case of separable wavelets. (**a**) compares the usual wavelet filter bank and the equivalent lifting scheme. (**b**) illustrates how the predict (P) filter is adapted: $P = \{p_0, \ldots, p_{N_p-1}\}$. The proposed solution trades off retrieval performance (in *green*) and good wavelet properties for compression purposes (in *red*). Update (U) filter adaptation is similar (Color figure online)

3.2.3 Localized Image Retrieval

The wavelet adaptation strategy above can also be used to adapt local image characterizations. We have presented a weakly supervised solution to identify pathological areas in images [39]. This solution relies on a search for image patches of various sizes whose characterizations are only found in pathological images, or in images with a given diagnosis. The problem is similar to what was presented above (Sect. 3.2.1), except that queries and retrieved objects are now image patches with fuzzy labels. This solution fits into the multiple-instance learning paradigm [2], that we also adopted for video retrieval, a task in which relevant information is not only lost in space, but also lost in time.

3.3 Our CBHRR Approach: Information Fusion and Data Mining

After characterizing each image in a health record individually, we tackled the problem of combining all pieces of evidence contained in health record to increase retrieval performance. In other words, we developed multimodal information retrieval solutions, a much more unexplored research area. We focused on the retrieval of medical records consisting of a varying length image series (each image being characterized as described above) together with a sparse vector or demographic and biological data. Demographic and biological data usually come as a (possibly sparse) structured list; if not, such a structured list can be extracted from texts (Sect. 2.3). Therefore, the challenge was to optimally combine elements in an incomplete list of multimodal information for health record retrieval.

3.3.1 Late Fusion Using Rules of Combination

First, a set of solutions based on late fusion was proposed: a retrieval engine is trained per health record element (i.e. per modality). Then, the outputs of these element-specific retrieval engines are fused using various probabilistic or evidential rules of combination. Bayesian networks were used as a probabilistic rule of combination [47]. The Dezert-Smarandache theory was used to derive evidential rules of combination [42, 47]. This approach is convenient when processing incomplete health records: data incompleteness does not need to be addressed specifically. In fact, rules of combinations do not need to be trained: they simply combine the sources of evidence that are available for a query. This is different from classifier-based late fusion (such as k-Nearest Neighbors or support-vector machines [30]), which may become invalid when inputs are incomplete. Another advantage of this approach is that the retrieval results can be refined progressively as new elements are added to the query health record.

The inclusion of prior knowledge (in the form of diagnosis rules derived from epidemiological studies) has also been investigated [46]. It did not improve retrieval performance: information obtained through machine learning, from the most similar health records, turned out to be more relevant, which tends to validate the CBR approach.

3.3.2 Early Fusion Using Data Mining

Second, a set of solutions based on early fusion was proposed: features extracted from all health record elements are put together to train a single retrieval engine. In this approach, correlations between health record elements are used in order to push performance further. As suggested above, not all machine learning algorithms are suitable for this task in the case of incomplete health records. The task of data mining algorithms consists in finding tuples of element values (e.g. age > 50 and sex = F) that are frequently observed in health records with similar diagnoses. If one element is missing in a health record, this kind of search is not invalidated.

Two types of data mining algorithms were investigated: (1) decision trees and their extensions (random forests and boosted decision trees) [40], which are frequently used in decisional databases, and (2) Apriori, a well-known algorithm for frequent item set mining and association rule learning [14].

3.4 Application 1: Diabetic Retinopathy Severity Assessment

As a toy example, the proposed methodologies were applied to DR severity assessment in a dataset collected at Brest University Hospital (France) [41]. DR severity in each health record was assessed by clinicians according to the Early Treatment Diabetic Retinopathy Study (ETDRS) scale: these manual gradings were used for both supervision and performance assessment. Results obtained for the proposed CBIR methods were encouraging: on average, given a query image, 56.5 % of the five most similar images belonged to the same DR severity stage as the query image (out of six severity levels). In a multimodal information retrieval context (when entire health records are used as queries), the performance went up to 81.8 %, which is very high, given the clinician's inter-observer variability. High performance was also achieved in a mammography dataset: a performance of 70.9 % or 86.9 % was measured when single images or entire health records were used as queries. The conclusion of these experiments is that context is important: large performance increases can be observed when the entire health record is used as query, rather than a single image.

3.5 Application 2: Retinal Pathology Screening

The present research is being conducted in the framework of TeleOphta[8] (*Telemedicine in Ophthalmology*), a consortium created in 2009 to develop automated and semi-automated tools for retinal pathology screening. TeleOphta aims primarily at automatically differentiating healthy patients from patients affected by at least one retinal pathology [14]. The goal is to reduce the number of health records collected in DR screening centers that should be read by an ophthalmologist. In a scenario where ophthalmologists are replaced by non-ophthalmologist readers, decision support tools will be available to help in analyzing the pathological cases. The TeleOphta system is illustrated in Fig. 2.

[8]http://teleophta.fr.

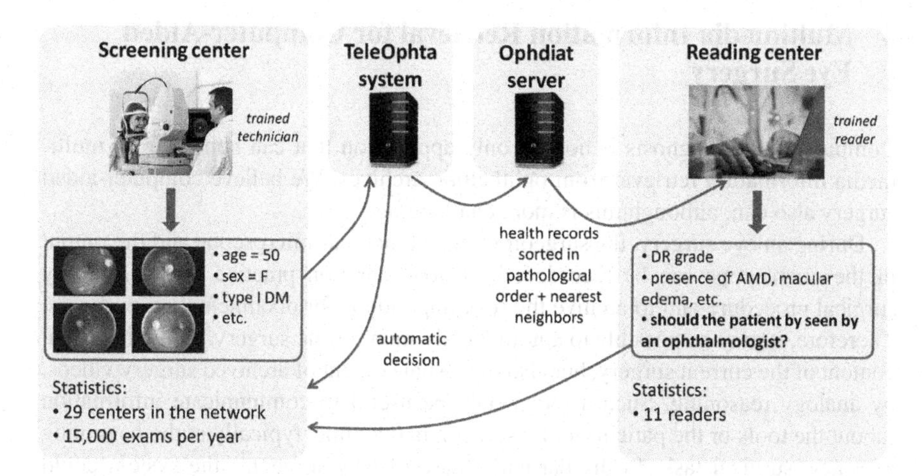

Fig. 2 The *TeleOphta* retinal pathology screening system. Multimodal eye fundus exams are performed by trained technicians in primary care centers. Health records are analyzed automatically by the TeleOphta software. Obviously healthy cases are sent back to the primary care center immediately. Others are archived on a server. Trained readers can access these health records, possibly sorted in pathological order, and grade them, possibly looking at the nearest neighbors to support their diagnosis. Their grades are then forwarded to the screening centers

The Ophdiat[9] DR screening network in Paris is involved in the TeleOphta consortium. They provided all screening records collected in the network during the years 2008 and 2009, namely 25,702 records, for training and assessing the TeleOphta system. A screening record consists of four fundus images on average (min: 1, max: 19), i.e. two per eye, nine general information fields (age, sex, weight, ethnicity, etc.) and 18 diabetes-related information fields (type, stability, cholesterol concentration, etc.). Many general and diabetes-related information fields are sparsely filled. In one out of four examinations, the patient was referred to an ophthalmologist.

Our image characterization (Sect. 3.2) and information fusion algorithms (Sect. 3.3) were combined with lesion detectors developed by MINES ParisTech and ourselves. The combined system was able to detect patients needing referral with a specificity of 70.0 % when the sensitivity is set to that of a second human reader (80.9 %). Although a second reader achieves a higher specificity (81.5 %), this automated decision is very useful: it can safely reduce the ophthalmologist workload by 50 % [14]. Future improvements in the TeleOphta system will be mostly about decision support, for the development of new screening networks, with less experienced readers.

[9]http://reseau-ophdiat.aphp.fr.

4 Multimedia Information Retrieval for Computer-Aided Eye Surgery

Computer-aided diagnosis is not the only application that can benefit from multimedia information retrieval from ophthalmic archives. We believe computer-aided surgery also can, although this is more challenging.

During an eye surgery, the surgeon wears a binocular microscope and the output of the microscope can be recorded. It is now common practice to record every surgical procedure and to archive the resulting videos for documentation purposes. Therefore, it may be possible to automatically monitor the surgery, using the visual content of the current surgery, but also the visual content of archived surgery videos, by analogy reasoning. Such a tool would be useful to communicate information (about the tools or the patient) to the surgeon in due time, typically at the beginning of a new surgical task. In the particular case of new surgeons, the system could also provide recommendations on how to best perform the current task, based on the experience of their peers in similar surgeries (similar patients, similar implants, etc.), as well as warnings if something wrong is detected. In line with our works on content-based image retrieval, we investigated the use of content-based video retrieval to provide this kind of feedback using digital surgical video archives. Note that, in the future, the same methodology could be used for the automatic monitoring of robotized eye surgeries.

The same idea can be used for surgical task recognition, recommendation generation and warning generation; this idea is illustrated in Fig. 3 in the case of warning generation. The following of this section describes how this idea has been implemented to temporally segment and categorize surgical tasks in real-time, the first step of the envisioned computer-aided eye surgery system.

4.1 Existing Surgical Task Segmentation and Recognition Systems

In recent years, a few systems were presented for the automatic recognition of surgical gestures, assuming a known temporal segmentation of these gestures. Two approaches were evaluated for the automatic classification of surgical gestures in video clips of minimally invasive surgery [20]. If visual cues are combined with kinematic data, classification performance is pushed further [67].

A second group of systems was presented for the automatic temporal segmentation of surgical tasks or gestures, given the full surgical video. Such a system was proposed for the automatic segmentation of surgical gestures in a laparoscopic video, using Hidden Markov Models (HMM) [59]. Another system was presented for the automatic segmentation of surgical tasks, also in laparoscopic videos, using Dynamic Time Warping (DTW) or an HMM [10]. A system was proposed for the automatic segmentation and recognition of surgical gestures using both visual

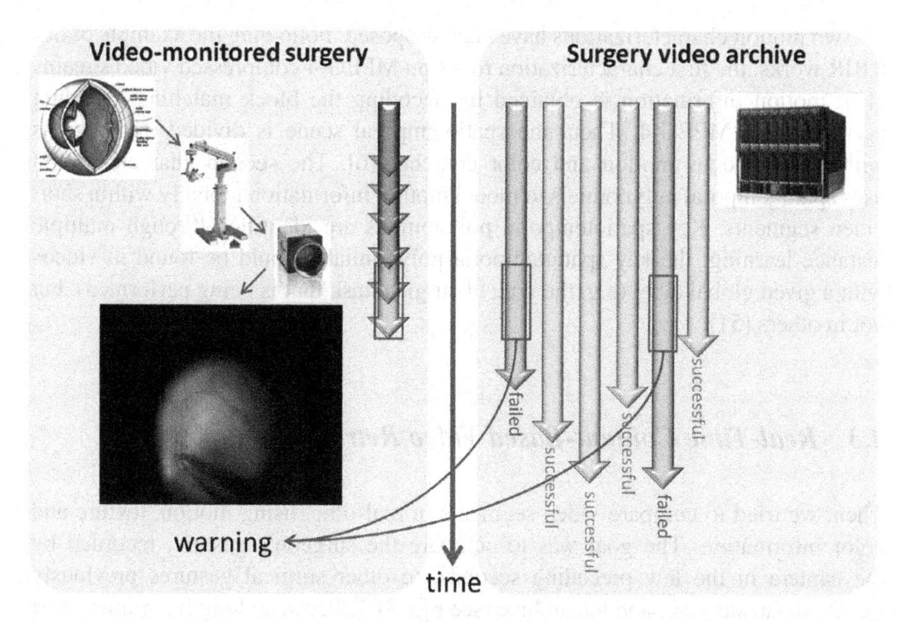

Fig. 3 Real-time video retrieval for computer-aided eye surgery. Short video segments are characterized in the current video-monitored eye surgery. Similar video segments are searched in a video archive. Based on expert annotations attached to the videos containing the nearest video segments, feedbacks can be given to the surgeon. In the case of warning generation, if the nearest neighbors are all found in surgical videos with complications, a warning should be generated

cues and kinematic data [60]. Finally, one system was presented for the automatic temporal segmentation of surgical phases in microscope videos using DTW or an HMM: the visual content of images is described by color histograms, Haar-based features and SIFT descriptors, among other features [26]. It was adapted for the automatic segmentation of cataract surgery videos into surgical phases, using visual features extracted inside the pupil only [26].

Unlike the first group of systems, we don't want to rely on a manual segmentation of the surgical tasks. Unlike the second group of methods, we don't want to wait for the surgery to be finished before processing the video: the segmentation has to be performed as the video is recorded.

4.2 *Motion Analysis*

Besides color and texture, which can be characterized like in CBIR systems, motion needs to be characterized in order to allow high-level analysis of the surgical scene. In order to characterize motion information reliably, surgical videos are normalized in order to compensate for eye motion and zoom level variations. In the case of anterior eye surgeries, such as cataract surgery, the pupil center is tracked and the zoom level is estimated using corneal reflections.

Two motion characterizations have been proposed. Following the example of our CBIR works, the first characterization relies on MPEG-4 compressed video streams [17]: motion information is obtained by decoding the block matching encoding performed by MPEG-4. Then, the spatiotemporal scene is divided into regions with homogeneous motion and color content [16]. The second characterization uses spatiotemporal polynomials to model motion information globally within short video segments. Key spatiotemporal polynomials are identified through multiple instance learning: the key spatiotemporal polynomials should be found in videos with a given global label (e.g. the type of surgical task that is being performed), but not in others [51].

4.3 Real-Time Content-Based Video Retrieval

Then, we tried to compare video segments in real-time, using motion, texture and color information. The goal was to compare the surgeon's gesture, recorded by the camera in the few preceding seconds, to other surgical gestures previously recorded and stored in a video archive (see Fig. 3). When searching for similar short video segments, and not simply video files as a whole, the number of items that should be compared to the query item explodes. And the proposed system needs to run in real-time. In order to meet the real-time constraint, a very fast similarity metric must therefore be used to compare video segments. In particular, the use of temporally flexible distance metrics such as DTW is prohibited for time reasons. An alternative solution is proposed: temporal flexibility is directly introduced in the way video segments are characterized. The idea is that video segments only need to be characterized once, whereas distances need to be computed every time the system processes a new segment, for as long as the video archive is used. So it is worth spending time computing a smart characterization for each video segment. The video segmentation characterization is obtained by concatenating and compressing video characterizations extracted from multiple subsets of the video segment (see Fig. 4). The subset selection and the distance metric are tuned through multiple instance learning [38].

4.4 Real-Time Surgical Task Segmentation and Recognition

Then, we tried to detect transitions between surgical tasks, in real-time, and to recognize the surgical tasks as their end is detected. This will allow sending recommendations to surgeons about the following task, as a function of preceding tasks. In order to detect transitions between surgical tasks, the distance metric above is tuned to optimally separate time instants when the surgeon interacts with the eye (corresponding to a surgical task or subtask) from time instants when only the eye moves (potentially corresponding to a transition between tasks). The proposed task

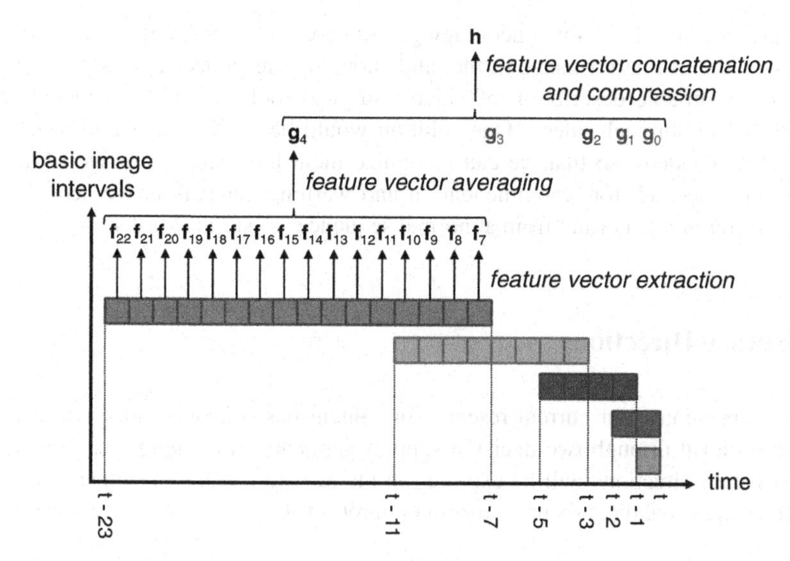

Fig. 4 Temporally-flexible video segment characterization. A feature vector is extracted from each image within a video segment. Then, feature vectors extracted from multiple subsets of the video segment (referred to as basic image intervals) are averaged. Finally, average feature vectors from all basic image intervals are concatenated and compressed. The idea is that two video segments are similar if similar actions happen in each basic image interval. But when exactly those actions happen within those basic image intervals does not matter. This makes comparisons between video segments temporally flexible

characterization method combines several sources of evidence indicating what the current task is. These sources of evidence rely on the visual content of the current task, the visual content of previous tasks or the duration of transitions between the current and previous tasks. They are combined using a conditional random field [48].

4.5 Application to Cataract Surgery

A dataset of 186 cataract surgery videos was collected at Brest University Hospital (Brest, France). In each video, a temporal segmentation was provided by one cataract expert for the nine main surgical tasks, plus miscellaneous tasks. For each surgical task, the expert indicated the date of first appearance of one tool related to this task into the field of view. Similarly, he indicated the date of last disappearance of one of these tools from the field of view. In those manually segmented videos, a good task recognition performance was achieved using the proposed flexible characterization (Sect. 4.3): an average area under the ROC curve $A_z = 0.794$ was achieved [38]. An even better performance was achieved for the joint segmentation and recognition of surgical tasks (i.e. when the system ignores the manual segmentation—Sect. 4.4): $A_z = 0.832$ on average.

These results are really encouraging. However, two challenges need to be addressed to push performance further and allow high-level decision support. First, for a more reliable description of videos, surgical tools should be detected and recognized in surgical videos. One solution would be to document tool usage in the archived videos, so that we can recognize them later through image retrieval techniques. Second, for recommendation and warning generation, we need to be able to differentiate normal from abnormal sequences of surgical tasks.

5 Future Directions

We have presented our current research in content-based image, video and health record retrieval through two decision support applications in ophthalmology. Two new research directions will be explored in the next few years: (1) expert-specific decision support and (2) secured information retrieval.

5.1 Expert-Specific Decision Support

Standard decision support systems (that rely on expert knowledge and usually focus on the detection of the most frequent pathological signs) tend to encode the cognitive processes of the average clinician. That is the reason why they have been mainly used for automatic decision (for simple tasks), not aided decision (for complex tasks). In CBR systems, two approaches have been considered to bridge the semantic gap between low-level features and the concept of clinical similarity: we used a machine learning approach; relevance feedback was used by others. With the machine learning strategy we adopted, chances are that retrieval is less successful when clinicians disagree with their peers, i.e. in difficult cases. This is because machine learning algorithms tend to generalize better for common situations to the detriment of rare situations. With relevance feedback, system adaptation only relies on the feedbacks of the current user. So, it is assumed that the user has enough experience for the case under study: because there is not experience pooling, inexperienced clinicians cannot take advantage of the experience of their peers. For the design of a personalized decision support system that can really help clinicians solve difficult cases (i.e. cases in which they have little experience), we will have to encode the differences between the personal cognitive processes of all clinicians [49]. Systems need to be trained to be complementary to human experts, not to mimic them. Mining cognitive processes (in patient records and human-system interactions) can help identify the difficult situations and support decisions in these situations.

5.2 *Secured Information Retrieval*

From the first hospital-wide Picture Archiving and Communication Systems (PACS), through regional PACS and telemedicine applications, we now come to the era of cloud computing and storage. Archives of medical data are progressively being outsourced from hospitals and will have to be processed remotely. To address security issues, particularly in terms of data privacy and confidentiality, the use of encryption mechanisms is essential and raises questions such as of how to exploit the protected data for data mining and information retrieval? This additional challenge is well worth the effort as it will give us access to much larger medical archives, and hopefully better decision support systems.

References

1. Abouelenien, M., Wan, Y., & Saudagar, A. (2012). Feature and decision level fusion for action recognition. In *Proceedings of International Conference on Computing, Communications and Networking Technologies (ICCCNT)* (pp. 1–7).
2. Amores, J. (2013). Multiple instance classification: Review, taxonomy and comparative study. *Artificial Intelligence, 201*, 81–105.
3. André, B., Vercauteren, T., Buchner, A. M., Shahid, M. W., Wallace, M. B., & Ayache, N. (2010). An image retrieval approach to setup difficulty levels in training systems for endomicroscopy diagnosis. In *Proceedings of Medical Image Computing and Computer Assisted Interventions (MICCAI)* (pp. 480–487).
4. André, B., Vercauteren, T., Buchner, A. M., Wallace, M. B., & Ayache, N. (2012). Learning semantic and visual similarity for endomicroscopy video retrieval. *IEEE Transactions on Medical Imaging, 31*(6), 1276–1288.
5. André, B., Vercauteren, T., Wallace, M. B., Buchner, A. M., & Ayache, N. (2010). Endomicroscopic video retrieval using mosaicing and visual words. In *Proceedings of IEEE International Symposium on Biomedical Imaging (ISBI)* (pp. 1419–1422).
6. Avni, U., Greenspan, H., Konen, E., Sharon, M., Goldberger, J. (2011). X-ray categorization and retrieval on the organ and pathology level, using patch-based visual words. *IEEE Transactions on Medical Imaging, 30*(3), 733–746.
7. Bettadapura, V., Schindler, G., Ploetz, T., & Essa, I. (2013). Augmenting bag-of-words: Data-driven discovery of temporal and structural information for activity recognition. In *Proceedings of IEEE Computer Vision and Pattern Recognition (CVPR)* (pp. 2619–2626).
8. Bichindaritz, I. (2006). Mémoire: A framework for semantic interoperability of case-based reasoning systems in biology and medicine. *Artificial Intelligence in Medicine, 36*(2), 177–192.
9. Bichindaritz, I., & Marling, C. (2006). Case-based reasoning in the health sciences: What's next? *Artificial Intelligence in Medicine, 36*(2), 127–135.
10. Blum, T., Feussner, H., & Navab, N. (2010). Modeling and segmentation of surgical workflow from laparoscopic video. In *Proceedings of Medical Image Computing and Computer Assisted Interventions (MICCAI)* (pp. 400–407).
11. Bruno, E., Moenne-Loccoz, N., & Marchand-Maillet, S. (2008). Design of multimodal dissimilarity spaces for retrieval of video documents. *IEEE Transactions on Pattern Analysis and Machine Intelligence, 30*(9), 1520–1533.
12. Cauvin, J. M., Le Guillou, C., Solaiman, B., Robaszkiewicz, M., Le Beux, P., & Roux, C. (2003). Computer-assisted diagnosis system in digestive endoscopy. *IEEE Transactions on Information Technology, 7*(4), 256–262.

13. Chatzichristofis, S. A., Iakovidou, C., Boutalis, Y., Marques, O. (2013). Co.Vi.Wo.: Color visual words based on non-predefined size codebooks. *IEEE Transactions on Cybernetics, 43*(1), 192–205.
14. Decencière, E., Cazuguel, G., Zhang , X., et al. (2013). TeleOphta: Machine learning and image processing methods for teleophthalmology. *IRBM, 34*(2), 196–203.
15. Douze, M., Jégou, H., Schmid, C. (2010). An image-based approach to video copy detection with spatio-temporal post-filtering. *IEEE Transactions on Multimedia, 12*(4), 257–266.
16. Droueche, Z., Lamard, M., Cazuguel, G., Quellec, G., Roux, C., & Cochener, B. (2011). Content-based medical video retrieval based on region motion trajectories. In *Proceedings of International Federation for Medical and Biological Engineering (IFMBE)* (pp. 622–625).
17. Droueche, Z., Lamard, M., Cazuguel, G., Quellec, G., Roux, C., & Cochener, B. (2012). Motion-based video retrieval with application to computer-assisted retinal surgery. In *Proceedings of IEEE Engineering in Medicine and Biology Society (EMBS)* (pp. 4962–4965).
18. Dyana, A., Subramanian, M. P., & Das, S. (2009). Combining features for shape and motion trajectory of video objects for efficient content based video retrieval. In *Proceedings of International Conference on Advances in Pattern Recognition (ICAPR)* (pp. 113–116).
19. Gao, H. P., & Yang, Z. Q. (2010). Content based video retrieval using spatiotemporal salient objects. In *Proceedings of International Petroleum Technology Conference (IPTC)* (pp. 689–692).
20. Haro, B. B., Zappella, L., & Vidal, R. (2012). Surgical gesture classification from video data. In *Proceedings of Medical Image Computing and Computer Assisted Interventions (MICCAI)* (pp. 34–41).
21. Haux, R. (2006). Health information system: Past, present, and future. *International Journal of Medical Informatics, 75*(3–4), 268–281.
22. Hoi, S. C. H., & Lyu, M. R. (2007). A multimodal and multilevel ranking framework for content-based video retrieval. In *Proceedings of International Conference on Acoustics, Speech and Signal Processing (ICASSP)* (pp. 1225–1228)
23. Hu, W., Xie, D., Fu, Z., Zeng, W., & Maybank, S. (2007). Semantic-based surveillance video retrieval. *IEEE Transactions on Image Processing, 16*(4), 1168–1181.
24. Ji, R., Duan, L. Y., Chen, J., Xie, L., Yao, H., & Gao, W. (2013). Learning to distribute vocabulary indexing for scalable visual search. *IEEE Transactions on Multimedia, 15*(1), 153–166.
25. Juan, K., & Cuiying, H. (2010). Content-based video retrieval system research. In *Proceedings of International Conference on Computer Science and Information Technology (ICCSIT)* (pp. 701–704).
26. Lalys, F., Riffaud, L., Bouget, D., & Jannin, P. (2012). A framework for the recognition of high-level surgical tasks from video images for cataract surgeries. *IEEE Transactions on Biomedical Engineering, 59*(4), 966–976.
27. Liu, Z., Li, H., Zhou, W., Zhao, R., & Tian, Q. (2014). Contextual hashing for large-scale image search. *IEEE Transactions on Image Processing, 23*(4), 1606–1614.
28. Mansencal, B., Benois-Pineau, J., Vieux, R., & Domenger, J. (2012). Search of objects of interest in videos. In *Proceedings of Content-Based Multimedia Indexing (CBMI)* (pp. 1–6).
29. Müller, H., Michoux, N., Bandon, D., & Geissbuhler, A. (2004). A review of content-based image retrieval systems in medical applications—clinical benefits and future directions. *International Journal of Medical Informatics, 73*(1), 1–23.
30. Müller, H., Seco de Herrera, A. G., Kalpathy-Cramer, J., Fushman, D. D., Antani, S., & Eggel, I. (2012). Overview of the ImageCLEF 2012 medical image retrieval and classification tasks. In *Conference and Labs of the Evaluation Forum (CLEF) 2012* working notes.
31. Naturel, X., & Gros, P. (2008). Detecting repeats for video structuring. *Multimedia Tools and Applications, 38*(2), 233–252.
32. Niemeijer, M., van Ginneken, B., Cree, M. J., et al. (2010). Retinopathy online challenge: Automatic detection of microaneurysms in digital color fundus photographs. *IEEE Transactions on Medical Imaging, 29*(1), 185–195.

33. Pan, W., Coatrieux, G., Cuppens, N., Cuppens, F., & Roux, C. (2010). An additive and lossless watermarking method based on invariant image approximation and haar wavelet transform. In *Proceedings of IEEE Engineering in Medicine and Biology Society (EMBS)* (pp. 4740–4743).
34. Patel, B. V., Deorankar, A. V., & Meshram, B. B. (2010). Content based video retrieval using entropy, edge detection, black and white color features. In *Proceedings of International Conference on Chemical Engineering and Technology (ICCET)* (pp. 272–276).
35. Perner, P. (Ed.). (2008). *Case-based reasoning on images and signals*. Studies in Computational Intelligence (Vol. 73). Heidelberg: Springer.
36. Pires, R., Jelinek, H. F., Wainer, J., Goldenstein, S., Valle, E., & Rocha, A. (2013). Assessing the need for referral in automatic diabetic retinopathy detection. *IEEE Transactions on Biomedical Engineering, 60*(12), 3391–3398.
37. Quantin, C., Cohen, O., Riandey, B., & Allaert, F. A. (2007). Unique patient concept: A key choice for european epidemiology. *International Journal of Medical Informatics, 76*(5–6), 419–426.
38. Quellec, G., Charrière, K., Lamard, M., Droueche, Z., Roux, C., & Cochener, B. (2014). Real-time recognition of surgical tasks in eye surgery videos. *Medical Image Analysis, 18*(3), 579–590.
39. Quellec, G., Lamard, M., Abràmoff, M. D., Decencière, E., Lay, B., & Erginay, A. (2012). A multiple-instance learning framework for diabetic retinopathy screening. *Medical Image Analysis, 16*(6), 1228–1240.
40. Quellec, G., Lamard, M., Bekri, L., Cazuguel, G., Roux, C., & Cochener, B. (2010). Medical case retrieval from a committee of decision trees. *IEEE Transactions on Information Technology in Biomedicine,14*(5), 1227–1235.
41. Quellec, G., Lamard, M., Cazuguel, G., Bekri, L., Daccache, W., & Roux, C. (2011). Automated assessment of diabetic retinopathy severity using content-based image retrieval in multimodal fundus photographs. *Investigative Ophthalmology and Visual Science, 52*(11), 8342–8348.
42. Quellec, G., Lamard, M., Cazuguel, G., Cochener, B., & Roux, C. (2009). Multimodal information retrieval based on DSmT. Application to computer aided medical diagnosis. In F. Smarandache & J. Dezert (Eds.), *Advances and applications of DSmT for information fusion III*, chap. 18 (pp. 471–502). Ann Harbor: American Research Press.
43. Quellec, G., Lamard, M., Cazuguel, G., Cochener, B., & Roux, C. (2010). Adaptive nonseparable wavelet transform via lifting and its application to content-based image retrieval. *IEEE Transactions on Image Processing, 19*(1), 25–35.
44. Quellec, G., Lamard, M., Cazuguel, G., Cochener, B., & Roux, C. (2010). Wavelet optimization for content-based image retrieval in medical databases. *Medical Image Analysis, 14*(2), 227–241.
45. Quellec, G., Lamard, M., Cazuguel, G., Cochener, B., & Roux, C. (2012). Fast wavelet-based image characterization for highly adaptive image retrieval. *IEEE Transactions on Image Processing, 21*(4), 1613–1623.
46. Quellec, G., Lamard, M., Cazuguel, G., Roux, C., & Cochener, B. (2008). Multimodal medical case retrieval using dezert-smarandache theory with a priori knowledge. In *Proceedings of International Federation for Medical and Biological Engineering (IFMBE)* (pp. 716–719).
47. Quellec, G., Lamard, M., Cazuguel, G., Roux, C., & Cochener, B. (2011). Case retrieval in medical databases by fusing heterogeneous information. *IEEE Transactions on Medical Imaging, 30*(1), 108–118.
48. Quellec, G. , Lamard, M., Cochener, B., & Cazuguel, G. (2014). Real-time segmentation and recognition of surgical tasks in cataract surgery videos. *IEEE Trans Med Imaging, 33*(12), 2352–2360.
49. Quellec, G., Lamard, M., Cochener, B., Droueche, Z., Lay, B., & Chabouis, A. et al. (2012). Studying disagreements among retinal experts through image analysis. In *Proceedings of IEEE Engineering in Medicine and Biology Society (EMBS)* (pp. 5959–5962).

50. Quellec, G., Lamard, M., Cochener, B., Roux, C., & Cazuguel, G. (2012). Comprehensive wavelet-based image characterization for content-based image retrieval. In *Proceedings of the Conference on Content-Based Multimedia Indexing (CBMI)*.
51. Quellec, G., Lamard, M., Droueche, Z., Cochener, B., Roux, C., & Cazuguel, G. (2013). A polynomial model of surgical gestures for real-time retrieval of surgery videos. In *Lecture Notes in Computer Science: Vol. 7723. Proceedings MCBR-CDS* (pp. 10–20).
52. Ren, R., & Collomosse, J. (2012). Visual sentences for pose retrieval over low-resolution cross-media dance collections. *IEEE Transactions on Multimedia, 14*(6), 1652–1661.
53. Rublee, E., Rabaud, V., Konolige, K., & Bradski, G. R. (2011). ORB: An efficient alternative to SIFT or SURF. In *Proceedings of IEEE International Conference on Computer Vision (ICCV)* (pp. 2564–2571).
54. Safran, C., Bloomrosen, M., Hammond, W. E., et al. (2007). Toward a national framework for the secondary use of health data: An american medical informatics association white paper. *Journal of the American Medical Informatics Association, 14*(1), 1–9.
55. Sivic, J., Russell, B. C., Efros, A. A., Zisserman, A., & Freeman, W. T. (2005). Discovering objects and their location in images. In *Proceedings of IEEE International Conference on Computer Vision (ICCV)* (pp. 370–377).
56. Strat, S. T., Benoit, A., & Lambert, P. (2013). Retina enhanced SIFT descriptors for video indexing. In *Proceedings of the Conference on Content-Based Multimedia Indexing (CBMI)* (pp. 201–206).
57. Sweldens, W. (1998). The lifting scheme: A construction of second generation wavelets. *SIAM Journal on Mathematical Analysis, 29*(2), 511–546.
58. Syeda-Mahmood, T., Ponceleon, D., & Yang, J. (2005). Validating cardiac echo diagnosis through video similarity. In *Proceedings of ACM Multimedia* (pp. 527–530).
59. Tao, L., Elhamifar, E., Khudanpur, S., Hager, G. D., & Vidal, R. (2012). Sparse hidden markov models for surgical gesture classification and skill evaluation. In *Proceedings of Information Processing in Computer-Assisted Interventions (IPCAI)* (pp. 167–177).
60. Tao, L., Zappella, L., Hager, G. D., & Vidal, R. (2013). Surgical gesture segmentation and recognition. In *Lecture Notes in Computer Science: Vol. 8151* (pp. 339–46).
61. Tsikrika, T., Kludas, J., & Popescu, A. (2012). Building reliable and reusable test collections for image retrieval: The Wikipedia task at ImageCLEF. *IEEE Multimedia, 19*(3), 24–33.
62. Tutac, A. E., Cretu, V. I., & Racoceanu, D. (2010). Spatial representation and reasoning in breast cancer grading ontology. In *Proceedings of International Joint Conference on Computational Cybernetics and Technical Informatics (ICCC-CONTI)* (pp. 89–94).
63. Vieux, R., Benois-Pineau, J., Domenger, J. P. (2012). Content based image retrieval using bag of regions. In *Proceedings of Multimedia Modeling (MMM)* (pp. 507–517).
64. Xu, D., & Chang, S. F. (2008). Video event recognition using kernel methods with multilevel temporal alignment. *IEEE Transactions on Pattern Analysis and Machine Intelligence, 30*(11), 1985–1997.
65. Yang, Y., & Newsam, S. (2013). Geographic image retrieval using local invariant features. *IEEE Transactions on Geoscience Remote Sensing, 51*(2), 818–832.
66. Yuan, C., Li, X., Hu, W., Ling, H., & Maybank, S. J. (2014) Modeling geometric-temporal context with directional pyramid co-occurrence for action recognition. *IEEE Transactions on Image Processing, 23*(2), 658–672.
67. Zappella, L., Béjar, B., Hager, G., & Vidal, R. (2013). Surgical gesture classification from video and kinematic data. *Medical Image Analysis, 17*(7), 732–745.
68. Zheng, L., & Wang, S. (2013). Visual phraselet: refining spatial constraints for large scale image search. *IEEE Signal Processing Letters, 20*(4), 391–394.

Characterisation of Data Quality in Electronic Healthcare Records

Sheena Dungey, Natalia Beloff, Rachael Williams, Tim Williams, Shivani Puri, and A. Rosemary Tate

1 Introduction

The use of electronic healthcare systems for recording patient treatment history is well established across the UK healthcare sector, the potential benefits of using such systems being numerous. Within the primary care setting, electronic healthcare records (EHR) can provide a near complete picture of patient care over time. This not only affords the opportunity to improve patient care directly through effective monitoring and identification of care requirements but also offers a unique platform for both clinical and service-model research [1] essential to the longer term development of the health service. The potential for using routinely collected patient records for research purposes has been steadily increasing [2] with recent advances and diminishing technical barriers in data storage and information processing. There are, however, significant challenges in using EHRs effectively in the research setting and in ensuring the quality of data recorded for this purpose. Incorrect or missing data can render records as useless or indeed misleading such that conclusions drawn from the data could have a negative impact.

The aim of this chapter is to outline both the key challenges to the management and assessment of data quality in EHRs and the key considerations for meeting these challenges. The Clinical Practice Research Datalink database CPRD GOLD,

S. Dungey (✉)
Department of Informatics, University of Sussex, Brighton BN1 9QJ, UK

MHRA, 151 Buckingham Palace Road, London SW1W 9SZ, UK
e-mail: sheena.dungey@mhra.gsi.gov.uk

N. Beloff • A.R. Tate
Department of Informatics, University of Sussex, Brighton BN1 9QJ, UK

R. Williams • T. Williams • S. Puri
MHRA, 151 Buckingham Palace Road, London SW1W 9SZ, UK

A. Briassouli et al. (eds.), *Health Monitoring and Personalized Feedback using Multimedia Data*, DOI 10.1007/978-3-319-17963-6_7

globally recognised as being one of the largest and most detailed sources of electronic patient data, will be used as an example throughout. In Sect. 2, the concept of data quality is presented within the setting of primary care databases and a framework for its assessment is set out, based on findings of an investigation carried out on CPRD GOLD. In Sect. 3, the importance of understanding data quality of an individual source of data in relation to alternative sources, both intra- and internationally, is examined, posing the emerging challenges to the future use EHRs for research. Finally Sect. 4 investigates data quality requirements from the perspective of a range of stakeholders through discussion of a day-long CPRD-led data quality workshop and we consider the way forward to a more comprehensive approach to tackling issues of data quality in EHRs.

2 Developing a Data Quality Framework at the Clinical Practice Research Datalink (CPRD)

2.1 Defining Data Quality Within the Context of Electronic Healthcare Records

The widely accepted conceptualization of data quality is that it is defined through "fitness for use" [3] i.e. the ability of the data to meet the requirement of the user. To be able to use data to attain information, it must be complete, consistent and accurate and so forth. Actually defining what can be considered as complete or consistent, thus determining the quality of the data, will reflect the required use of the data and will therefore vary across purpose.

The primary purpose for recording patient data within the General Practice (GP) setting is to facilitate patient care and to assess and optimise the care of the practice population as well as to provide documentation for administrative and legal purposes. Guidelines are provided [4] to promote good recording practice including the use of codes to express clinical information and standard procedures for capturing information from outside the practice and with regards to sharing information. In 2004 the Quality and Outcomes Framework (QOF) was introduced with the provision of large financial incentives based on practice achievement on a range of quality of care indicators over 22 clinical areas [5]. This had a major impact on the use of clinical recording software and the development of a more standardised approach to data recording, particularly in coding of key disease areas [6] and recording of key lifestyle measures such as smoking status [7]. However, despite efforts to ensure quality and consistency in data recording, using GP records for research remains challenging [8].

The primary care setting is complex and constantly changing and this reflected in similarly complex and transient recording mechanisms. Coding systems, such as the Read code system predominantly used to categorise clinical events in UK primary care databases, including CPRD GOLD, are a prime example. Introduced to curb the vast number of ways a clinical concept can be described, the number of codes has grown massively increasing risk of inconsistency in use and necessitating staff training [9, 10]. Further still, not all data is coded instead being entered as free text; such information is challenging to extract at the research stage, particularly amid growing concerns over record anonymisation [11, 12].

Additionally, some level of bias in data collection is hard to avoid. Whilst QOF has reaped many benefits to recording quality, these are tied to the clinical areas covered by QOF [13, 14]. Stigmatization of certain conditions is believed to lead to under-recording [15, 16] and the way data is recorded may depend on the type of staff entering the data and when it's entered relative to actual consultation [10]. And this is to name but a few examples.

Ultimately, the effort-benefit balance for detail of recording sits differently for patient care, for which GPs are striving to provide a face-to-face consultation in a time-pressurised environment, and for research, where meticulous and consistent recording is crucial.

Whilst the importance of addressing data quality in the reuse of EHR for research is widely acknowledged, with various frameworks having been put forward [17–19], there is no commonly recognised methodology for undertaking an assessment of data quality in this setting. Hitherto, data quality has been mainly addressed via one-off validation studies [20]. The evolution of a unified approach has no doubt been hampered by the vast number of possible measures and the variability in importance of measures between studies leading to data quality considerations derived for a given study being isolated to that study. A recent Clinical Practice Research Datalink (CPRD) sponsored project entitled "Methods to characterise and monitor data quality in the Clinical Practice Research Datalink" led by the University of Sussex, has addressed the need for standardisation and facilitation in data quality assessment. The rest of this section recounts the development of an approach for characterising data quality in primary care databases [6] based on investigative work carried out on the CPRD database.

2.2 *Introduction to the Clinical Practice Research Datalink*

The Clinical Practice Research Datalink (CPRD) GP OnLine Database (GOLD) contains diagnostic, demographic and prescribing information for over 14 million patients, broadly representative of the UK, providing a significant potential resource for public health and epidemiological research—its usage has led to over 1,500 published research studies and conference abstracts to date [21].

What is now known as CPRD was initially developed by an Essex general practitioner, Dr Alan Dean, to facilitate day-to-day management of his own general practice. This was so successful that a venture capital company was set up in 1987 named VAMP (Value Added Medical Products Ltd) to recruit other practices and form an information base. In late 1993, the company was taken over by Reuters and the database was offered to the Department of Health as independent custodian to supervise access to the information for the benefit of public health. Early in 1994, the Office of Population Censuses and Statistics took over maintenance and running of the information resource (which was then renamed to General Practice Research Database), until 1999 at which point the Medicines Control Agency took over. Throughout this period, data collection and validation was maintained without loss of information from individual practices. This agency became the Medicines and Healthcare Products Regulatory Agency (MHRA) in 2003 following a merger with the Medical Devices Agency. Since then, use of the database expanded within the UK and overseas. In March 2011, the UK Government launched its "Plan for Growth" [22] which detailed steps needed to enable the British economy to become more internationally competitive. As part of this initiative the Government pledged to build a consensus on using e-health record data to create a unique position for the UK in health research. Under this motivation, CPRD was launched in April 2012 co-funded by the National Institute for Health Research (NIHR) and the MHRA.

CPRD is aiming at providing capability, products and services across a number of areas including secure integrated and linked data collection and provision, as well as advanced observational and innovative interventional research services [21, 23]. That the data is of high quality and can be validated as being so is paramount for all work carried out at CPRD.

2.3 CPRD GOLD Data Quality: Developing a Methodological Approach for Characterising Data Quality in Primary Care Research Databases

In order to ensure high quality data, CPRD historically has constructed a set of internal data quality measurements, at both patient and practice level. The practice level quality assessment is manifested by an 'up-to-standard' (UTS) date derived using a CPRD algorithm that looks primarily at practice death recording and gaps in the data. At patient level, records are labelled as 'acceptable' for use in research by a process that identifies and excludes patients with non-contiguous follow up or patients with poor data recording that raises suspicion as to the validity of that patient's record. However, these checks are limited in scope, and with the expansion and increasing use of the database a more comprehensive approach is needed whereby CPRD undertake data quality assessment for these data both individually and jointly as linked data sets.

Work carried out under the current CPRD/University of Sussex data quality project commenced with a comprehensive examination of CPRD data quality and correlations between different measures, with a view to reducing the effective number of variables needed to characterise data quality [6]. This study was carried out as part of a wider project funded by the UK Technology Strategy Board and incorporated input from a user group consisting of representatives of pharmaceutical companies and clinical research organisations [23].

Several frameworks for data quality have been suggested in the literature. In our opinion, the use of different frameworks may not in itself be a major problem if clear definitions and examples are provided and all important aspects are considered. Ultimately, use of a framework is advantageous in encouraging a consistent and comprehensive approach to data quality assessment and hence, after a review of the literature, a suitable framework for describing dimensions of data quality was proposed. The dimensions include accuracy, validity, reliability, timeliness, relevance, completeness and integrity with full definition and examples given in [6]. Measures were then identified according to the framework and also as either basic or study-specific measures. Here, the definition as basic pertains to general measures such as recording of height and weight, duplicate records or missing values for fields such as staff ID. Definition as condition-specific pertains to measures characterising the coding of specific conditions. Note, all Clinical, Referral and Test event records have an associated Read code, as described above.

Table 1 shows the correlation matrix for selected basic measures. For most of the variables examined representing different aspects of patient records, correlations were very weak with (Spearman) correlation coefficients typically below 0.2 (absolute value). Most practices that were "bad" at recording one thing were almost always fine at recording all others. However, correlations between variables representing the same aspect were much higher. For example, percentages representing completeness of patient's height, weight, smoking and alcohol status were found to be highly correlated (Pearson coefficient ≥ 0.79). The same was found to be true for study-specific measures for selected groups of patients, e.g. diabetes patients.

Additionally, the quality of coded data for research purposes (such as specificity and consistency of coding) in CPRD GOLD was found to be reasonably high for most of the criteria that we measured, especially in more recent years. Recording of most of the data elements that were investigated improved significantly between 2000 and 2010 with a noticeable improvement in 2004 for measures (such as those related to diabetes recording) that are included the Quality Outcomes Framework (QOF) introduced in that year.

2.4 *Proposed Approach for Assessing Data Quality for Research*

The fact that correlations between dissimilar variables are weak, representing the variability in recording for different criteria within each practice, leads to the necessity of an approach in which most of the data quality metrics are tailored to the intended use of the data. This approach was supported by the user group who agreed that some variables will be much more relevant to them than others, for example, the variables relating to the study-specific patient selection criteria. Additionally, study-specific variables are more likely to be intercorrelated and aggregation of variables into data quality summary scores becomes more feasible.

Table 1 Correlation matrix for general measures no time element—coefficients calculated using Stata 11 (StataCorp. 2009. Stata Statistical Software: Release 11. College Station, TX: StataCorp LP) [9] extracted from 528 practices contributing data to CPRD over time period 2000–2010

Framework category	Variable	Valid reg. date	Valid age	Duplicates	Consultation date	Clinical event date	Referral date	Weight	Smoking	Alcohol
Validity	**Valid reg. date**	1.00								
Validity	**Valid age**	0.03	1.00							
Accuracy	**Duplicates**	0.01	0.03	1.00						
Validity	**Cons. date**	0.04	0.01	0.43	1.00					
Validity	**Cli. event date**	0.02	0.01	0.37	0.84	1.00				
Validity	**Ref. date**	−0.02	−0.09	−0.04	−0.19	−0.13	1.00			
Accuracy	**Weight**	−0.04	0.03	0.10	0.25	0.27	−0.03	1.00		
Accuracy	**Smoking**	−0.02	0.04	0.11	0.25	0.26	−0.02	0.83	1.00	
Accuracy	**Alcohol**	−0.05	0.08	0.11	0.21	0.24	−0.05	0.88	0.79	1.00

The disadvantage of this approach is that it may be necessary to measure data quality dynamically on a study-by-study basis (however, many criteria will be common across studies, e.g. completeness of recording of registration and life-style measures). To address this, computational methods for facilitating the dynamic calculation of study-specific measures are being investigated as part of the on-going work of the CPRD/University of Sussex data quality project.

It is proposed that basic checks are always carried out first for consistency of data elements between tables, duplicate values, missing values etc., before checking more complex elements. While this may seem obvious, in our experience these are often overlooked and even if the checks are carried out they are not often reported. It is also very important to investigate completeness and correctness of elements, such as dates and gender, as more complex elements will depend on these—for example if the registration dates of many patients are invalid then the incident rates will be flawed. Once these basic checks have been carried out, data quality measures based upon the intended use of the data can be derived via the following steps:

1. List all data elements required to define the cohort for the particular study, including all elements that these are dependent upon, e.g. registration and transfer out dates and specificity of coding of condition(s) of interest.
2. List all other elements that will be needed for the study, e.g. test results, smoking status, type of consultation.
3. According to the framework, determine data quality measures associated with each data element, specifying any conditions which must apply for a given data quality measure to be relevant. For example, in validating the coding pertaining to a condition, one can utilise the framework, working through the different components such as: **Accuracy**: are there coding errors? (e.g. a type 1 code for a patient diagnosed as type 2) and is there the coverage of expected associated tests? (e.g. HbA1c for diabetes). **Timeliness**: is the coding consistent over time? (e.g. consistent coding indicating the severity of a condition over time). **Relevance**: is the coding specific (e.g. the type of diabetes is given) . . . and so on.
4. It is proposed that the vast array of measures can be calculated using a contained set of core, input-driven computational routines. The underlying computations are identified as assessments of missing and implausible data, tested across the different natural structures of the database, namely within a given consultation record or over time. The inputs are a Read code list for a given condition or set of events and likewise for a set of associated events (e.g. diabetes and HbA1c test); the location of the entities of interest within the database (for HbA1c this would be the test table); a description of expected relationships between the data entities (here, as a gold standard, it is expected that three HbA1c should be recorded within 1 year). The proportion of patients failing each check can then be calculated at practice level.
5. Calculate incidence and prevalence rates for condition and check that these agree with data from the published literature and other sources. This step could be skipped if published validation studies exist.

6. Construct a set of indicators or scores for each practice. These could be the values of the practice based variables (i.e. the total number of fails of each measure for a given practice), or a combination of them (combination of data quality variables, as discussed above, is most likely possible for measures relating to a specific condition). The most appropriate method for combining variables into scores will depend on their intercorrelations and the intended use of the data. For more basic measures, simple thresholding could be applied for acceptable values. It is emphasised that scores should be used as a guide for further investigation rather than a hard and fast method of eliminating poor quality data. Exploring correlations and combining measures, thresholding measures and profiling patient sets (e.g. looking for underlying trends in poor quality such as patient age or particular ranges of test results) are essential components of an investigation into data quality.

2.5 *Points for Further Consideration*

In this study we investigated only coded data; however, additional information can be recorded in the free text. Free text is not widely used due to the cost and governance of anonymisation and wider difficulties of information extraction. However, free text could be highly valuable for validating coded information and for finding missing information that has not been coded [10, 24, 25], free text in relation to perceived data quality will be an important area of focus for future research. An interesting question being whether the use of free text could itself be used as an indicator of data quality relating to the completeness of recording.

Moving forward, it will be crucial to gauge the actual implications of poor quality to research outcomes to derive truly meaningful measures of quality. Whilst a measure of poor quality can indeed be defined as a shortcoming of the data in meeting a gold standard in data recording, this is not an absolute concept and does not necessarily predict consequences of including the poor quality data on the study outcomes derived from the data. This issue is currently being investigated by comparing outcomes from different data sets subject to a range of data quality constraints.

There is also a distinct need to understand the quality of the database compared with other sources of data. Although based on an investigation of a primary care database, much of our proposed approach would be equally applicable to other health care databases that are used for research, such as hospital records or registries, and also to linked data sets. The broader challenges involved in assessing multi-system data quality are the focus of the next section.

3 Emerging Challenges of Data Quality: Combing Data from Disparate Sources

3.1 Introduction

The potential advantages to research outcomes of inferring information from multiple sources, therefore extending the breadth and density of information available, are considerable [2]. However, if different sources of the same data are not comparable it indicates a quality issue intrinsic to one or both of the systems as a whole. If the limitations of data quality in different sources are not understood and addressed, poor quality may be amplified as a course of the linking process, manifesting as bias in conclusions drawn [26].

In this section, data quality issues relating to the combination of data from different national healthcare databases, from different domains of the UK healthcare system and from different data recording software systems are addressed, pursuing CPRD GOLD as an example, to pose what are widely perceived to be the key emerging challenges to the future use of EHRs for research.

3.2 *TRANSFoRM and the NIVEL Data Quality Framework*

The TRANSFoRm (Translational Research and Patient Safety in Europe) project [27] is an EU funded collaboration intended as a milestone project in the use of primary care EHRs for research. The aim is to provide interoperability between primary care databases (including CPRD GOLD) across Europe in order to facilitate research across resources, requiring common standards for data integration, data presentation, recording, scalability, and security. An extensive body of work has been carried out under the project; including a component to develop methodology for assessing and comparing primary care EHR data quality within different European databases, led by NIVEL (Netherlands Institute for Health Services Research) and in collaboration with CPRD [28].

Here we briefly present an example from the application of the NIVEL framework to the TRANSFoRm diabetes use case [29], the aim of which is to create a database of patients with type-2 diabetes containing genetic and phenotypic information compiled from genetic and primary care data sources (repositories). The combination of data from CPRD GOLD and NPCD[1] and derivation of a set of data quality measures were carried out according to the following steps. The study purpose and population as determined by the study purpose were defined for each database (via the individual national coding systems)—in this case the population of

[1]The Netherlands national primary care database (NPCD), hosted by NIVEL, holds information from about 1.5 million patients (approximately 10 % of the total population).

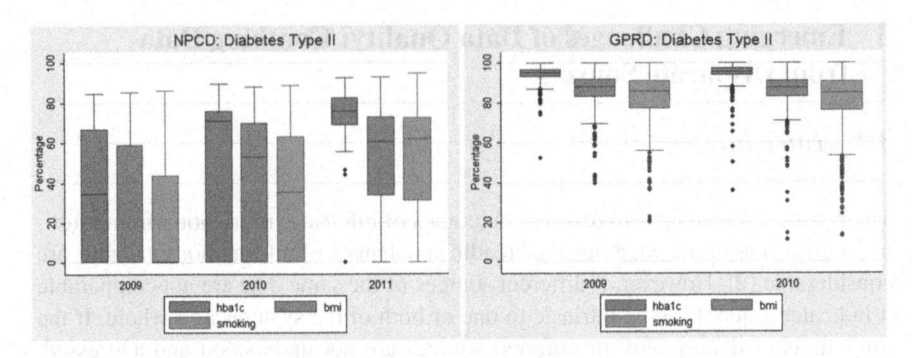

Fig. 1 Percentage of patients with type-2 diabetes at a given practice having at least one measurement of a given type within a given year for NPCR (300 practices) and CPRD (650 practices)

patients with type-2 diabetes. The data elements of interest to the researcher, such as related coded diagnoses and test results, were then set out. Finally, a quality assessment was made in terms of an evaluation of completeness, accuracy and correctness characteristics for each data element. An example is given in Fig. 1 of the completeness metric of three required data elements: measurements for hbA1c, smoking and weight [28]. Clearly, the degree of recording of these entities differs greatly between the two databases.

On comparison and integration of data derived from different national healthcare systems, data quality may be compromised by structural and operational differences in the healthcare systems. For one, there is variation across Europe in the extent that patient treatment is referred by the GP (largely the case in the UK). Whilst the Netherlands healthcare system is also based on the "Gatekeeper" model, patients in Holland are commonly seen by diabeticians rather than GPs which could account for the lack of measures for many patients. Differences in the data coding systems used present further challenges. The UK Read code system is greatly more complex than the ICPC coding system used across much of Europe, for example, for diabetes diagnosis there are only 2 standard ICPC codes compared to over 250 UK Read codes. Additionally the use of both coding systems has been shown to be variable within a given country and to be dependent on software package used to record data [30].

A wealth of other factors such as the age, set-up and management of a given database could be consequential in the recording of such events, as in Fig. 1, leading to patient information which is not, in the first instance, comparable across different databases. In the example of the diabetes use case, comparing prevalence between the two databases was problematic as the NIVEL database was much newer and thus some patients who had been diagnosed earlier may not be found. Key themes pertaining to data quality issues are summarised in Table 2 at the end of this section.

Ultimately this case study demonstrates the significant challenges faced on comparing just two European databases and the importance of ensuring a formalised, unified approach to assessing data quality for future linking of medical datasets across Europe.

Table 2 Factors affecting data quality, organised by stage of data usage from occurrence of the event to end-use of data

Data stage	Data stage factors	Potential effect on data quality
Occurrence of event	Organisational aspects of the health care system	For example, in a non-gatekeeping system (where patient treatment is referred through GP), an event may take place outside primary care, resulting in fewer event readings in primary care settings.
	Financial incentives in the health care system	The reimbursement system in one country may stimulate event readings—as occurs with the UK QOF system.
	Quality of care guidelines	For example, if a guideline says an event reading should be done every year, it will be more likely that such a measurement takes place.
	Practice workload	Practice workload may have a negative effect on the carrying out of clinical events.
Recording of event	Computerisation and EHR software	Studies [24] have shown there to be considerable differences between software packages in the way episodes of care are recorded.
	Strategic recording behaviour	Separate reimbursement schemes for patients with chronic illness will stimulate GPs to diagnose patients with chronic disease.
	Recording guidelines	Absence of recording guidelines will lead to less accurate, less complete and less correct data.
	The use of coding systems and free text	For example, the variety of ICPC codes is much smaller than Read codes or Snomed coding system.
	Knowledge	Software packages and coding systems may enable GPs to record effectively, but if a GP is not trained, this may be of no use.
	Practice workload	Shortage of time in a consultation will be detrimental recording behaviour.

(continued)

Table 2 (continued)

Data stage	Data stage factors	Potential effect on data quality
Data extraction	Extraction software	Extraction software that determines which data elements will be extracted.
	Governance issues	Some patients/practice will opt out of data sharing.
Data import	Capacity of database to capture data extract	Semantic interoperability may not be achievable across different software packages.
	Pre data entry quality control measures	This may for example mean that data that are incomplete are not entered into the database.
Generation of a research data file	Selection of data	Quality checks/filters may be employed such that not all data that is in a repository will go into the data file that is used by a researcher.
	Linkage studies	Where data is linked, the resulting database will may hold only data on the common population.
	Governance of repositories	There may be regulations restricting use of a certain repository for a certain purpose affecting completeness of data.
Data analysis	Choice of analysis method	Different methods of analysis as used by different researchers may render different results.

3.3 Linked Data Sources

Linking primary care data with data from different areas of UK healthcare is an evolving CPRD specialist service with data volume and coverage, as well as methodological expertise, seeing significant recent expansion. The need for a likewise development in classification and methodology for assessing data quality issues associated with linked data, and ensuring these considerations are a core component of the process of combining data from disparate sources, is now presented considering two CPRD linkage case studies.

3.3.1 Cancer Recording and Mortality in the General Practice Research Database[2] and Linked Cancer Registries

Boggon et al. [31] carried out an investigation into the completeness of case ascertainment in UK cancer registries collated into the National Cancer Data Repository (NCDR) by comparing information held within the NCDR to that of GPRD. UK cancer survival has been reported as being poor in international comparisons, however, it has been suggested that inaccuracies in cancer registration could invalidate international comparisons of cancer survival [32]. In particular, if patients with a good prognosis are missed or if patients are detected initially through death certification and if tracing backwards is inadequate, reported survival may be poorer than true survival.

At present, cancer registration is largely based on information supplied by hospitals and from death certification (via the Office for National Statistics). GPs in England do not routinely receive information directly from cancer registries and there has previously been little investigation comparing data from cancer registries and GP records [33, 34].

Boggon et al. found that on comparison of the two databases, levels of concordance between cancer registries and GPRD were reasonably high; however, numbers of patients known only to one dataset were non-trivial and levels of concordance in recording were observed to vary by cancer type. Overall survival rates were found to be higher in GPRD cancer cases. Ultimately the root of the recording disparities revealed in this study lies in the differences in how data is recorded for the two depositories such as when the data is recorded in relation to the event date and where the data is being generated in the first place.

[2]Study conducted before the 2012 transition from GPRD to CPRD GOLD.

3.3.2 Completeness and Diagnostic Validity of Recording Acute Myocardial Infarction Events in Primary Care, Hospital Care, Disease Registry, and National Mortality Records: Cohort Study

Herrett et al. [35] compared the incidence, recording, agreement of dates and codes, risk factors, and all-cause mortality of acute myocardial infarction recorded in four national health record sources encompassing primary care, hospital care, the national acute coronary syndrome registry, and the national death registry.

In over 2,000 patients, each data source missed a substantial proportion (25–50 %) of myocardial events. Again, missing data could be related to non-random features of the data life-span, particular to each data source. CPRD GOLD was the single most complete source of non-fatal myocardial infarction demonstrating the importance of incorporating primary care data in order to reduce biased estimates of incidence. Also highlighted here was the need for research into how electronic health record data are coded and how this can be improved. Additionally, it was concluded that more extensive cross referencing is required against additional sources of information on myocardial infarction, including investigation of electronic free text recorded by general practitioners (for example, diagnoses that are not recorded using a Read code).

Both the above examples demonstrate how data representing the same set of events but recorded in different settings can lead to different conclusions being drawn about those events. The value of linking these EHR databases is not that it will identify the gold standard or a superior database. Rather, the linkage will complement the information provided in each data source. An assessment of data quality is necessary to identify valid information within different sources and to understand why information may not be valid.

Conversely, the potential for comparison that is made possible through linkage enables greater insight into data quality issues. It may not be apparent that data is missing or incorrect until different sources of data are compared. Hence the increased utilisation of linkages will be important for development of more comprehensive and more fit-for-purpose data quality assessment procedures.

3.4 Comparing Data and Data Quality for Different GP Software Systems

Whist CPRD GOLD data is currently collected from practices using a single software system (Vision), planned CPRD expansion will encompass data collection from additional data recording systems. In this scenario the reason for recording data is the same whichever source data is obtained from, unlike the examples discussed in Sect. 3.3, and is recorded under the same national system unlike the case discussed in Sect. 3.2. However, it would be naïve to dismiss the potential of the mechanism for recording of a given software system to affect recording behaviour. UK systems are based around a problem-oriented model [36] i.e. events such as referrals and prescriptions are linked to a patient-problem, however, system-functionality can vary considerably under this premise [37]. Indeed [38] reported differences in the provision of primary care, as determined by levels of achievement on QOF indicators, dependent on the choice of software system. That variation is observed in the primary setting for data recording, without doubt warrants the investigation into the possible implications to the research setting.

To date there has not been a great deal of investigation into the impact of software system choice on research outcomes. Recently Reeves et al. [39] reported encouraging findings comparing the use of two primary care databases, CPRD (Vision software) and QResearch (EMIS software) [40], to carry out a study into the use of statins in patients with ischaemic heart disease. Whilst certain data characteristics were reported to vary between the different systems, the impact on the research findings was observed to be minor. However, as has been stressed throughout the course of this chapter, quality considerations vary greatly from study to study and more comprehensive examination is required. Understanding differences at this level may additionally aid a more fundamental understanding of the recording process at the point of care and help identify areas for focus in quality assessment.

3.5 Summary of Key Factors Affecting Data Quality

Key themes of the preceding sections are summarised in Table 2, based on the NIVEL/TRANSFoRm stepwise approach to data quality [28].

4 Report on 2013 Workshop on Data Quality

Throughout the preceding sections of this chapter, the scope of data quality in EHRs has been presented as wide-ranging, multi-faceted and dependent on use. Ultimately, the development of an approach for systematically assessing data quality for research needs to incorporate the expertise of the broad range of EHR users and contributors. In 2013, CPRD sponsored a one-day workshop entitled "Towards a common protocol for measuring and monitoring data quality in European primary care research databases" [41]. Researchers, clinicians and database experts, including representatives from primary care databases in the UK [6], Catalonia [42], Norway [43] and Spain [44] and from the Primary Care Information Services (PRIMIS) [45], were invited to give their perspectives on data quality and to exchange ideas on which data quality metrics should be made available to researchers.

The key findings from the discussions which took place throughout the day are summarised against the workshop objectives as follows:

1. Share experiences of assessing data quality in electronic health records (EHRs).
   The expectation was that there would be a diverse set of perspectives from stakeholders. However, throughout the course of the day's discussions, there was a surprising amount of consensus, in particular concerning the characteristics that were important (particularly completeness, reliability and validity). Additionally presentations given on the day covered:
   - Approaches for assessing data quality for different national primary care databases and quality improvement through feedback mechanisms.
   - The impacts of poor data quality on research outcomes, the difficulty of revealing hidden quality issues and the time-consuming nature of data quality assessment and correction.
2. Discuss the issues and challenges involved with measuring data quality in EHRs for epidemiological and clinical research.
   - Clinicians highlighted that teams that generate personal health data are focused on clinical care rather than research.
   - Bias arises in areas that directly impact on reimbursement such as QOF [14]. Another potential influence is the role of the GP as patient advocate, where they might emphasise certain clinical findings to justify an investigation or referral.
   - Database managers highlighted the need for transparency of methods used to calculate variables and difficulties resulting when a data item can be recorded in multiple ways.
   - Those other than clinicians highlighted the need for a study-specific specific approach and that this makes a standard approach to assessment harder to achieve.
   - Data users stressed the importance of understanding unstructured/un-coded data and the ability to link data.
3. Work towards development of an approach to ensure compatibility of data quality measures for different European primary and secondary care databases.
   - Some technical proposals were put forward. A key discussion point for all groups was the communication of data quality metrics.
   - The importance of the publication of data quality work to inform the understanding of third parties, including future users of the data, regardless of whether this was the primary focus of a particular research study, or an early phase of data exploration. This should include how different parties handle the various aspects of data quality, the algorithms used for the identification of outcomes including code lists.
   - Most stakeholders agreed that data should be made available "warts and all" so users can make the decision on whether or not and how to use the data.
   - All agreed that it is important to have transparency on how the data is collected, and to understand the processes involved.

- Clinicians focussed on more specific examples such as test results and prescribing records.
- Data experts discussed the benefits to quality that can be gained by being able to discuss data with clinicians.

4. Discuss how to help data contributors improve data quality (for both clinical care and research) at source.

- Providing practices feedback on their data quality, as pioneered by PRIMIS [45] in the UK and NOKLUS in Norway, has made an important contribution in motivating practices in high quality recording [46].
- Data experts stressed the importance of providing recording guidelines to clinicians and feeding back data quality metrics directly to them.
- It was generally agreed that incentivising GP's to produce higher quality data is key, either by feedback loops or by demonstrating how the data could be used to benefit their own patients. Clinicians are unlikely to prioritise data quality unless it benefits patient care or it affects their payments.
- A critical success factor for the future will be to ensure that good quality data delivers value to those individuals who capture them, for example through decision support, alerts, charts of trends etc.
- It may also be influential if clinical effort investments in data quality can be perceived as beneficial by patients themselves.
- Further work is needed to understand the costs and benefits of improving data quality.

Based on the results of this workshop our suggestions for the moving forward are summarised below:

1. Data providers

- Provide meta-data and practice-based data quality scores to users (bearing in mind the concept of fitness for use).
- Be transparent about how data is handled providing as much information as possible on the processing steps.
- Provide information/training on how data is recorded at source.
- Explore ways to incentivise GPs to record better e.g. feedback data quality information or enabling database access for patient treatment.

2. Data users/experts

- Communicate impact of data quality on primary care data research to clinicians.
- Be aware of the limitations and impact of poor quality when carrying out research.
- Document or publish operational definitions so that researchers can easily validate research.

3. Clinicians
   - Encourage training of staff within general practice to record data using coding as much as possible.
4. All
   - Set up a network to continue the discussions of the workshop in order to develop a unified approach for measuring and improving data quality in Primary Care (and linked data) research databases.

Although the workshop did not result in a proposed overall approach for measuring data quality, many of the participants indicated that they would be interested in joining a data quality network to discuss these issues further. The network has recently been launched and will act as an international forum for discussion, aiming, through participation from different user groups, to develop a comprehensive, robust, integrated and widely used approach to measuring and delivering data quality across all aspects of EHRs (the reader is encourage to contact the authors or further information about the network).

5 Conclusions

The issue of understanding data quality could not be more pertinent given the context of 'Big' healthcare data. Within England CPRD is the data service for healthcare research and has access to 10 % of the UK population at the primary care data level. It sits within a broader governmental project to make the entire health care data set available for use for the improvement of public health understanding and clinical research and delivery as well as surveillance. The principle component of this activity is a project from NHS England to gain access to primary care records from all General Practices in England, known as Care.Data [47]. Given this and increasing access to linkable data sets a key and full understanding of data quality within data sets from different platforms and collected for different purposes using different paradigms is central to use of this data for research and allied activities.

Furthermore, in the epidemiology and pharmacoepidemiology arena a growing number of projects are being developed to utilise big data from different global settings to answer key questions on drug safety for example. Projects such as OMOP (Observational Medical Outcomes Partnership) [48], OHDSI (Observational Health Data Sciences and Informatics) [49] and IMEDS (Innovation in Medical Evidence Development and Surveillance) [50] utilise disparate data sets in a common data model. The appropriateness of this approach needs to carefully consider the relative data quality aspects of each component data source.

Another focus of this chapter has been to convey the importance of collaboration in mapping out such an extensive and dynamic field, pursued here with the launch of a data quality network to facilitate the sharing of ideas. Understanding the requirements of and the demands upon clinicians must be central to addressing quality issues in EHRs so as to be able to implement procedures that can realistically support data recording to facilitate patient management directly as well as to increase the capacity for research into cutting edge therapeutic and general patient care supporting longer term sustainability of the healthcare system.

References

1. Lawrenson, R., Williams, T., & Farmer, R. (1999). Clinical information for research; The use of general practice databases. *Journal of Public Health Medicine, 21*, 299–304.
2. Williams, T., van Staa, T., Puri, S., & Eaton, S. (2012). Recent advances in the utility and use of the General Practice Research Database as an example of a UK Primary Care Data resource. *Therapeutic Advances in Drug Safety, 3*, 89–99.
3. Juran, J. M. (1988). *Juran's quality control handbook* (4th ed.). TX: McGraw-Hill.
4. The good practice guidelines for GP electronic patient records, v4. (2011). Provided by NHS/connecting for health.
5. National Institute for Health and Care Excellence. (2014). *Quality and outcomes framework*. Available via NICE http://www.nice.org.uk/aboutnice/qof/indicators.jsp. Accessed May 2014.
6. Tate, A. R., Beloff, N., Padmanabhan, S., Dungey, S., Williams, R., & Williams, T., et al. (2015). Developing a methodological approach for characterising data quality in primary care research databases (in press).
7. Taggar, J. S., Coleman, T., Lewis, S., & Szatkowski, L. (2012). The impact of the Quality and Outcomes Framework (QOF) on the recording of smoking targets in primary care medical records: Cross-sectional analyses from The Health Improvement Network (THIN) database. *BMC Public Health, 12*, 329–340.
8. de Lusignan, S., & van Weel, C. (2006). The use of routinely collected computer data for research in primary care: Opportunities and challenges. *Family Practice, 23*, 253–263.
9. de Lusignan, S. (2005). Codes, classifications, terminologies and nomenclatures: Definition, development and application in practice. *Informatics in Primary Care, 13*, 65–70.
10. Porcheret, M., Hughes, R., Evans, D., Jordan, K., Whitehurst, T., Ogden, H., et al. (2004). Data quality of general practice electronic health records: The impact of a program of assessments, feedback, and training. *Journal of the American Medical Informatics Association, 11*, 78–86.
11. Nicholson, A., Ford, E., Davies, K. A., Smith, H. E., Rait, G., Tate, A. R., et al. (2013). Optimising use of electronic health records to describe the presentation of rheumatoid arthritis in primary care: A strategy for developing code lists. *PLoS One, 8*(2), e54878.
12. Tate, A. R., Martin, A. G. R., Ali, A., & Cassell, J. A. (2011). Using free text information to explore how and when GPs code a diagnosis of ovarian cancer. Observational study using the General Practice Research database. *BMJ Open, 1*, 1–9.
13. Bhaskaran, K., Forbes, H. J., Douglas, I., Leon, D. A., & Smeeth, L. (2013). Representativeness and optimal use of body mass index (BMI) in the UK Clinical Practice Research Datalink (CPRD). *BMJ Open, 3*(9), e003669.
14. Mannion, R., & Braithwaite, J. (2012). Unintended consequences of performance measurement in healthcare: 20 salutary lessons from the English National Health Service. *Internal Medicine, 42*, 569–574.
15. Rait, G., Walters, K., Griffin, M., Buszewicz, M., Petersen, I., & Nazareth, I. (2009). Recent trends in the incidence of recorded depression in primary. *British Journal of Psychiatry, 195*, 520–524.
16. Salomon, R. M., Urbano Blackford, J., Rosenbloom, S. T., Seidel, S., Wright Clayton, E., Dilts, D. M., et al. (2010). Research paper: Openness of patients' reporting with use of electronic records: Psychiatric clinicians' views. *Journal of the American Medical Informatics Association, 17*, 54–60.

17. Gray Weiskopf, N., & Weng, C. (2013). Methods and dimensions of electronic health record data quality assessment: Enabling reuse for clinical research. *Journal of the American Medical Informatics Association, 20*, 144–151.
18. Salati, M., Brunelli, A., Dahan, M., Rocco, G., Van Raemdonck, D. E., & Varela, G. (2011). Task-independent metrics to assess the data quality of medical registries using the European Society of Thoracic Surgeons (ESTS) Database. *European Journal of Cardio-Thoracic Surgery, 40*, 91–98.
19. Kahn, M. G., Raebel, M. A., Glanz, J. M., Riedlinger, K., & Steiner, J. F. (2012). A pragmatic framework for single-site and multisite data quality assessment in electronic health record-based clinical research. *Medical Care, 50*, S21–S29.
20. Herrett, E., Thomas, S. L., Schoonen, W. M., Smeeth, L., & Hall, A. J. (2010). Validation and validity of diagnoses in the General Practice Research Database: A systematic review. *British Journal of Clinical Pharmacology, 69*, 4–14.
21. Clinical Practice Research Datalink. http://www.cprd.com. Accessed May 2014.
22. HM Treasury. (2011). *Plan for growth.* Available via GOV.UK. https://www.gov.uk/government/uploads/system/uploads/attachment_data/file/221514/2011budget_growth.pdf. Accessed May 2014.
23. Tate, A. R., Beloff, N., Al-Radwan, B., Wickson, J., Puri, S., Williams, T., et al. (2014). Exploiting the potential of large databases of electronic health records for research using rapid search algorithms and an intuitive query interface. *Journal of the American Medical Informatics Association, 21*, 292–298.
24. Woods, C. (2001). Impact of different definitions on estimates of accuracy of the diagnosis data in a clinical database. *Journal of Clinical Epidemiology, 54*, 782–788.
25. Wurst, K. E., Ephross, S. A., Loehr, J., Clark, D. W., & Guess, H. A. (2007). The utility of the general practice research database to examine selected congenital heart defects: A validation study. *Pharmacoepidemiology and Drug Safety, 16*, 867–877.
26. Harron, K., Wade, A., Gilbert, R., Muller-Pebody, B., & Goldstein, H. (2014). Evaluating bias due to data linkage error in electronic healthcare records. *BMC Medical Research Methodology, 14*, 36.
27. TRANSFoRm. (2010). Website available at: http://www.transformproject.eu. Accessed May 2014.
28. Khan, N. A., McGilchrist, M., Padmanabhan, S., van Staa, T., & Verheij, R. A. (2013). *Deliverable 5.1: Data quality tool.* NIVEL, University of Dundee, CPRD. Available via TRANSFoRm. http://transformproject.eu/Deliverables.html. Accessed May 2014.
29. Leysen, P., Bastiaens, H., Van Royen, P., Agreus, L., & Andreasson, A. N. (2011). *Development of use cases*. University of Antwerp, Karolinska Institutet. Available via TRANSFoRm. http://transformproject.eu/D1.1Deliverable_List_files/DetailedUsecases_V2.1-2.pdf. Accessed May 2014.
30. de Lusignan, S., Minmagh, C., Kennedy, J., Zeimet, M., Bommezijn, H., & Bryant, J. (2001). A survey to identify the clinical coding and classification systems currently in use across Europe. *Studies in Health Technology and Informatics, 84*, 86–89.
31. Boggon, R., Van Staa, T., Chapman, M., Gallagher, A. M., Hammad, T. A., & Richards, M. A. (2012). Cancer recording and mortality in the General Practice Research Database and linked cancer registries. *Pharmacoepidemiology and Drug Safety, 22*, 168–175.
32. Beral, V., & Peto, R. (2010). UK cancer survival statistics are misleading and make survival worse than it is. *British Medical Journal, 341*, c4112.
33. Berkel, J. (1990). General practitioners and completeness of cancer registry. *Journal of Epidemiology and Community Health, 44*, 121–124.
34. Schouten, L. J., Höppener, P., van den Brandt, P. A., Knottnerus, J. A., & Jager, J. J. (1993). Completeness of cancer registration in Limburg, The Netherlands. *International Journal of Epidemiology, 22*, 369–376.

35. Herret, E., Dinesh Shah, A., Boggon, R., Denaxas, S., Smeeth, L., Van Staa, T., et al. (2013). Completeness and diagnostic validity of recording acute myocardial infarction events in primary care, hospital care, disease registry, and national mortality records: Cohort study. *British Medical Journal, 346*, f2350.
36. Weed, L. (1968). Medical records that guide and teach. *New England Journal of Medicine, 278*, 593–600.
37. Bossen, C. (2007). Evaluation of a computerized problem-oriented medical record in a hospital department: Does it support daily clinical practice? *International Journal of Medical Informatics, 76*, 592–600.
38. Kontopantelis, E., Buchan, I., Reeves, D., Checkland, K., & Doran, T. (2013). Relationship between quality of care and choice of clinical computing system: Retrospective analysis of family practice performance under the UK's quality and outcomes framework. *BMJ Open, 3*, e003190.
39. Reeves, D., Springate, D. A., Ashcroft, D. M., Ryan, R., Doran, T., Morris, R., et al. (2014). Can analyses of electronic patient records be independently and externally validated? The effect of statins on the mortality of patients with ischaemic heart disease: A cohort study with nested case–control analysis. *BMJ Open, 4*, e004952.
40. QResearch. Website available at: http://www.qresearch.org/. Accessed May 2014.
41. Tate, A. R., Kalra, D., Boggon, R., Beloff, N., Puri, S., & Dungey, S., et al. (2014). *Data quality in European primary care research databases*. Report of a workshop sponsored by the CPRD in London September 2013. IEEE-EMBS International Conference on Biomedical and Health Informatics (BHI), 2014, IEEE.
42. García-Gil Mdel, M., Hermosilla, E., Prieto-Alhambra, D., Fina, F., Rosell, M., Ramos, R., et al. (2011). Construction and validation of a scoring system for the selection of high-quality data in a Spanish population primary care database (SIDIAP). *Informatics in Primary Care, 19*, 135–145.
43. Bellika, J. G., Hasvold, T., & Hartviysen, G. (2007). Propagation of program control: A tool for distributed disease surveillance. *International Journal of Medical Informatics, 76*, 313–329.
44. Sáez, C., Martinez-Miranda, J., Robles, M., & Garcia-Gomez, J. M. (2012). Organizing data quality assessment of shifting biomedical data. *Studies in Health Technology and Informatics, 180*, 721–725.
45. PRIMIS (Primary Care Information Services). Website available at: http://www.nottingham.ac.uk/primis/index.aspx. Accessed May 2014.
46. Lagerqvist, B., James, S. K., Stenestrand, U., Lindbäck, J., Nilsson, T., & Wallentin, L. (2007). Long-term outcomes with drug-eluting stents versus bare-metal stents in Sweden. *New England Journal of Medicine, 356*, 1009–1019.
47. Care.Data. Website available at: http://www.nhs.uk/nhsengland/thenhs/records/healthrecords/pages/care-data.aspx. Accessed May 2014.
48. Observational Medical Outcomes Partnership. Website available at: http://omop.org/. Accessed May 2014.
49. Observational Health Data Sciences and Informatics. Website available at: http://www.ohdsi.org/. Accessed May 2014.
50. Innovation in Medical Evidence Development and Surveillance. Website available at: http://imeds.reaganudall.org/. Accessed May 2014.

Part II
Multimedia Event and Activity Detection and Recognition for Health-Related Monitoring

Activity Detection and Recognition of Daily Living Events

Konstantinos Avgerinakis, Alexia Briassouli, and Ioannis Kompatsiaris

1 Introduction

Ambient assisted living (AAL) focuses on the development of tools that can help people with chronic degenerative conditions continue living independently for as long as they can. They aim to provide continuous, unobtrusive monitoring, to ensure their safety in case of an emergency, and help build an accurate activity, lifestyle and behavioral profile, to detect changes in their condition. The results of the monitoring and profiling provided by such systems can then be used as input for appropriate feedback, both for the people being monitored, as well as for their carers.

Existing assisted living solutions usually employ physiological and environmental sensors that are relatively simple, like accelerometers and contact or pressure sensors. Attention has recently turned to the inclusion of more sophisticated technologies, such as audiovisual monitoring. In this work we focus on the use of video for remote monitoring of people with conditions like dementia, living home alone. For effective video-based monitoring, the recognition of Activities of Daily Living (ADLs) is central, and also the focus of this work. Activity detection and recognition from video for assisted living is based on unobtrusive ambient sensors, namely static video cameras, which do not disturb people in their daily life.

In this work, we expand our previous research [1], which focused on ADL recognition, and introduce a novel approach for activity detection, which is a necessary precursor to recognition in real applications. While studying the related work on activity detection, we noticed that some issues have been overlooked until now. One issue is that in the literature it is common to use overlapping temporal sliding windows in order to detect activities, however no spatial localization

K. Avgerinakis (✉) • A. Briassouli • I. Kompatsiaris
Centre for Research and Technology Hellas (CERTH), Thessaloniki, Greece
e-mail: koafgeri@iti.gr; abria@iti.gr; ikom@iti.gr

A. Briassouli et al. (eds.), *Health Monitoring and Personalized Feedback using Multimedia Data*, DOI 10.1007/978-3-319-17963-6_8

is provided. We overcome this limitation by proposing a novel algorithm that localizes ADLs both in time and space by extracting Activity Areas (AA) [1] to find where in the video frame an activity is taking place.

Another disadvantage of some of the existing activity detection methods is the great computational cost needed to analyze an entire video. Sliding window techniques tend to filter all frames, many of which do not contain any useful information. In this work, we reduce the duration of the video analysis to almost half of real time by applying temporal segmentation before analyzing the video frames. This is achieved by segmenting video frames based on the lifetime of the trajectories in the AAs, which indicate where different activities occur.

A hybrid descriptor is proposed for boosting the ADL representation, as opposed to related techniques in the literature that usually focus on only one representation category (Sect. 2). Our motivation for this is that there are strong indications showing that the recognition of ADLs in constrained environments (labs, nursing homes) can usually be enhanced by including their spatial context in their description.

Finally, we haven't found any work in the literature that deals with the ambiguous time intervals often present between two pre-defined ADLs. We cover this gap and extract "ambiguous activity intervals" between recognized ADLs by deploying multi-class SVMs with one-against-one comparisons. A voting index is proposed in order to accumulate scores and also indicate ambiguous temporal intervals in the case of multiple draws. In concluding, we contribute in the literature of activity detection with the following:

- Activity Areas (AA) for spatial localization of ADLs inside long videos that include ambiguous intervals ("noise").
- Dense trajectories extracted in AAs for temporal activity boundary detection and automatic ADL video segment extraction.
- A hybrid ADL representation in order to enhance recognition rates, created by adding global information to a local descriptor.
- Detection of ambiguous time intervals between ADLs, based on an one-against-one multi-class SVM and a voting procedure.

This chapter is organized as follows: Sect. 2 analyzes related work and State-of-the-Art (SoA) activity detection and recognition methods. Section 3 follows with the adopted ADL representation in this work, while Sect. 4 presents the detection and classification of ADLs in videos. Experiments on data-sets from real scenarios are provided in Sect. 5, in order to evaluate the overall system.

2 Related Work

Activity representation has been extensively studied during the last decade, leading to continually improving recognition rates. The methods that have been proposed until today can be separated into two major categories: (1) holistic algorithms that gather information from the entire person and aggregate them in a common descriptor and (2) local approaches that accumulate visual cues around spatio-temporal interest points and describe them in a Bag-of-Words (BoW), Fisher, and VLAD(Vector of Locally Aggregated Descriptors) global representations [10, 23].

Most notable holistic approaches include Motion History Volumes [29], space-time shapes [9] and temporal templates [4], while local approaches use 3D local patches [13, 16, 26–28], either based on the extension of local patches over time (i.e. SIFT3D, HOG3D, SURF3D in [13, 26, 28]) or on the construction of motion histograms around sampled interest points (i.e. HOF (Histograms of Oriented Optical Flow), MBH(Motion Boundary Histograms) in [16, 27]). Interest points can either be sampled in a sparse manner, as in [7, 15, 28], or densely [27], with the latter providing better recognition rates than the former. Further attention has recently been given to the tracking of such interest points, which lead to the construction of trajectory descriptors [20, 21, 27], and are useful for retaining relations between them.

The **Activity detection** literature can also be separated into two categories, based on the action representation used, namely into holistic and local methods.

Holistic activity detection approaches tend to represent actions as a sequence of states which inherit their features from the whole actor, e.g. poses, spatio-temporal templates, and use dynamic probabilistic models to detect transitions within activities. Distinctive examples include methods based on the construction of coupled Hidden Markov Models (HMMs) [22] and dynamic Bayesian Networks [18]. Another holistic activity detection approach uses exemplars in conjunction with Dynamic Time Warping (DTW) [25] among action and template sequences in order to compute alignment scores and thus detect which activities in the exemplars are present. Other holistic approaches entail the computation of spatio-temporal volumes to represent activities and compare them using Tensor Canonical Correlation Analysis [12]. Sliding window with part-based template matching using pictorial structures is used in [11]. More recently, action banks have been deployed in [24], in conjunction with spatio-temporal oriented energy templates [6] and have achieved SoA results.

Local-based approaches, on the other hand, mainly focus on detecting specific action-poses in movies (e.g. the pose corresponding to drinking), either by using a spatio-temporal video block classifier, or by training specialized space-time cubes, guided by activity localization hypotheses [17]. Similarly, in [14], a human detector is used, based on the HOG tracker, combined with a spatio-temporal sliding window for recognizing the kind of activity that happens at each instance. The most recent sliding window technique was proposed in [8], leading to SoA detection accuracy. The authors introduced the idea of decomposing actions into atomic action units (actoms) and used an Actom Sequence Model (ASM) in order to evaluate the appearance of an action in a video interval and its activity boundaries.

3 Activity Representation

In the first step of our algorithm, we use the activity representation framework of our previous work [1], which is specifically designed for deployment in real-life scenarios at a low computational cost. This framework involves the construction of a trajectory structure over Activity Areas by tracking densely sampled interest points (Sect. 3.1) and computing spatio-temporal cuboids around them. Hybrid cuboids are specially designed, to encapsulate motion and visual characteristics via HOGHOF (aka Histograms of Oriented Gradients and Histograms of Oriented Optical Flow) descriptors. Trajectory descriptors are also added to this structure, so as to incorporate spatial location information (Sect. 3.2). The overall procedure for the construction of our ADL descriptor is depicted in Fig. 1.

3.1 Dense Trajectories Over Activity Areas

Interest point sampling entails the construction of a motion segmentation algorithm that detects regions of interest (RoI) where motion undergoes change. This algorithm relies on kurtosis-based Activity Areas [5], for which optical flow values [30] are analyzed statistically over successive frames, leading to the real-time

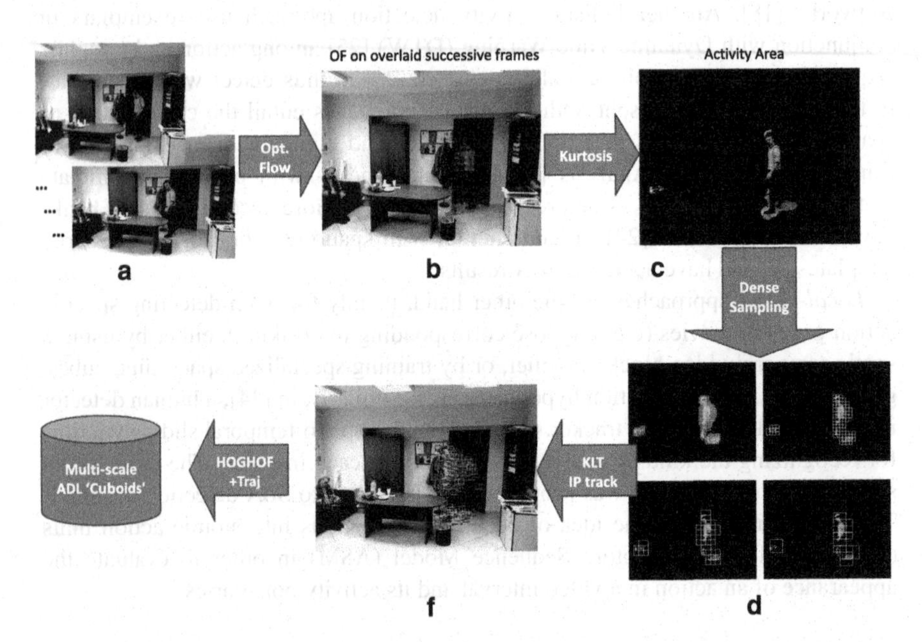

Fig. 1 Block diagram of the ADL representation schema. Kurtosis analysis on OF values results in Activity Areas, where dense sampling is performed. Trajectory and local histograms (HOGHOF) are constructed around interest points on multiple scales, to be tracked using the KLT tracker algorithm leading to trajectory and HOGHOF structures which constitute the final ADL descriptor

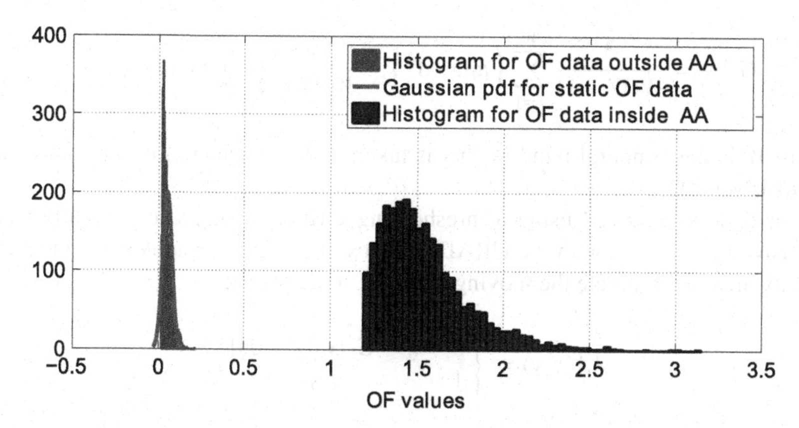

Fig. 2 Optical flow histograms for static (*green*) and moving (*blue*) pixels. In *red* the Gaussian distribution that models the static data (Color figure online)

localization of regions containing motion. From these regions, we sample candidate interest points and track them with a boosted KLT tracker until they become motionless, after which the video subsequence is considered to end, and we build a trajectory structure in it.

Activity Areas separate moving pixels from static ones based on the assumption that the static pixels' noise-induced optical flow follows a Gaussian distribution over time. Thus, for the construction of Activity Areas, optical flow values are monitored over successive frames and a kurtosis-based test of Gaussianity is applied on them. In order to determine if static video data can indeed be modelled by a Gaussian distribution, we applied the Kolmogorov–Smirnov test [19] on ten training videos from the University of Rochester Activities of Daily Living (URADL) dataset [20], which is a commonly used benchmark dataset for human activity recognition. From the histograms of aggregated optical flow values in Fig. 2, it can be seen that the optical flow of static pixels can indeed be modelled by a Gaussian distribution, unlike that of moving pixels. Thus, the optical flow values of static and moving pixels respectively, are expressed by the following two hypotheses:

$$H_0 : u_t^0(x, y) = z_t(x, y)$$
$$H_1 : u_t^1(x, y) = u_t(x, y) + z_t(x, y),$$

where $u_t(x, y)$ denotes optical flow at pixel (x, y) at time t, and $z_t(x, y)$ represents the additive noise in that location. A fast and robust way to separate this data is by computing the Kurtosis for each set of flow values over time and binarizing the resulting flow map. Recent studies in statistics introduce a novel technique for accurately estimating the empirical value of the kurtosis [3] in an unbiased manner, by approximating excess Kurtosis from the fourth-order cumulant estimator as:

$$G_2[y] = \frac{3}{W(W-1)} \sum_{i=1}^{W} \left(u_i(x,y)^4 \right) - \frac{W+2}{W(W-1)} \sum_{i=1}^{W} \left(u_i(x,y)^2 \right)^2 ,$$

where W is the temporal window that is taken under consideration for computing Kurtosis values.

Our data is binarized using a thresholding formula acquired empirically from extensive experiments with the URADL videos, in order to accurately segment the activity area, i.e. separate the moving from the static pixels:

$$AA(x,y) = \begin{cases} 0 & \textit{if} \quad G_2[x,y] < 0.02 \\ 1 & \quad \textit{else} \end{cases}$$

A spatial grid is formed in each video frame, creating a number of blocks; the central pixel (x, y) of each block is considered as a candidate interest point only when more than 50 % of the block belongs to the moving pixels of the AAs. All candidate interest points are tracked using KLT, boosted by a homography test which uses a RANSAC (RANdom SAmple Consensus) estimator to validate the interest point correspondences. Interest points that pass the test are used to track moving objects and form a trajectory descriptor (see Sect. 3.2).

3.2 Hybrid Spatio-Temporal Descriptor

In order to benefit from the advantages of both local and global SoA activity representations, we designed a hybrid descriptor, aiming at increased recognition rates at a lower computational cost. Thus, we use a local approach for describing appearance and motion characteristics of the activities, and a holistic one for capturing global spatial information. The sampled interest points, already extracted for activity detection, provide us with very accurate trajectory vectors. Regions around interest points are described on four spatial scales to ensure scale invariance in our activity descriptor. We use one of the fastest methodologies for extracting an activity descriptor, namely the HOGHOF [16], where HOGs describe appearance information, and HOFs capture motion characteristics. The spatiotemporal descriptor is formed by concatenating all HOGs and HOFs that belong to the same trajectory. Each descriptor is then subdivided into a $n_x \times n_y \times n_t$ ($n_x = n_y = 2, n_t = 3$) grid of cuboids and, for each cuboid, histograms are averaged over time and normalized.

The construction of the average descriptor is depicted in Fig. 3: subsequences of length $N \geq 15$ are divided into $n_t = 3$ groups and HOGHOF descriptors are extracted in each of them. They are then averaged for each scale and normalized, giving more concise but informative appearance and motion descriptors. Trajectory coordinates are also added to the vector to include global spatial information in our descriptor (i.e. HOGHOF+Traj), creating a hybrid local-global descriptor, which is shown to improve recognition rates, as demonstrated in the experiments of Sect. 5.

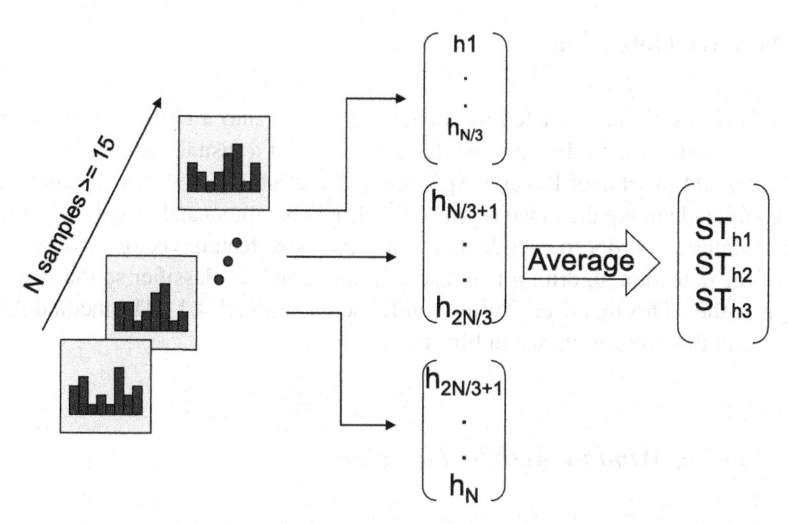

Fig. 3 Construction of average HOGHOF descriptor by dividing video subsequences into three groups over time and averaging the corresponding HOGHOFs

Let $HOG(x, y, t)$ and $HOF(x, y, t)$ be the histograms extracted inside a trajectory block $B_{sc} = (x, y, w_{sc}, h_{sc})$, around an interest point with coordinates (x, y), with sampling of size 8 for each scale: $(w_{sc}, h_{sc}) = (8, 8)(16, 16)(24, 24)(32, 32)$, where w_{sc} and h_{sc} are the width and height respectively, and the index sc denotes the different sizes of the block, corresponding to scale size. The resulting spatio-temporal descriptor around each interest point (x, y) is the L_2 normalized concatenation of the averaged histograms within each temporal sub-volume that is formed by dividing the initial descriptor by n_t:

$$ST_{desc} = \left[\underset{j=1}{\overset{n_t}{concat}} \left\{ \left\| \frac{\sum_{t_i=(j-1)*(N/n_t)+1}^{j*N/n_t} HOG(B_{sc}, t_i)}{N/n_t} \right\|_2 \cdots \left\| \frac{\sum_{t_i=(j-1)*(N/n_t)+1}^{j*N/n_t} HOF(B_{sc}, t_i)}{N/n_t} \right\|_2 \right\} (x_t, y_t) \right],$$

where ST_{desc} denotes the final feature vector formed by the concatenation, denoted here as *concat*, of the spatio-temporal volumes, used for the representation of each activity. $HOG(B_{sc}, t_i)$ and $HOF(B_{sc}, t_i)$ are both represented below as *hist*. Thus each block has each weight to each histogram and is computed by:

$$hist_{block}(B_{sc}, t_i) = \left\| \underset{i,j}{accumulate} \left(hist \left(x + i * \frac{w_{sc}}{n_x}, \ldots, y + j * \frac{h_{sc}}{n_y}, \frac{w_{sc}}{n_x}, \frac{h_{sc}}{n_y} \right) \right) \right\|_2,$$

where $(i, j) = \{(1, 1)(1, -1)(-1, 1)(-1, -1)\}$ and *hist* returns the spatial block histogram around each trajectory interest point. *hist* is the accumulation, here denoted as *accumulate*, of its four cell histograms.

4 Activity Detection

The *hybrid* descriptor extracted as above is imported into a clustering algorithm, either K-Means or GMM in this work, so as to build a visual vocabulary for our ADLs. VLAD (Vector of Locally Aggregated Descriptors) or Fisher distances are then computed among the video segments' hybrid descriptors and visual vocabulary cluster centers, in order to encode them into fixed size feature vectors. The central idea of the detection algorithm is to use a sliding window classifier so that we can recognize the ADLs that occur during a video segment (Sect. 4.1). The methodology followed in this work is shown in Fig. 4.

4.1 Sliding Window Activity Detection

Temporal sliding windowing is a widespread technique that is used to recognize activities that may exist within an unsegmented video sample. Before the application of the sliding window, a global representation needs to be constructed for each training video segment, to model the appropriate classifiers. A clustering algorithm (i.e. K-Means, GMM) is deployed in order to partition the ADL feature space and acquire the corresponding visual vocabulary. Afterwards, appropriate encoding

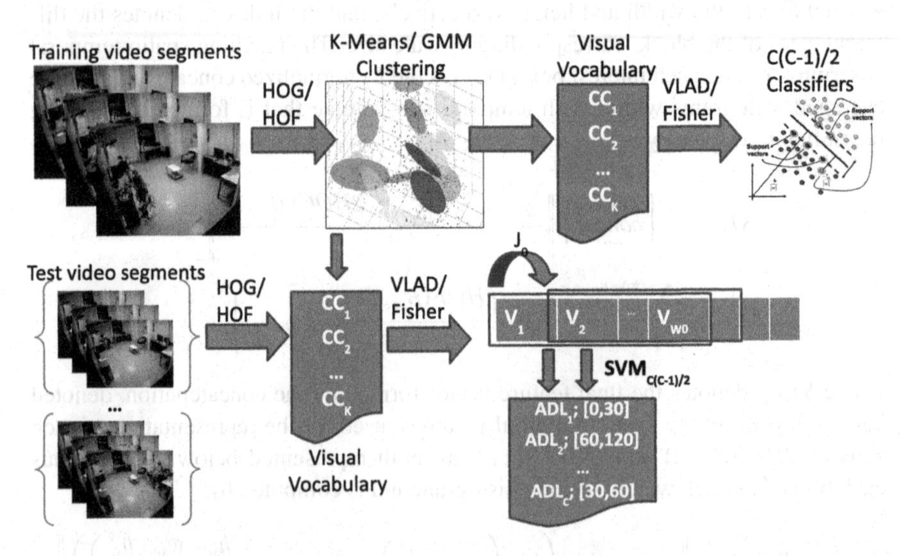

Fig. 4 Block diagram of ADL detection and recognition. Training pre-segmented samples are used to build visual vocabularies (i.e. cluster centers or means) and all data is encoded into a single size feature vector based on the VLAD/Fisher framework. $C(C-1)/2$ multi-class SVM classifiers are built and a sliding window filters the unsegmented test data, in order to detect the ADLs and ambiguous temporal segments that occur within them

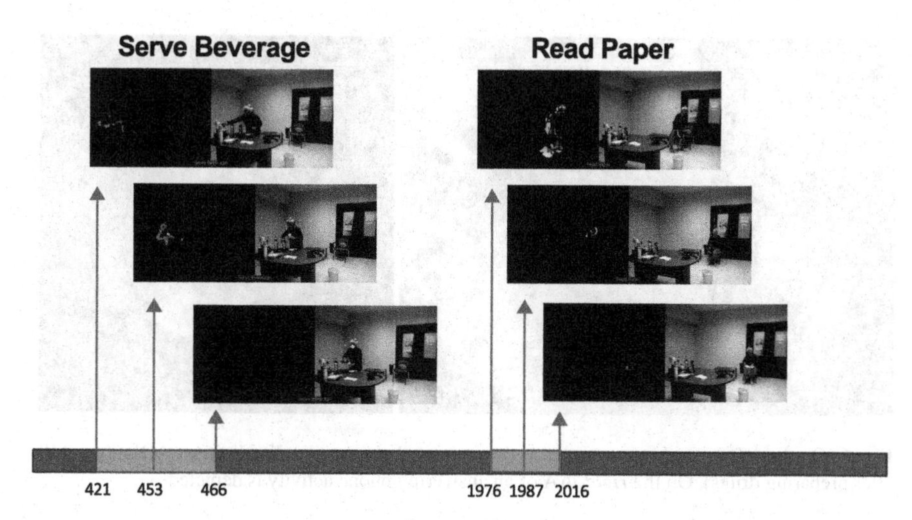

Fig. 5 Activity areas on the *left* and trajectories on the *right* of each frame depict the start and end of a video segment

(e.g. VLAD [10], Fisher [23]) takes place to characterize each video segment by a fixed size feature vector. The feature vectors obtained from training videos are used to train $C(C-1)/2$ SVM linear classifiers, where C is the number of ADL classes, for the global representation of the training video classes.

Test video samples, on the other hand, are segmented using an activity detection algorithm and classified using a temporal sliding window. For activity detection, an activity is considered to start when the first trajectory is sampled from an AA, while the termination of a video segment is denoted when no other trajectories are found in that AA. Figure 5 depicts an example where activity detection locates the start, the end frame and also shows an in-between frame of a video segment.

These video segments are then filtered by a temporal sliding window of size W_0, with a sampling step J_0. For $J_0 < W_0$, there is temporal overlap among the predictions, and thus better localization of activity boundaries (i.e. the start and end of an activity) is achieved.

One-against-one SVM classification is used to predict each windowed video segment's class, and a voting schema is applied, in order to declare the prominent ADL recognized by the classifiers. The class with the highest number of votes always wins, while in the case of a draw, a second round among the 1st and 2nd classifiers is performed, so that the one that wins among these two is the detected activity. In case of triple or higher draws, we announce ambiguity among classes (a scenario that usually occurs during a transition between clearly detected activities) and no recognition data is stored. Results are accumulated in a classification index, which is used to find the number of activities that exist in the video segment, their activity boundaries and ambiguous regions between successive detected activities. Figure 6 visualizes ambiguous activity states and an AA sample.

Fig. 6 On the *left* we ambiguous status is demonstrated among two discriminant ADLs (pay bill after preparing drugs). On the *right*, AA of an answering phone activity is depicted

5 Experiments

In this experimental section, we evaluate the activity recognition performance of our descriptor on the pre-segmented videos of ADLs in the benchmark URADL action dataset. ADL detection is tested on realistic videos recorded for the EU project Dem@Care in home-like environments in the Centre Hospitalier Universitaire de Nice (CHUN) in Nice, France, and the day center of the Greek Association for Alzheimer's Disease and Related Disorders (GAADRD) in Thessaloniki, Greece. For ADL recognition, we estimate average accuracy over all classes, while for temporal localization we consider the OV20 evaluation criterion [8], which requires the Jaccard coefficient (intersection over union) to be over 20 % among groundtruth and detected activities, in order to consider a detected activity to be a true positive.

5.1 URADL Dataset

URADL [20] is a well-known benchmark dataset for recognizing ADLs and is used mostly for evaluation purposes. In this dataset, five different actors perform ten different activities, three times each, in a kitchen environment. A serious disadvantage of this dataset is that it lacks significant anthropometric variance, the environmental conditions are quite simple, the activities take place in the same location, with the same illumination, a static camera, no environmental noise and no occlusions. The environment is also quite uncluttered, and the actors are standing directly in front of the camera, while they carry out the ADLs under examination. For evaluating our algorithm, we choose to use leave-one-subject-out testing. Thus, we initialized the recognition procedure five times, so that we can recognize the activity of each human subject independently. The names of the activities of this

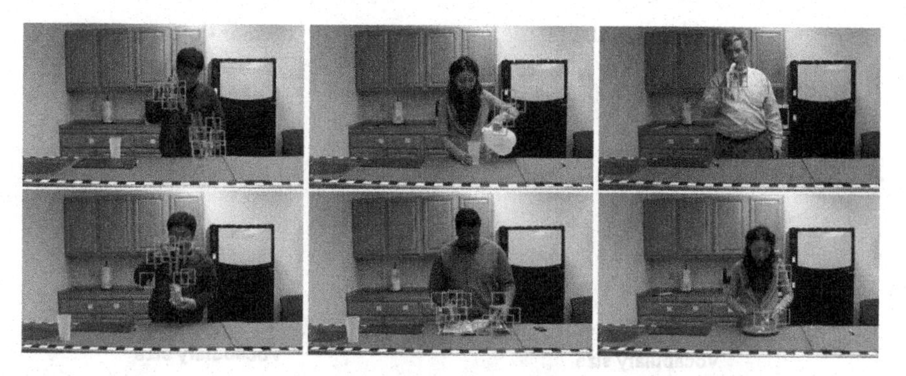

Fig. 7 Characteristic URADL video frames. *Top row*: answer phone (AP), drink water (DW) and eat banana (EB). *Bottom row*: Eat snack (ES), lookup in phonebook (LiP), use silverware (US)

Table 1 Average accuracy over all classes for URADL action dataset

Encoding	Vocabulary size	Descriptor	Average Accuracy (%)
VLAD	32	HOGHOF	76,00
		HOGHOF+Traj	63,33
	64	HOGHOF	82,00
		HOGHOF+Traj	68,00
	128	HOGHOF	81,33
		HOGHOF+Traj	71,33
	256	HOGHOF	80,67
		HOGHOF+Traj	75,33
Fisher	32	HOGHOF	88,67
		HOGHOF+Traj	92,67
	64	HOGHOF	92,67
		HOGHOF+Traj	**94,67**
	128	HOGHOF	92,67
		HOGHOF+Traj	94,00
	256	HOGHOF	92,00
		HOGHOF+Traj	94,00

dataset are encoded in the tables as AP = Answer Phone, CB = Chop Banana, ES = Eat Snack, DP = Dial Phone, DW = Drink Water, EB = Eat Banana, LiP = Look up in Phonebook, PB = Peel Banana, US = Use Silverware, WoW = Write on Whiteboard. Figure 7 depicts characteristic URADL activities in some sample video frames, with their trajectories drawn in the corresponding local blocks.

Table 1 and Fig. 8 show that adding trajectory information to the HOGHOF descriptor, combined with Fisher encoding, improves recognition rates. Even small vocabulary sizes can provide very accurate recognition rates in that case, due to

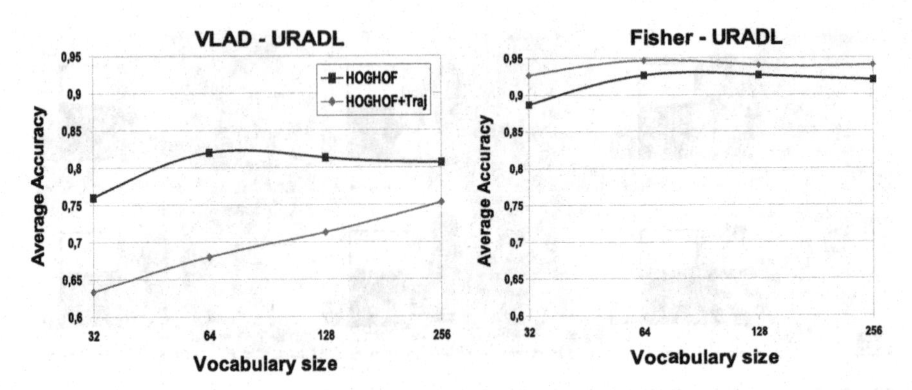

Fig. 8 HOGHOF and HOGHOF+Traj comparison when VLAD (*left*) and Fisher (*right*) encoding is applied to the URADL action dataset

Table 2 The best recognition results on URADL action dataset, when HOGHOF representation is combined with the GMM & Fisher recognition schema

	AP	CB	DP	DW	EB	ES	LiP	PB	US	WoW
AP	66,7%		33,4%							
CB		100,0%								
DP	20,0%		80,0%							
DW				100,0 %						
EB					100,0%					
ES						100,0%				
LiP							100,0%			
PB								100,0%		
US									100,0%	
WoW										100,0%
Av.Acc	**94, 67%**									

the meaningful information that is provided by this descriptor and its encoding. On the other hand, VLAD performs worse when trajectory information is added, showing that it cannot exploit the additional information provided by the trajectory coordinates.

In Table 2 we present the confusion matrix with the highest recognition rates achieved with this schema. We observe that our algorithm leads to very accurate activity recognition for most ADLs. Only "answer phone" was recognized with low accuracy, as it was confused with "dialing phone" and vice versa, which makes sense, since these two ADLs are very similar to each other.

5.2 *CHUN Action Dataset*

The proposed method was also applied on two real-life datasets that were recorded during the Dem@Care project at CHUN and GAADRD [2]. The CHUN dataset consists of 15 h and 10 min recordings of 64 PwD that perform ADLs in a Lab environment. The camera viewpoint is such that it monitors the whole room where the PwD is performing semi-directed ADLs (i.e. they followed instructions for performing a set of ADLs listed on a paper). The recording frequency is at 8 fps and our algorithm performance achieved 3.075–4.035 fps for video analysis, which is a near real time process achievement considering that for 2 video frames of the actual video, one is processed by our ADL detection algorithm. The ADLs observed are: (AP) answering phone and (DP) dialing phone, (LoM) look on map, (PB) pay bill, (PD) prepare drugs, (PT) prepare tea, (RP) read paper, (WP) water plant and (WtV) watch TV. They included large anthropometric variations and activity performance styles, while severe occlusions introduced great difficulty in discriminating actions. Figure 9 depicts some activities with their trajectory and HOGHOF rectangles. Despite these challenges, Fig. 10 shows that high accuracy results were achieved, proving the applicability of our technique in real applications.

Table 3 shows that the inclusion of trajectory coordinates in the HOGHOF descriptor improves recognition performance, with both the VLAD and Fisher encoding schemas. The differences between HOGHOF and HOGHOF+Traj are

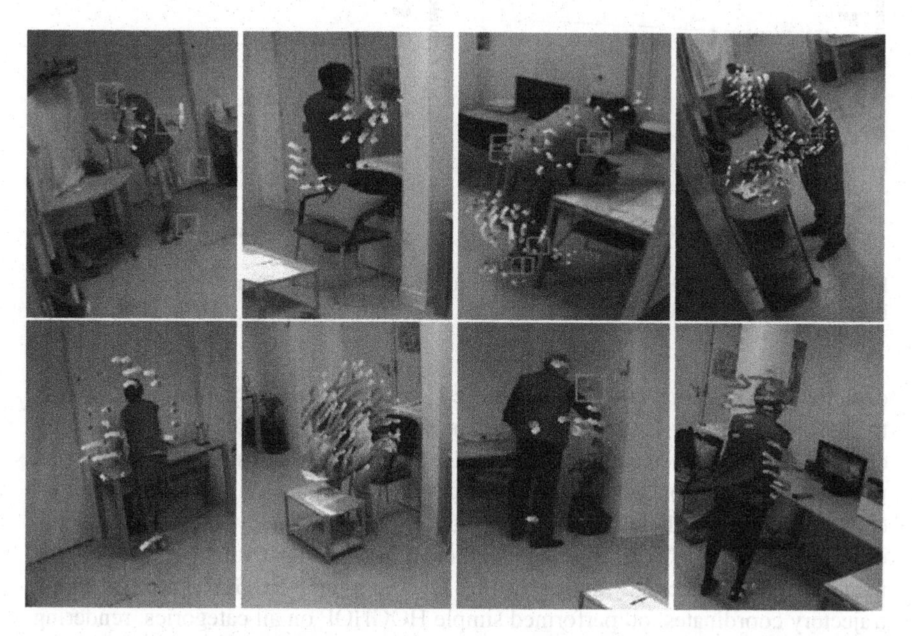

Fig. 9 CHUN dataset sample ADLs. From *left to right* and *top to bottom* we have: answer phone, look on map, pay bill, prepare drugs, prepare tea, read paper, water plant and watch TV

Table 3 Average accuracy over all classes for the CHUN dataset

Encoding	Vocabulary size	Descriptor	Average Accuracy (%)
VLAD	32	HOGHOF	90,65
		HOGHOF+Traj	96,43
	64	HOGHOF	91,66
		HOGHOF+Traj	96,43
	128	HOGHOF	91,32
		HOGHOF+Traj	96,66
	256	HOGHOF	93,47
		HOGHOF+Traj	96,78
Fisher	32	HOGHOF	92,12
		HOGHOF+Traj	95,96
	64	HOGHOF	93,34
		HOGHOF+Traj	95,75
	128	HOGHOF	92,36
		HOGHOF+Traj	95,55
	256	HOGHOF	91,57
		HOGHOF+Traj	95,35

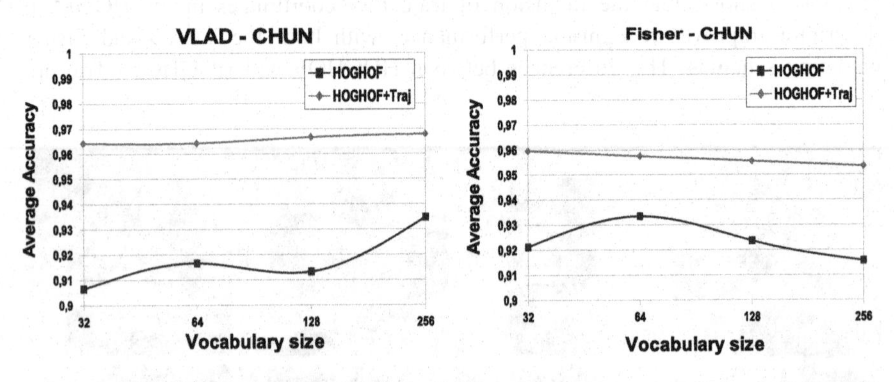

Fig. 10 HOGHOF and HOGHOF+Traj comparison when VLAD (*left*) and Fisher (*right*) encoding is applied on the CHUN action dataset

depicted clearly in Fig. 10. VLAD achieved very accurate recognition rates, similar to those achieved by Fisher encoding, but at a lower computational and memory cost (i.e. fewer distance computations for the same vocabulary size). This motivated us to test our detection schema on the CHUN dataset with a VLAD descriptor of 64 cluster centers, which achieves a fair balance among accuracy and computational cost, and used the OV20 Jaccard coefficient as a comparison metric, with the resulting average accuracy shown in Fig. 11. The hybrid HOGHOF, i.e. including trajectory coordinates, outperformed simple HOGHOF on all categories, rendering it more appropriate for this challenging, real-life dataset.

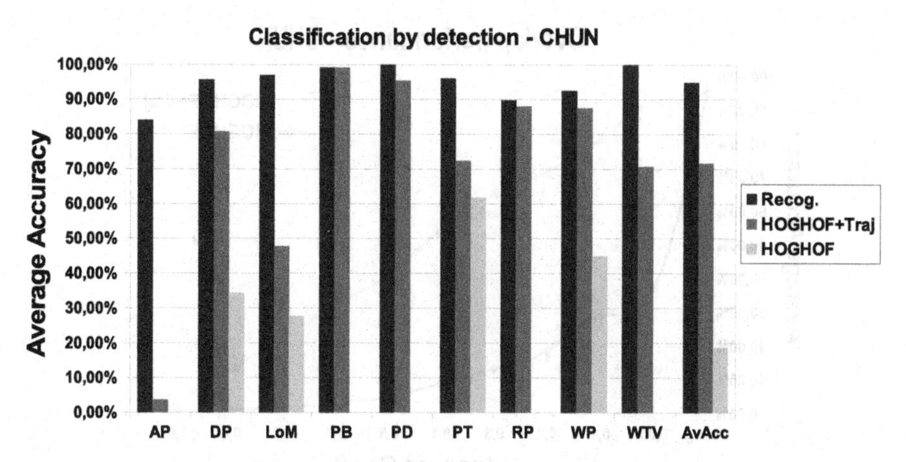

Fig. 11 Classification by detection results. Average accuracy for HOGHOF and HOGHOF+Traj descriptors are provided for OV20 on the CHUN dataset

Table 4 ADL recognition, on average accuracy means, on the CHUN dataset

	AP	DP	LoM	PB	PD	PT	RP	WP	WtV
AP	89,09 %	9,09 %				1,82 %			
DP	2,99 %	97,01 %							
LoM			100,0 %						
PB		0,86 %		98,28 %		0,86 %			
PD					100,0 %				
PT						98,67 %		1,33 %	
RP				1,72 %		3,45 %	94,83 %		
WP	2,5 %					2,5 %		95,00 %	
WtV				1,72 %					98,28 %
AA	**96,79 %**								

Table 4 depicts the recognition rates achieved when leave-one-Subject-out was used to classify the CHUN videos. The robustness of our recognition algorithm is obvious, as all activities, except for AP (which, as before, is confused with the very similar DP), are very accurately classified.

Figure 12 shows the performance of the HOGHOF and HOGHOF+Traj descriptors when varying the Jaccard Coefficient. For all percentages of overlap, we have a consistent improvement in accuracy with the inclusion of trajectory coordinates in the HOGHOF descriptor.

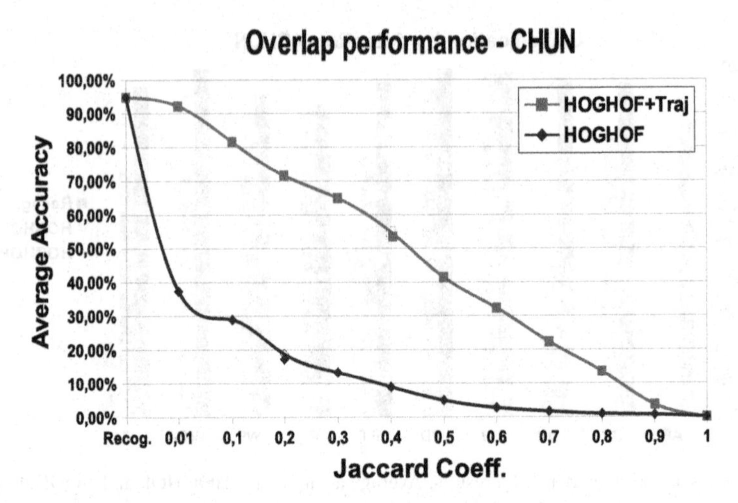

Fig. 12 Average accuracy for different overlap ratios when HOGHOF and HOGHOF+Traj are applied on the CHUN dataset

5.3 GAADRD Action Dataset

The GAADRD action dataset consists of 1 h and 52 min recordings of 32 PwDs that perform ADLs in a home-like environment. The camera viewpoint is in front of the person while they perform directed ADLs (i.e. activities dictated by a clinician). The recording frequency is at 8 fps and our algorithm performance achieved 3.27–3.85 fps for video analysis, which is a near real time process achievement considering that for 1.5–2 video frames of the actual video, one frame is processed by our ADL detection algorithm. The ADLs observed include: Cleaning Up (CU), Drink Beverage (DB), End Phonecall (EP), Enter Room (ER), Eat Snack (ES), Hand-Shake (HS), Prepare Snack (PS), Read Paper (RP), Serve Beverage (SB), Start Phonecall (SP) and Talk to Visitor (TV). They included large anthropometric differences and activity performance styles, while continuity among the ADLs introduced difficulty and increased the computational cost needed for activity detection. Characteristic video frames for GAADRD action dataset are provided in Fig. 13.

Table 5 aggregates the average accuracy rates over all ADL classes when our representation scheme was used in a one-subject-against-all scenario. Figure 14 shows the classification performance of HOGHOF and HOGHOF+Traj. It is obvious that the Fisher encoding schema surpasses the VLAD recognition rates and the hybrid HOGHOF+Traj descriptor outperforms the local HOGHOF, even for small vocabularies. With VLAD, on the other hand, the inclusion of trajectory information does not significantly improve recognition rates.

Fig. 13 Characteristic video frames for enter room (ER), handshake (HS) and serve beverage (SB) ADLs on the *top row*. Drink Beverage (DB), prepare snack (PS) and read paper (RP) on the *bottom row*, taken from URADL dataset from *left to right*

Table 5 Average accuracy over all classes for GAADRD action dataset

Encoding	Vocabulary size	Descriptor	Average Accuracy (%)
VLAD	32	HOGHOF	75,90
		HOGHOF+Traj	76,02
	64	HOGHOF	80,69
		HOGHOF+Traj	77,13
	128	HOGHOF	80,66
		HOGHOF+Traj	78,15
	256	HOGHOF	81,75
		HOGHOF+Traj	80,54
Fisher	32	HOGHOF	83,20
		HOGHOF+Traj	85,10
	64	HOGHOF	81,80
		HOGHOF+Traj	88,10
	128	HOGHOF	83,02
		HOGHOF+Traj	87,60
	256	HOGHOF	81,83
		HOGHOF+Traj	87,20

Table 6 depicts the recognition rates achieved when leave-one-Subject-out was used for classifying the GAADRD videos under a Fisher encoding framework with a vocabulary size of 64. Most ADLs are distinguished quite clearly except from some that are very similar, such as ES and PS which are confused with DB and SB (i.e. same location-similar action).

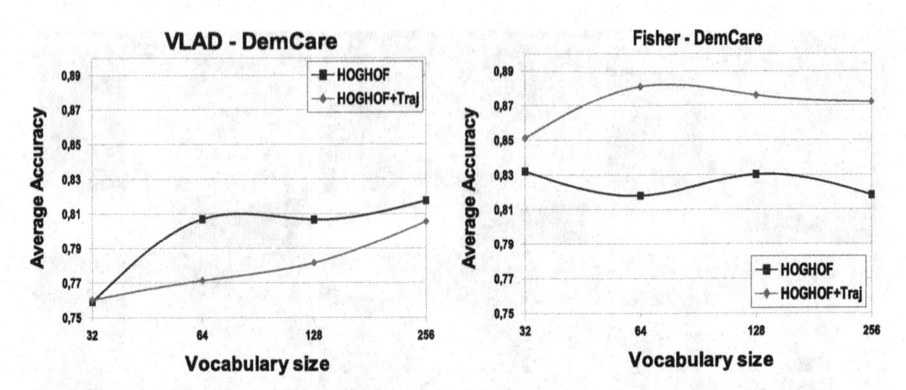

Fig. 14 HOGHOF and HOGHOF+Traj comparison when VLAD (*left*) and Fisher (*right*) encoding is applied on GAADRD action dataset

Table 6 Confusion matrix for ADL recognition, using average accuracy (% scores) over all classes, on the GAADRD dataset

	CU	DB	EP	ER	ES	HS	PS	RP	SB	SP	TV
CU	85,3 %		2,9 %		5,9 %		2,9 %			2,9 %	
DB	2,0 %	93,9 %			4,1 %						
EP	3,1 %		84,4 %		3,1 %					9,4 %	
ER				100 %							
ES		24,4 %	2,2 %		71,1 %			2,2 %			
HS						90,6 %					9,4 %
PS		2,9 %			8,6 %		71,4 %		17,1 %		
RP	3,1 %			3,1 %				93,8 %			
SB					2,9 %		5,9 %		91,2 %		
SP			6,1 %		6,1 %					87,9 %	
TV											100 %
AA	**88.1 %**										

Figure 15 shows the performance of the HOGHOF and HOGHOF+Traj descriptors for a varying Jaccard Coefficient. Again it is obvious that trajectory coordinates boost HOGHOF descriptor and lead to better classification rates by detection than when using simple HOGHOF.

Figure 16 shows classification by detection over all classes for the HOGHOF and HOGHOF+Traj descriptors applied to the GAADRD action dataset. The OV20 Jaccard coefficient was used as a comparison metric and a Fisher descriptor with 64 cluster centers was picked as a vocabulary size. The results are quite similar in distinct classes (see ER/DB/ES), while great differences are spotted in static ADLs, where trajectory coordinates make the difference (see TV/RP/HS) and HOGHOF+Traj outerperforms HOGHOF. Other activities (classes), such as

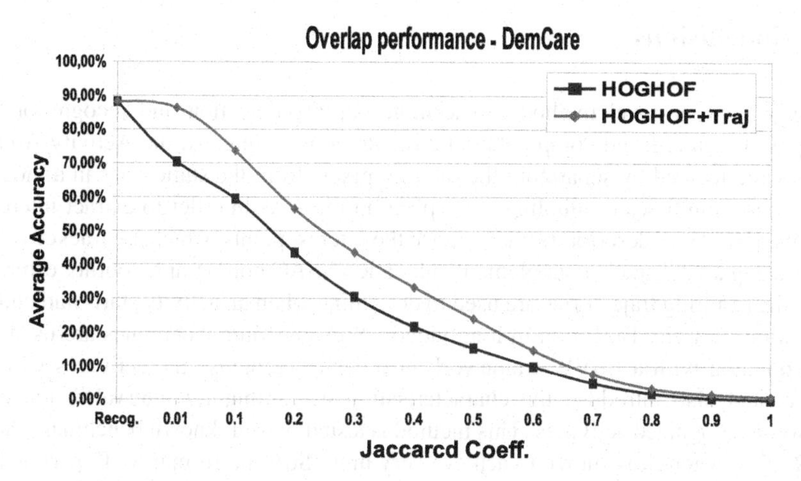

Fig. 15 Average accuracy for different overlap ratios when HOGHOF and HOGHOF+Traj applied on DemCare dataset

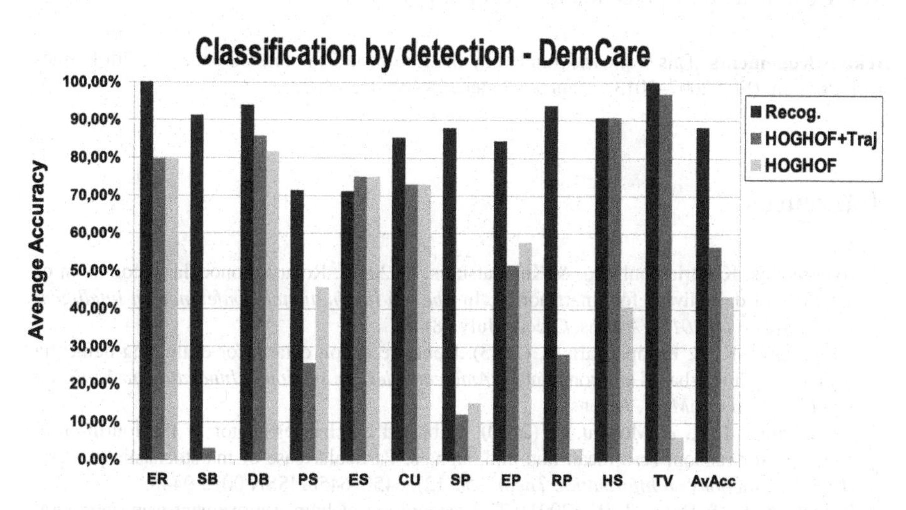

Fig. 16 Classification by detection results. Average accuracy for HOGHOF and HOGHOF+Traj descriptors are provided for OV20 on the GAADRD dataset

SB/PS and SP/EP, located in the same region and including very similar actions were usually confused and performed quite poorly with both descriptors. Generally, HOGHOF+Traj performs better for most ADLs and can be considered a better option than simple HOGHOF.

6 Conclusions

In this work, a novel method for accurate activity detection and recognition is proposed, at a reduced computational cost. Regions of interest, the Activity Areas (AAs) are located by separating the moving pixels from the static ones in a video, and dense, multi-scale sampling takes place in the AAs in order to extract interest points. HOGHOF descriptors characterise the interest points, which are tracked over time using a KLT tracker supplemented by a RANSAC homography outlier estimator. The resulting trajectories are used to determine when an activity starts and ends, providing Activity Detection in long videos. SoA encoding techniques are used in combination with a BoVW framework in order to recognise the activities taking place, and also introduce the characterisation of "ambiguity intervals", located between recognized activities. This method is tested on well known benchmark data (URADL) where it is shown to achieve very high, SoA, performance. Experiments also take place with home-like environments in more challenging, real life datasets from CHUN and GAADRD, where it can be seen that the proposed method leads to very accurate activity detection and recognition rates.

Acknowledgements This work was funded by the European Commission under the 7th Framework Program (FP7 2007–2013), grant agreement 288199 Dem@Care.

References

1. Avgerinakis, K., Briassouli, A., & Kompatsiaris, I. (2013). Robust monocular recognition of activities of daily living for smart homes. In *The 9th International Conference on Intelligent Environments (IE2013)*, Athens, Greece, July 18–19.
2. Avgerinakis, K., & Kompatsiaris, I. (2013). Demcare action dataset for evaluating dementia patients in a home-based environment. In *Ambient TeleCare session of Innovation in Medicine and Healthcare (InMed)*, Athens.
3. Blagouchine, I. V., & Moreau, E. (2010). Unbiased efficient estimator of the fourth-order cumulant for random zero-mean non-i.i.d. signals: Particular case of ma stochastic process. *IEEE Transactions on Information Theory, 56*(12), 6450–6458. ISSN 0018-9448.
4. Bobick, A. F., & Davis, J. W. (2001). The recognition of human movement using temporal templates. *IEEE Transactions on Pattern Analysis and Machine Intelligence, 23*, 257–267.
5. Briassouli, A., & Kompatsiaris, I. (2009). Robust temporal activity templates using higher order statistics. *IEEE Transactions on Image Processing, 18*(12), 2756–2768.
6. Derpanis, K. G., Sizintsev, M., Cannons, K., & Wildes, R. P. (2010). Efficient action spotting based on a spacetime oriented structure representation. In *IEEE Conference on Computer Vision and Pattern Recognition (CVPR)*.
7. Dollar, P., Rabaud, V., Cottrell, G., & Belongie, S. (2005). Behavior recognition via sparse spatio-temporal features. In *IEEE International Workshop on Visual Surveillance and Performance Evaluation of Tracking and Surveillance* (pp. 65–72).
8. Gaidon, A., Harchaoui, Z., & Schmid, C. (2013). Temporal localization of actions with actoms. *IEEE Transactions on Pattern Analysis and Machince Intelligence, 35*(11), 2782–2795.

9. Gorelick, L., Blank, M., Shechtman, E., Irani, M., & Basri, R. (2007). Actions as space-time shapes. In *IEEE Transaction on Pattern Analysis and Machine Intelligence (TPAMI)* (pp. 1395–1402).
10. Jegou, H., Perronnin, F., Douze, M., Sanchez, J., Perez, P., & Schmid, C. (2012). Aggregating local image descriptors into compact codes. *IEEE Transactions on Pattern Analysis and Machine Intelligence, 34*(9), 1704–1716. ISSN 0162-8828.
11. Ke, Y., Sukthankar, R., & Hebert, M. (2010). Volumetric features for video event detection. *International Journal of Computer Vision, 88*(3), 339–362.
12. Kim, T. K., & Cipolla, R. (2009). Canonical correlation analysis of video volume tensors for action categorization and detection. *IEEE Transactions on Pattern Analysis and Machine Intelligence, 31*, 1415–1428.
13. Klaser, A., Marszalek, M., & Schmid, C. (2008). A spatio-temporal descriptor based on 3d-gradients. In *In British Machine Vision Conference (BMVC).*
14. Klaser, A., Marszałek, M., Schmid, C., & Zisserman, A. (2010). Human focused action localization in video. In K. N. Kutulakos (Ed.), *Lecture Notes in Computer Science: Vol. 6553. IEEE European Conference on Computer Vision (ECCV Workshops)* (pp. 219–233). Berlin: Springer. ISBN 978-3-642-35748-0.
15. Laptev, I., & Lindeberg, T. (2003). Space-time interest points. In *IEEE International Conference on Computer Vision (ICCV)* (Vol. 1, pp. 432–439).
16. Laptev, I., Marszałek, M., Schmid, C., & Rozenfeld, B. (2008). Learning realistic human actions from movies. In *IEEE Conference on Computer Vision & Pattern Recognition (CVPR).*
17. Laptev, I., & Perez, P. (2007). Retrieving actions in movies. *IEEE International Conference on Computer Vision (ICCV).*
18. Laxton, B., Lim, J., & Kriegman, D. (2007). Leveraging temporal, contextual and ordering constraints for recognizing complex activities in video. In *Computer Vision and Pattern Recognition (CVPR).*
19. Marsaglia, G., Tsang, W. W., & Wang, J. (2003). Evaluating kolmogorov's distribution. *Journal of Statistical Software, 8*(18), 1–4, 11. ISSN 1548-7660. URL http://www.jstatsoft.org/v08/i18.
20. Messing, R., Pal, C., & Kautz, H. (2009). Activity recognition using the velocity histories of tracked keypoints. In *IEEE International Conference on Computer Vision (ICCV).* Washington, DC.: IEEE Computer Society.
21. Matikainen, P., Hebert, M., & Sukthankar, R. (2009). Trajectons: Action recognition through the motion analysis of tracked features. In *International Conference on Computer Vision Workshop on Video-Oriented Object and Event Classification (ICCV Workshop).*
22. Oliver, N. M., Rosario, B., & Pentland, A. P. (2000). A Bayesian computer vision system for modeling human interactions. *IEEE Transactions on Pattern Analysis and Machine Intelligence, 22*(8), 831–843.
23. Oneata, D., Verbeek, J., & Schmid, C. (2013). Action and event recognition with fisher vectors on a compact feature set. In *IEEE International Conference in Computer Vision (ICCV).*
24. Sadanand, S., & Corso, J. J. (2012). Action bank: A high-level representation of activity in video. In *IEEE Conference on Computer Vision and Pattern Recognition (CVPR).*
25. Sakoe, H., & Chiba, S. (1978). Dynamic programming algorithm optimization for spoken word recognition. *Transactions on Acoustics, Speech and Signal Processing, 26*(1), 43–49.
26. Scovanner, P., Ali, S., & Shah, M. (2007). A 3-dimensional sift descriptor and its application to action recognition. In *Proceedings of the 15th International Conference on Multimedia*, MULTIMEDIA '07 (pp. 357–360). New York: ACM. ISBN 978-1-59593-702-5. URL http://doi.acm.org/10.1145/1291233.1291311.
27. Wang, H., Klaser, A., Schmid, C., & Liu, C.-L. (2011). Action recognition by dense trajectories. In *IEEE Conference on Computer Vision & Pattern Recognition (CVPR)* (pp. 3169–3176). Colorado Springs. URL http://hal.inria.fr/inria-00583818/en.

28. Willems, G., Tuytelaars, T., & Gool, L. (2008). An efficient dense and scale-invariant spatio-temporal interest point detector. In *Proceedings of the 10th European Conference on Computer Vision: Part II* (pp. 650–663). Berlin/Heidelberg: IEEE European Conference on Computer Vision (ECCV). ISBN 978-3-540-88685-3.
29. Weinland, D., Ronfard, R., & Boyer, E. (2006). Free viewpoint action recognition using motion history volumes. In *Computer Vision and Image Understanding (CVIU)*.
30. Zach, C., Pock, T., & Bischof, H. (2007). A duality based approach for realtime tv-l1 optical flow. In *Annual Symposium of the German Association for Pattern Recognition* (pp. 214–223).

Recognition of Instrumental Activities of Daily Living in Egocentric Video for Activity Monitoring of Patients with Dementia

Iván González-Díaz, Vincent Buso, Jenny Benois-Pineau, Guillaume Bourmaud, Gaelle Usseglio, Rémi Mégret, Yann Gaestel, and Jean-François Dartigues

1 Introduction

The task of recognizing human activities in videos has become a fundamental challenge among the computer vision community [15]. In order to face the limited field of view and the difficulty of accessing all relevant information from fixed cameras, an alternative has been found in egocentric videos, recorded by cameras worn by subjects. Indeed, in addition to dealing with the previously listed drawbacks, wearable cameras represent a cheap and effective way to record users activity for scenarios such as telemedicine or life-logging.

In this chapter we focus on the problem of recognizing Instrumental Activities of Daily Living (IADL) for the assessment of the ability of patients suffering from Alzheimer disease and age-related dementia. Indeed, an objective assessment of a patient's capability to perform IADLs is a part of clinical protocol of dementia

I. González-Díaz
Department of Signal Theory and Communications, Universidad Carlos III de Madrid, Leganés, 28911, Madrid, Spain
e-mail: igonzalez@tsc.uc3m.es

V. Buso • J. Benois-Pineau (✉)
LaBRI, UMR 5800 CNRS/University of Bordeaux 1, Talence, France
e-mail: vbuso@labri.fr; benois-p@labri.fr

G. Bourmaud • G. Usseglio • R. Megret
IMS, UMR 5218 CNRS/University of Bordeaux 1, Talence, France
e-mail: guillaume.bourmaud@ims-bordeaux.fr; remi.megret@ims-bordeaux.fr

Y. Gaestel • J.-F. Dartigues
ISPED, U897 INSERM, Bordeaux, France
e-mail: yann.gaestel@isped.u-bordeaux2.fr; dartigues@isped.u-bordeaux2.fr

A. Briassouli et al. (eds.), *Health Monitoring and Personalized Feedback using Multimedia Data*, DOI 10.1007/978-3-319-17963-6_9

diagnostics and evaluation of efficacy of therapetical treatment [1]. Traditional ways of assessment with the help of questionnaires do not bring satisfaction as two kinds of errors have been observed, which do not allow a practitionner to fully trust the responses. The error of the first kind is that one commited by the patients. At the early stage of dementia, they cannot admit that they become less performant in their everyday activities and diminish their difficulties. The error of the second kind is commited by the caregives. They are permanently stressed watching their relatives' mental capacities to deteriorate. Hence they over estimate the difficulties of patients with dementia [17]. This is why the egocentric video has been first used for the recording of IADLs on patients with dementia in [22]. Later on, first results of recognition of IADLs in such recorded video were reported in [19]. Nevertheless, the recognition problem being very complex, efficient ways of solving it still remain an open research issue.

There has been a fair amount of work on recognizing everyday at home activities by analyzing egocentric videos, many of them based on the fact that manipulated objects represent a significant part of the actions. However, most of the studies were conducted under a constrained scenario, in which all the subjects wearing the cameras perform actions in the same room and, therefore, interact with the same objects: e.g. a hospital scenario in which the medical staff asks patients to perform several activities. Typical constrained scenarios allow to make assumptions on the objects or even to use instance-level visual recognition: the authors in [12] present a model for learning objects and actions with very little supervision, whereas in [28] a dynamic Bayesian network that infer activities from location, objects and interactions is proposed. The problem still open under such a scenario becomes even more complex, if an ecological observation is performed, i.e. at person's home. The individual environment varies, the objects of the same usage, e.g. a tea-pot or a coffee machine, can be of totaly different appearance. We call this scenario "unconstrained". In this case the recognition of activities in a wearable camera video has to be funded on the features of higher abstraction level than simple image and video descriptors computed from pixels.

It is only recently that the more challenging unconstrained scenario has been examined regarding activity recognition, such as in the work of [21], where the authors recognize ego-actions in outdoor environments using a stacked Dirichlet Process Mixture model. Pirsivash and Ramanan [23] propose to train classifiers for activities based on the output of the well-known deformable part model [14] using temporal pyramids. They demonstrate that performances are dramatically increased if one has knowledge of the object being interacted with. The approach making use of these "active" areas for ADL recognition has also been studied by Fathi and al. in [13] under a constrained scenario, where the authors enhanced their performances by defining visual saliency maps.

In the context of medical research on Alzheimer disease the unconstrained scenario means an epidemiological study of performances of patients in an ecological situation at their homes as it was done in [20]. Hence in this paper we model an activity as a combination of a meaningful object the person interacts with and the environment. The rationale here is quite straightforward. Indeed a reasonable assumption can be made that e.g. if a person is manipulating a tea pot in front of a kitchen table, than the activity consists in "making tea". If a TV set is observed in the camera view field and the person is in living room, then the activity would be "watching TV". Therefore, efficient recognition approaches have to be proposed for object recognition and localization of a person in its environment, and, more than that an efficient combination of results of these two detectors have to be designed in the activity recognition framework.

We therefore make the following contributions: (1) we further develop object recognition approach with psycho-visual weighting by saliency maps [5], (2) we show that analyzing the dynamics of a sequence of active objects + context by means of temporal pyramids [23] becomes a suitable paradigm for activity recognition in egocentric videos. However, in this optic we claim that context can be better described by the output of place recognition module rather by the outputs of many non-active object detectors as proposed in [23]. We provide experimental evaluation on a publicly available dataset of activities in egocentric videos.

The remainder of the paper is organized as follows: in Sect. 2 we describe the involved modules in our activity recognition approach. Section 3 assesses our model and compares it to the current state-of-the-art performances and Sect. 4 draws our main conclusions and introduces our further research.

2 The Approach

We aim to recognize IADLs by analyzing human-object interactions and as well as the contextual information surrounding them. Hence let us firstly introduce the notion of an *'active object'* (AO). An AO is an object which the subject/patient wearing camera interacts with. Here the interaction is understood as manipulation or observation. We claim that the analysis of this kind of objects becomes the main source of information for the activity recognition, and that the explicit recognition of 'non-active' objects as in [20] is not longer needed. We suppose that they can be efficiently encoded in a global descriptor of the scene/context. The activity model is therefore understand as the interaction with specific objects (AO) in a specific environment (context). In this particular work, we have considered that context can be successfully represented by identifying the place in which the user is performing the activity.

We propose a hierarchical approach with two connected processing layers (see Fig. 1). The first layer contains a set of *Active Object detectors* (Sect. 2.1) and a *Place Recognition* system (Sect. 2.2). Hence it allows for identification of the elements of our activity model. The second one addresses the activity recognition task on the basis of identified elements (Sect. 2.3).

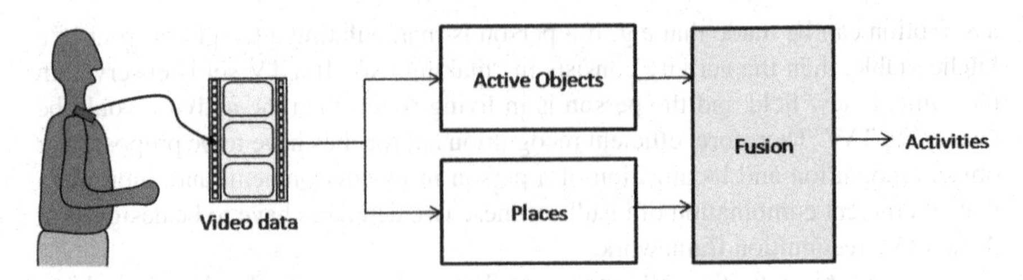

Fig. 1 Processing pipeline for the activity recognition

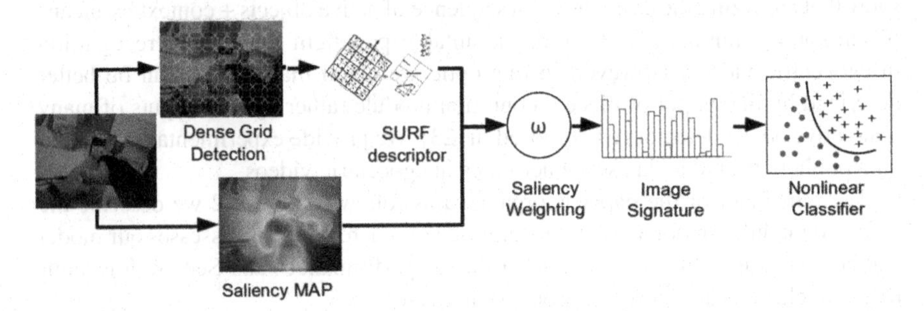

Fig. 2 Processing pipeline for the saliency-based object recognition in first-person camera videos

2.1 *Object Recognition*

As already mentioned, we aim to recognize activities under an unconstrained scenario in which each video is recorded at a different place. This is therefore a more difficult task than the recognition of specific objects instances. It remains an open problem for the computer vision community.

In general, we consider one individual detector for each object category although, as shown in the processing pipeline presented in Fig. 2, the nonlinear classification stage is the only step that is specific for each category. We have built our model on the well-known Bag-of-Words (BoW) paradigm [9] and proposed to add saliency masks as a way to provide spatial discrimination to the original Bag-of-Words approach. Hence, for each frame in a video sequence, we extract a set of N SURF descriptors d_n [3], using a dense grid of circular local patches. Next, each descriptor d_n is assigned to the most similar word $j = 1..V$ in a visual vocabulary by following a vector-quantization process. The visual vocabulary, computed using a k-means algorithm over a large set of descriptors in the training dataset (about 1M descriptors in our case), has a size of $V = 4{,}000$ visual words.

In parallel, our system generates a geometric-spatio-temporal saliency map S of the frame with the same dimensions of the image and values in the range [0,1] (the higher the more salient a pixel is, see Fig. 3). Details about the generation of saliency maps can be found in [5]. Here we briefly remind the key components for prediction a salient area in wearable video.

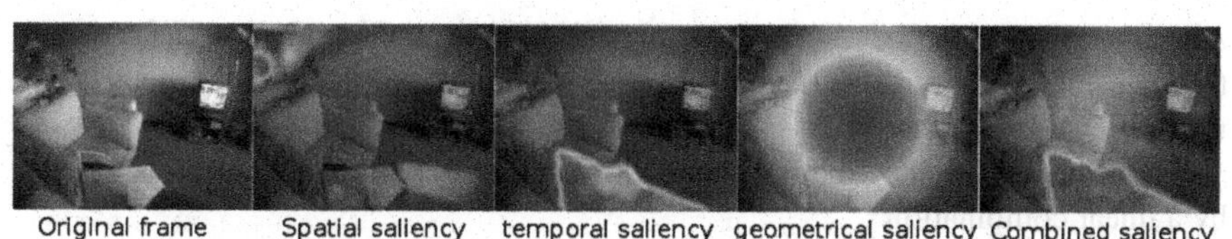

Fig. 3 Illustration of the different saliency cues (geometrical, spatial and temporal) composing the spatio-temporal saliency maps from [5]

First of all the saliency prediction approach we follow is a "bottom-up" one, or stimuli—driven. This means that the local characteristics of video frames are used to model attraction of Human Visual Attention (HVA) by the elements of visual scenes. Since the fundamental work by Itti [18] for stills and later developed models for video [4], bottom-up models for prediction of visual attention incorporate three cues: (i) spatial, (ii) temporal, (iii) geometric. Spatial cue stands for sensitivity of Human Visual System (HVS) to luminance and colour contrasts and orientation in image plane. Temporal cue expresses its sensitivity to motion. Finally, the geometrical cue usually expresses the so-called "central biais" hypothesis put forward by Buswell [6].

As for the spatial cue, various local filtering approaches have been proposed in order to compute local contrast and orientation. In our work as in [5] we used local operators allowing for computation of seven local contrast features in HSI colour domain: Contrast of Saturation, Contrast of Intensity, Contrast of Hue, Contrast of Opponents, Contrast of Warm and Cold Colors, Dominance of Warm Colors, and Dominance of Brightness and Saturation. These features proposed by Aziz and Mertsching [2] proved to be efficient for predicting sensitivity of HVS to colour contrasts. Then the spatial saliency map value for each pixel in a frame is a mean of these features.

Motion saliency map was built on the basis of non-linear sensitivity of HVS to motion magnitude, proposed by Daly [10] and expressing the fact that HVS is not sensitive to very low motion magnitude and to a very high motion magnitude neither. As the measure of motion magnitude we took the "residual motion" which is a local motion observed in image plane after compensation of camera motion according to affine motion model [5].

Finally, we devoted a specific study to the geometrical cue, which also is a non-trivial question in case of body-worn cameras. Indeed Buswell's hypothesis of central biais is not hold when the body-worn camera is not fixed on "symmetry axis" of human head as it was the case in the study [5]. Nevertheless, a reasonable assumption that the gaze direction coincides with the camera optical axis orientation can be made when the camera is fixed on the body such as on glaces or in a central position on the chest. (Note this is the case in the dataset we use in this chapter for experiments). In this case the geometric saliency map can be modelled by an isotropic gaussian with a spread $\sigma = 5$ visual degrees centered on image center.

All three maps: spatial $S_{sp}(x,y)$, temporal $S_t(x,y)$ and geometric $S_g(x,y)$ are normalized by their respective maximum, which is called "saliency peak". An illustration of these three cues is given in Fig. 3.

The resultant saliency map is obtained from the three normalized saliency maps by a linear combination

$$S(x,y) = \alpha * S_{sp}(x,y) + \beta * S_t(x,y) + \gamma * S_g(x,y) \quad (1)$$

The coefficients α,β,γ are estimated by linear regression with regard to ideal maps from a training set that is gaze fixation maps of subjects.

Coming back to the visual signature of a video frame, we use the resultant saliency map to weight the influence of each descriptor in the final image signature, so that each bin j of the BoW histogram H is computed following the next equation:

$$H_j = \sum_{n=1}^{N} \alpha_n w_{nj} \quad (2)$$

where the term $w_{nj} = 1$ if the descriptor or region n is quantized to the visual word j in the vocabulary and the weight α_n is defined as the maximum saliency value S found in the circular local region of the dense grid. Finally, the histogram H is L1-normalized in order to produce the final image signature.

Once each image is represented by its weighted histogram of visual words, we use a SVM classifier [8] with a nonlinear χ^2 kernel, which has shown good performance in visual recognition tasks working with normalized histograms as those ones used in the BoW paradigm [27]. Using the Platt approximation [24], we finally produce posterior probabilistic estimates O_k^t for the occurrence of the object of class k in the frame t.

2.2 *Place Recognition*

In this section we detail the place recognition module. Place recognition plays a role of *context recognition* in our overall approach for IADLs modeling and recognition.

The general framework can be decomposed into three steps. First of all, for each image, a global image descriptor is extracted. We choose the Composed Receptive Field Histograms (CRFH) [25] since it was proven to perform well for indoor localization estimation [11]. Then a non-linear dimensionality reduction method is employed. In our case, we use a Kernel Principal Component Analysis (KPCA) [26]. The purpose of this step is twofold: it reduces the size of the image descriptor which alleviates the computational burden of the rest of the framework, and it provides descriptors on which linear operations can be performed. Finally, based on these features, a linear Support Vector Machine (SVM) [8] is applied to perform the place recognition, and the result is regularized using temporal accumulation [11].

For the application considered in this paper, each video is taken in a different environment. Consequently, our module has to learn generic concepts instead of specific ones as it is usually the case [11]. In this context, we need to define concepts both relevant for action recognition and as constrained as possible to obtain better performances. Indeed, for example the concept 'stove' has probably less variability and may be more meaningful for action recognition than the concept 'kitchen'. This will be discussed in detail in Sect. 3.3.

Again, following the Platt approximation [24], the output of this module is then a vector P_j^t with the probability of a frame t representing the place j.

2.3 *Activity Recognition*

Our activity recognition module uses the temporal pyramid of features presented in [23], which allows to exploit the dynamics of user's behaviour in egocentric videos. However, rather than combining features for active/non-active objects, we represent activities as sequences of AOs and places (context). For instance, cooking may involve user's interaction with various utensils whereas cleaning the house might require a user to move around various places of the house.

In particular, for each frame t being analyzed, we consider a temporal neighborhood Ω_t corresponding to the interval $[t - \Delta/2, t + \Delta/2]$. This interval is then iteratively partitioned into two subsegments following a pyramid approach, so that at each level $l = 0 \ldots L - 1$ the pyramid contains 2^l subsegments. Hence, the final feature of a pyramid with L levels is defined as:

$$F_t = \left[F_t^{0,1} \ldots F_t^{l,1} \ldots F_t^{l,2^l} \ldots F_t^{L-1,2^{L-1}}\right] \tag{3}$$

where $F_t^{l,m}$ represents the feature associated to the subsegment m in the level l of the pyramid and is computed as:

$$F_t^{l,m} = \frac{2^l}{\Delta} \sum_{s \in \Omega_{tm}^l} f_s \tag{4}$$

where Ω_{tm}^l represents the m temporal neighborhood of the frame t in the level l of the pyramid and f_s is the feature computed at frame s in the video. In the experimental section, we will assess the performance of our approach using the outputs of K object detectors $\left[O_1^s \ldots O_K^s\right]$, the outputs of J place detectors $\left[P_1^s \ldots P_J^s\right]$, or the concatenation of both, as features f_s.

In this work, we have used a sliding window method with a fixed window of size Δ, parameter that is later studied in the Sect. 3, and a pyramid with $L = 2$. Finally, the temporal feature pyramid has been used as input for a linear multiclass SVM in charge of deciding the most likely action for each frame.

The complexity of the classifier system, being layered, precludes the easy interpretation of the results as probabilistic elements, as they are defined on an arbitrary axis that is suitable for deciding of a best class, but not to associate a probabilistic interpretation to it. Since automatic activity recognition from wearable camera is a difficult problem, it is very important to be able to assign confidence measures to these predictions, in order to monitor their validity and uncertainty for higher level inference. This problem corresponds to a calibration problem [16]. Even though the automatic detection of all possible events is not possible in all cases, computing confidences can mitigate this, by trusting the prediction only when the system is confident.

For a two-class classifier, each observation x_k (in our case the input fatures belonging to a multi-dimensional space) is associated a predicted binary label y_n in $\{0, 1\}$. In practice the prediction is based on the thresholding of the classifier score s_n, which is produced by the decision function as $s_n = f(x_n)$. The calibration problem consists in finding a transformation $p_n = g(s_n)$ of these scores into a value in the interval [0, 1] such that the result can be interpreted as the probability $p_k = P(y_k = 1|x_k)$ that of a true positive conditioned on the observed sample. The calibrated values have then reasonable properties to be used in a fusion approach with other sources of information.

In our work, we used the Platt approach [24], generalized to the one-to-one multi-class classification [29] and detailed in [7]. Each test sample is therefore associated with probabilistic confidence value $p_{kc} = P(L_k = c|x_k)$ that it belongs to class c, such that it is normalized by $\sum_c p_{kc}$.

The experimental part will evaluate both the raw recognition performance, using the classification strategy that assigns a sample to the class with higher probability, as well as the reliability of the estimated confidence value.

3 Experimental Section

3.1 Experimental Set-up

We have assessed our model in the ADL dataset, proposed by the authors of [23], that contains videos captured by a chest-mounted GoPro camera on 20 users performing various daily activities at their homes. This dataset was already annotated for 44 object-categories and 18 activities of interest (see Fig. 4) and we have additionally labeled 5 rooms and 7 places of interest.

This dataset is very challenging since both the environment and the object instances are completely different for each user, thus leading to an unconstrained scenario. Hence, and due to the hierarchical nature of the activity recognition process, we have trained every module following a leave-k-out procedure (k = 4 in our approach). This approach allows us to provide real testing results in object and place recognition for every user, so that the whole set can be later used for activity

Fig. 4 Overview of the 18 activities annotated in the ADL dataset

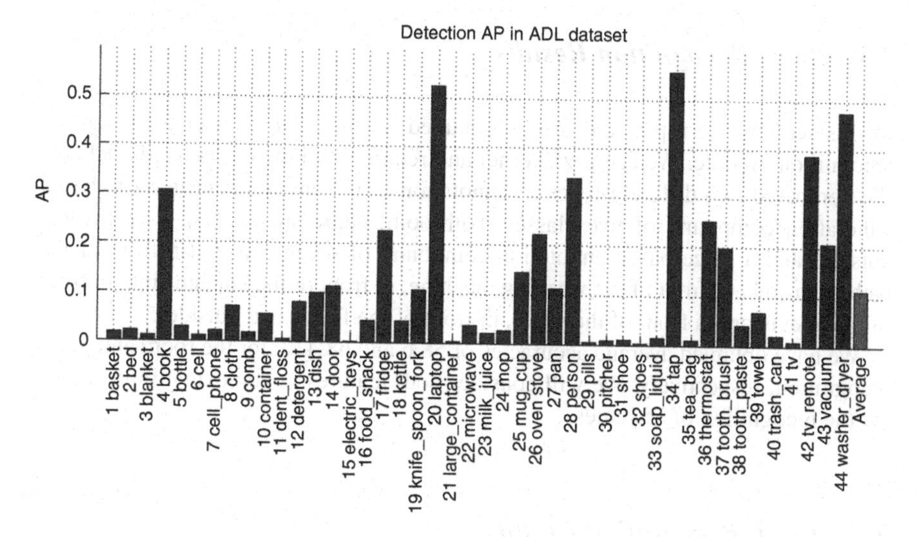

Fig. 5 Results in object detection

recognition. Furthermore, for activity recognition, the first six users have been taken to cross-validate the parameters of a linear SVM [8], whereas the remainder ones (7–20) have been used to train and test the models following a leave-1-out approach. The library libSVM [7] was used for the classification.

3.2 *Object Recognition Results*

Figure 5 shows the per-category and average results achieved by our active object detection approach in terms of Average Precision (AP). We have used this quality measure rather than accuracy due to the nature of the dataset, which is highly unbalanced for every category. The mean AP of our approach is 0.11 but, as can be noticed from the figure, the performance notably differs from one class to another. Main errors in classification are due to various reasons: (a) a high degree of intra-class variation between instances of objects found at different homes, what leads to poor recognition rates (e.g. bed clothes or shoes show large variations in their appearance), (b) some objects are too small to be correctly detected (dent floss, pills, etc.), and (c) for some objects that theoretically show a lower degree of intra-class variation (TV, microwave), performance is lower than expected since it is very hard for a detector to distinguish when they can be considered as 'active' in the scene (e.g. a user just faces a 'tv remote' or a 'laptop' when using them, whereas the TV or the microwave are more likely to appear in the field of view even when they are not 'active' for the user).

3.3 Place Recognition Results

In this section, we report the results obtained on the ADL dataset for the place recognition module. We use a χ^2 kernel and retain 500 dimensions for the KPCA. We compared two different types of annotation of the environment: a room based annotation compound of five classes (bathroom, bedroom, kitchen, living room, outside) and a place based annotation compound of seven classes (in front of the bathroom sink, in front of the washing machine, in front of the kitchen sink, in front of the television, in front of the stove, in front of the fridge and outside).

We have obtained average accuracies of 58.6 and 68.4 %, for the room and place recognition, respectively. We will consider both features as contextual information for the recognition of activities.

3.4 IADL Recognition Results

In this section we show our results in IADL recognition in egocentric videos. As already mentioned, our system identifies the activity at every frame of the video using a sliding window. The performance is evaluated using the accuracy at frame level, which is defined as the number of correctetly estimated frames divided by the total number of frames. For that end, we have also included a new class 'no activity' associated to frames that are not showing any activity of interest. It is also worth noting that the global performance is computed by averaging the particular accuracies for each class (rather than simply counting the number of correct decisions) and, thus, adapts better to highly unbalanced sets as the one being used (where most of the time there is no activity of interest).

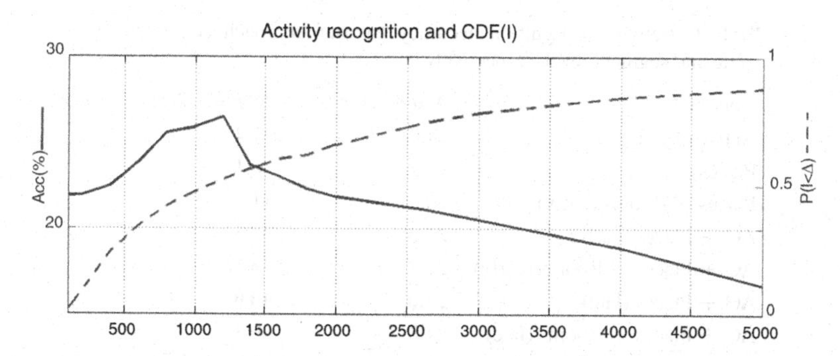

Fig. 6 Activity recognition accuracy with respect to the window size Δ (*blue solid line*) and cumulative distribution of activity lengths (*green dotted line*) (Color figure online)

3.4.1 Window Size

In our first experiment, we have studied the influence of the window size Δ defined in Sect. 2.3. Based on the results shown in Fig. 6 (blue line), we can draw interesting conclusions: on the one hand, too short windows do not model the dynamics of an activity, understood in our case as sequences of different active objects or places. Oppositely, too long windows may contain video segments showing various activities. Although, from our point of view, this fact might help to detect several strongly related activities by reinforcing the knowledge about one activity by the presence of the other (e.g. washing hands/face and drying hands/hair are activities that usually occur following the same temporal sequence), it might also lead to features containing too many active objects and places. These features would therefore make these frames difficult to assign to a particular activity. In our case, the value that best fits the activities in ADL dataset is $\Delta = 1{,}200$ frames, which corresponds to approximately 47 s of video footage. In fact, looking at the cumulative distribution of the activities length in the dataset (green line in Fig. 6), we have found this value is close to the median value which yields approximately 1,100 frames, thereby being consistent with the intuition that the window size should be chosen to be representative of typical activities length.

3.4.2 Recognition Performance

In the first column of Table 1, we show the results of our approach using either just active object or place detectors, and using an early combination of both of them by feature concatenation. As one can notice from the results, the active objects using sliency alone achieves slightly better performance than the approach of [23]. The place and room information alone yield lower performance, possibly being less informative to discriminate the activities. Combining objects and their context (the place where they are located) notably improves the performance achieved by simply

Table 1 Activity recognition accuracy for our approach computed at frame and segment level, respectively

Approach	Avg Fr. Acc (%)	Avg Seg. Acc (%)
Active objects (AO)	24.0	37.4
Places	18.5	6.1
Places + Rooms (early)	20.0	11.1
AO + Places (early)	**27.3**	38.5
AO + Places + Rooms (early)	26.3	36.5
AO + Places (late)	25.0	**40.0**
AO + Places + Rooms (late)	24.8	39.3
Pirsiavash et al. [23]	23.0	36.9

Bold values state the best performing approach in each scenario.

using the object detectors. Let us note that we have also tested several late fusion schemes (linear combinations, multiplicative, logarithmic, etc.) that did not lead to improvements in the system performance.

Furthermore, for comparison, we also include the results obtained with the software provided by the authors of [23]. This approach uses the outputs of various detectors of active and non-active objects implemented using the Deformable Part Models (DPM) [14]. Let us note that, as mentioned by the authors in the software, results differ from the ones reported in [23] due to changes in the dataset. From the results, and due to the similar classification pipeline of both methods, we can conclude that our features are more suitable for the activity recognition problem.

Finally, as made in [23], we additionally include results of a segment based evaluation in which ground truth time segmentations of the video are available in both training and testing steps. Hence, this case simplifies the activity recognition from a category segmentation problem to a simple classification problem for each segment. This case lacks the 'no activity' class, so that only video intervals showing activities of interest are taken into account. Combining objects and context provides the best performance, which is again superior to the one obtained by Pirsiavash and Ramanan [23].

In order to analyze these results in more details, Fig. 7 shows the Average Precision for each class separately. Performance is shown for several approaches, either using each mid-level feature alone, or using early or late fusion. It is clear from the results that several profiles appear for different kind of activities: some activities (Watching TV, Using the computer) are better recognized from object detection alone, while others (laundry, washing dishes, making coffee) are more linked to places. Their fusion tend to improve the mean performance, although the best way to do so depends on the activity category.

Overall, these results lead us to conclude that, recognizing activities in egocentric video does not require identifying every object in a scene, but simply detect the presence of 'active' objects and provide a compact representation of the object context. This context has been implemented in this work by means of a global classifier of the place. Future work could consider additional complementary features.

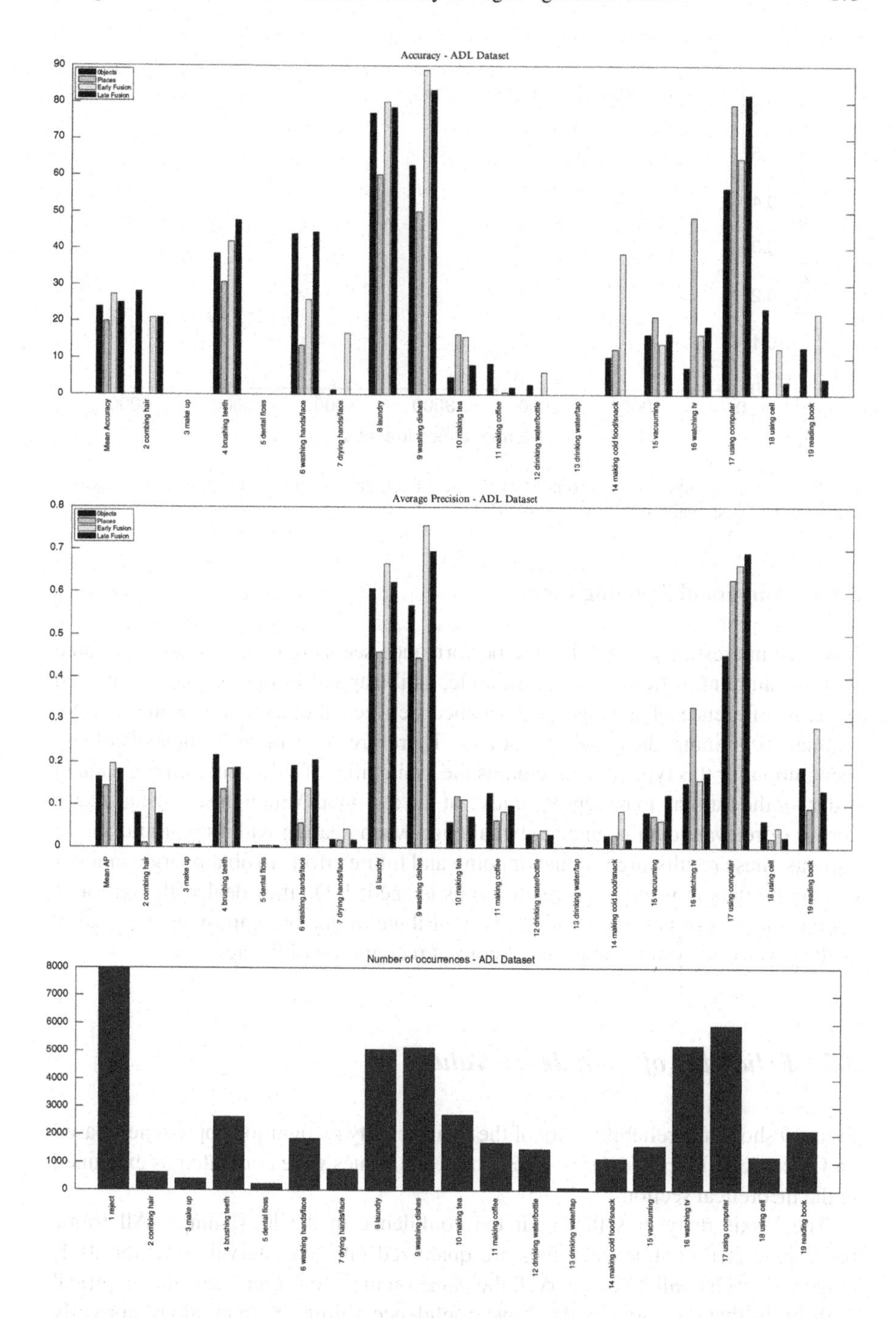

Fig. 7 (*Top*) Accuracy and (*Middle*) Average precision for activity recognition for various strategies: active object alone, places alone, early fusion, late fusion of active objects and places. (*Bottom*) Number of occurrences in the dataset

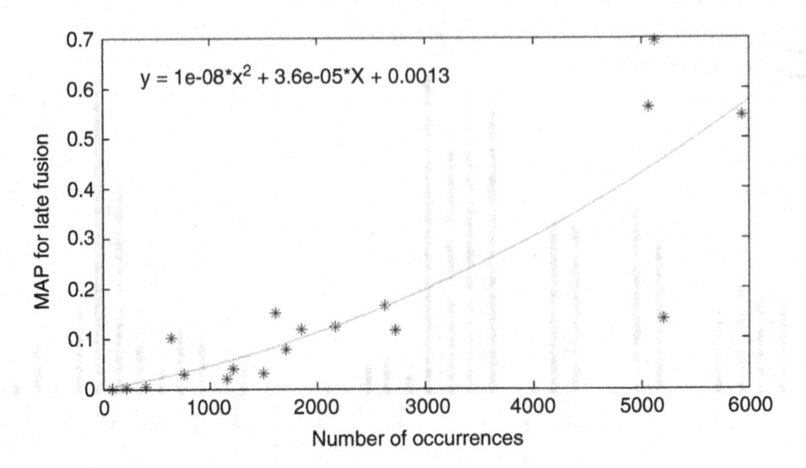

Fig. 8 MAP vs number of occurrences in dataset. Each star of the scatter plot represents one category. The best quadratic fitting is shown

3.4.3 Amount of Training Data

It is very interesting to note that the performance seems to be positively correlated with the amount of training data available, as illustrated in Fig. 8. There is indeed a sharp difference of average performance between the categories with a larger amount of training data and the others. Therefore, one main bottleneck of the recognition for this type of data remains the availability of sufficient training data, in order for the training to be representative of the test data. Although acquiring a large corpus of relevant data is an actual challenge when dealing with the monitoring of patients, these results suggest that ongoing and future efforts to obtain larger amount of training data in wearable camera setups is needed. Do they deal with control or patient subjects or not, they will likely contribute in notable improving the quality of the developed systems in terms of correct recognition of the activities.

3.5 Reliability of Confidence Values

Figure 9 shows the reliability plot of the main activity recognition approaches, based on Objects and Places features. The confidence values were computed as explained in the theoretical section.

The *x*-axis represents the predicted confidence in the [0, 1] range. All frame based predicted confidence values are quantized into ten intervals over the [0, 1] range. For each confidence interval, the value on the *y*-axis represents the empirical probability that the samples that have confidence within the interval are correctly classified. Therefore, an unperfect classifier with perfect calibration is called reliable and should ideally produce confidence values that match the empirical

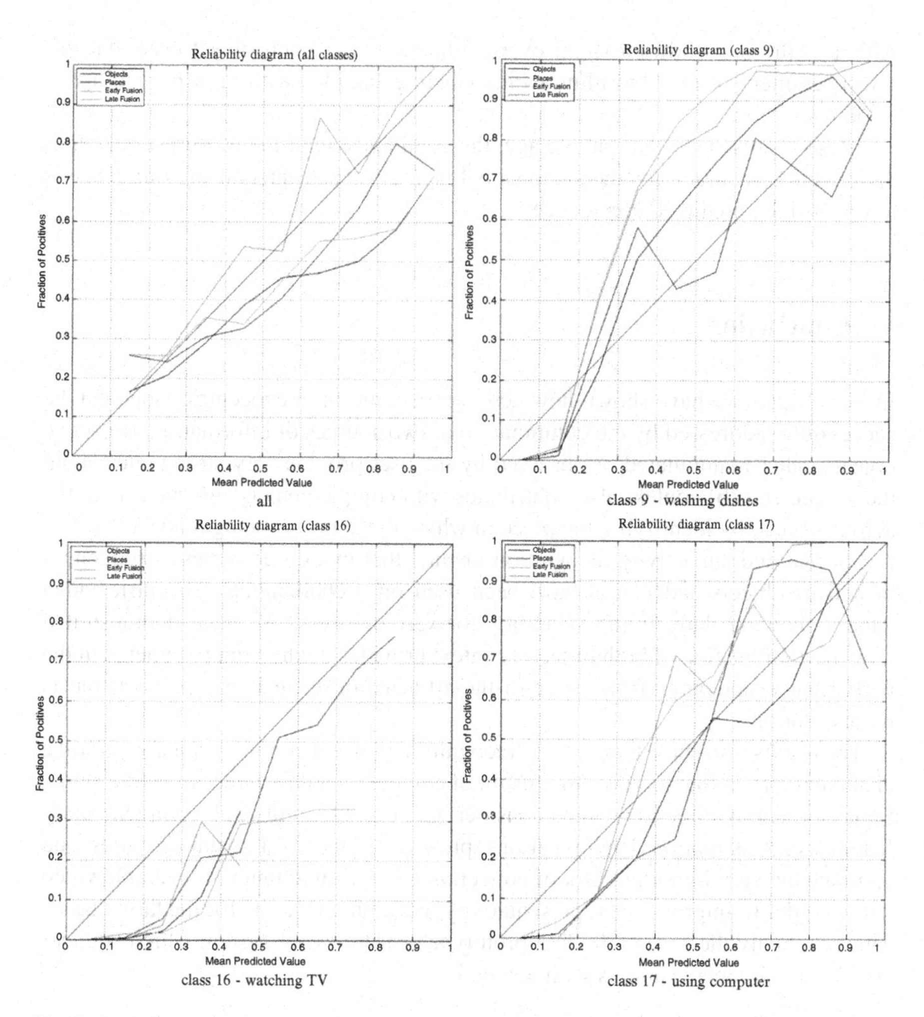

Fig. 9 Reliability diagrams of estimated confidence values

probability: points on the $(0, 0) - (1, 1)$ diagonal. Points over the diagonal show an underestimation of the quality of the classifier; point under the diagonal are overconfident.

Overall, for all classes, the reliability of all shown approaches follows approximately the diagonal. We have also shown the reliability plots of the categories with the three largest amounts of training data. Class 9, which has the best AP performance overall, has a reliability that is slightly under-confident: the estimates are actually better that predicted. Class 16 has a quite low AP performance, although a large amount of training data is available; this is reflected in the truncated curves, since no predictions on the test data was produced with high confidence.

Although the plot shows a slight over-confidence, it is interesting to see that this self-assessment of the algorithm avoids labeling samples with high probability in error.

These results show that on average, the confidence tend to be overestimated, as the effective accuracy for the samples belonging to each interval of confidence is lower than the predicted confidence.

4 Conclusion

In this chapter we have shown how activity recognition in egocentric video can be successfully addressed by the combination of two sources of information: (a) active objects either manipulated or observed by the user provide very strong cues about the action, and (b) context also contributes with complementary information to the active objects, by identifying the place in which the action is being made.

For that end, an activity recognition method that models activities as sequences of actives objects and places have been used on a challenging egocentric video dataset showing daily living scenarios for various users. We have demonstrated how the combination of both objects+context provides notable improvements in the performance, and outperforms state-of-the-art methods using active+passive objects representations.

The results also show that activity recognition in unconstrained scenarios is still a challenging task, that requires the fusion of complementary sources of information. Future research directions may consider the use of additional complementary features such as motion, hand positions, presence of faces for social activities, and continue the very important task of collecting significant amount of wearable video data in order to improve the representativity of training datasets for the target tasks. This is actually the case in the first prototype of Dem@care system which is under tests with volunteers patients with dementia.

Acknowledgements This research is supported by the EU FP7 PI Dem@Care project under grant agreement #288199. The authors would like to thank Olalla Rodríguez López for her valuable help.

References

1. Amieva, H., Goff, M. L., Millet, X., Orgogozo, J. M. M., Pérès, K., Barberger-Gateau, P., et al. (2008). Prodromal alzheimer's disease: Successive emergence of the clinical symptoms. *Annals of Neurology, 64*(5), 492–498.
2. Aziz, M. Z., & Mertsching, B. (2008). Fast and robust generation of feature maps for region-based visual attention. *IEEE Transactions on Image Processing, 17*(5), 633–644.
3. Bay, H., Ess, A., Tuytelaars, T., & Van Gool, L. (2008). Speeded-up robust features (surf). *Computer Vision and Image Understanding, 110*, 346–359.

4. Borji, A., & Itti, L. (2013). State-of-the-art in visual attention modeling. *IEEE Transactions on Pattern Analysis and Machine Intelligence, 35*(1), 185–207.
5. Boujut, H., Benois-Pineau, J., & Megret, R. (2012). Fusion of multiple visual cues for visual saliency extraction from wearable camera settings with strong motion. In *European Conference on Computer Vision - Workshops*, ECCV'12, (pp. 436–445).
6. Buswell, G. T. (1935). *How people look at pictures*. Chicago, IL: The University of Chicago Press.
7. Chang, C.-C., & Lin, C.-J. (2011). LIBSVM: A library for support vector machines. *ACM Transactions on Intelligent Systems and Technology, 2*, 27:1–27:27. Software available at http://www.csie.ntu.edu.tw/~cjlin/libsvm.
8. Cortes, C., & Vapnik, V. (1995). Support-vector networks. *Machine Learning, 20*, 273–297.
9. Csurka, G., Dance, C. R., Fan, L., Willamowski, J., & Bray, C. (2004). Visual categorization with bags of keypoints. In *Workshop on Statistical Learning in Computer Vision, ECCV* (pp. 1–22).
10. Daly, S. J. (1998). Engineering observations from spatiovelocity and spatiotemporal visual models. In *IS&T/SPIE Conference on Human Vision and Electronic Imaging III*, (Vol. 1).
11. Dovgalecs, V., Mégret, R., & Berthoumieu, Y. (2013). Multiple feature fusion based on co-training approach and time regularization for place classification in wearable video. *Advances in Multimedia, 2013*, 22 pp. doi:10.1155/2013/175064. Article ID 175064.
12. Fathi, A., Farhadi, A., & Rehg, J. M. (2011). Understanding egocentric activities. In *International Conference on Computer Vision, 2011*, ICCV '11 (pp. 407–414). Washington, DC, USA.
13. Fathi, A., Li, Y., & Rehg, J. M. (2012). Learning to recognize daily actions using gaze. In *Proceedings of the 12th European conference on Computer Vision - Volume Part I*, ECCV'12 (pp. 314–327). Berlin/Heidelberg: Springer.
14. Felzenszwalb, P. F., Girshick, R. B., McAllester, D. A., & Ramanan, D. (2010). Object detection with discriminatively trained part-based models. *IEEE Transactions on Pattern Analyisis and Machine Intelligence, 32*(9), 1627–1645.
15. Gaidon, A., Marszalek, M., & Schmid, C. (2009). Mining visual actions from movies. In A. Cavallaro, S. Prince & D. Alexander (Eds.), *British machine vision conference* (pp. 125.1–125.11). Londres, United Kingdom: British Machine Vision Association BMVA Press. Page web de l'article : http://lear.inrialpes.fr/pubs/2009/GMS09/.
16. Gebel, M. (2009). *Multivariate calibration of classifier scores into probability space*. Saarbrücken, Germany: VDM Publishing.
17. Helmer, C., Peres, K., Letenneur, L., Guttierez-Robledo, L. M., Ramaroson, H., Barberger-Gateau, P., et al. (2006). Measuring the objectness of image windows. *Geriatric Cognitive Disorders, 1*(22), 87–94.
18. Itti, L., Koch, C., & Niebur, E. (1998). A model of saliency-based visual attention for rapid scene analysis. *IEEE Transactions on Pattern Analysis and Machine Intelligence, 20*(11), 1254–1259.
19. Karaman, S., Benois-Pineau, J., Dovgalecs, V., Mégret, R., Pinquier, J., André-Obrecht, R., et al. (2014). Hierarchical hidden markov model in detecting activities of daily living in wearable videos for studies of dementia. *Multimedia Tools and Applications, 69*(3), 743–771.
20. Karaman, S., Benois-Pineau, J., Mégret, R., Dovgalecs, V., Dartigues, J.-F., & Gaëstel, Y. (2010). Human daily activities indexing in videos from wearable cameras for monitoring of patients with dementia diseases. In *International Conference on Pattern Recognition (ICPR), 2010* (pp. 4113–4116).
21. Kitani, K. M., Okabe, T., Sato, Y., & Sugimoto, A. (2011). Fast unsupervised ego-action learning for first-person sports videos. In *2011 IEEE Conference on Computer Vision and Pattern Recognition (CVPR)* (pp. 3241–3248).
22. Mégret, R., Dovgalecs, V., Wannous, H., Karaman, S., Benois-Pineau, J., Khoury, E. E., et al. (2010). The immed project: Wearable video monitoring of people with age dementia. In *Proceedings of the International Conference on Multimedia*, MM '10 (pp. 1299–1302). New York: ACM.

23. Pirsiavash, H., & Ramanan, D. (2012). Detecting activities of daily living in first-person camera views. In *2012 IEEE Conference on Computer Vision and Pattern Recognition (CVPR)*. IEEE.
24. Platt, J. C. (1999). Probabilistic outputs for support vector machines and comparisons to regularized likelihood methods. In *Advances in large margin classifiers* (pp. 61–74). Cambridge: MIT Press.
25. Pronobis, A., Mozos, O. M., Caputo, B., & Jensfelt, P. (2010). Multi-modal semantic place classification. *The International Journal of Robotics Research (IJRR), 29*(2–3), 298–320.
26. Schölkopf, B., Smola, A., & Müller, K.-R. (1998). Nonlinear component analysis as a kernel eigenvalue problem. *Neural Computing, 10*(5), 1299–1319.
27. Sreekanth, V., Vedaldi, A., Jawahar, C. V., & Zisserman, A. (2010). Generalized RBF feature maps for efficient detection. In *Proceedings of the British Machine Vision Conference (BMVC)*.
28. Sundaram, S., & Cuevas, W. W. M. (2009). High level activity recognition using low resolution wearable vision. In *Conference on Computer Vision and Pattern Recognition, CVPR Workshops 2009* (pp. 25–32).
29. Wu, T.-F., Lin, C.-J., & Weng, R. C. (2004). Probability estimates for multi-class classification by pairwise coupling. *Journal of Machine Learning Research, 5*, 975–1005.

Combining Multiple Sensors for Event Detection of Older People

Carlos F. Crispim-Junior, Qiao Ma, Baptiste Fosty, Rim Romdhane, Francois Bremond, and Monique Thonnat

1 Introduction

Human Behavior (or event) monitoring has experienced continuous advances since last decade promoted by Computer Vision, Wearable and Ubiquitous Computing fields. Examples of applications range from security field, such as video surveillance, crime prevention, and older people monitoring at home, to tools to support objective assessment of emerging symptoms of diseases (medical diagnosis), and even as a part of human-machine interfaces for game entertainment.

Wearable and Pervasive Computing communities have proposed multimodal event monitoring based on sensors such as, wearable inertial sensors, passive infrared presence sensors, change of state sensors, microphones. For instance, Gao et al. [9] and Rong and Ming [19] have demonstrated thes fusion of wearable inertial

C.F. Crispim-Junior • B. Fosty • R. Romdhane • F. Bremond • M. Thonnat (✉)
Inria Sophia Antipolis - Méditerranée, 2004, route des Lucioles - BP 93,
06902 Sophia Antipolis Cedex
e-mail: carlos-fernando.crispim_junior@inria.fr; baptiste.fosty@inria.fr; rim.romdhane@inria.fr; francois.bremond@inria.fr; monique.thonnat@inria.fr

Q. Ma
Inria Sophia Antipolis - Méditerranée, 2004, route des Lucioles - BP 93,
06902 Sophia Antipolis Cedex

Ecole Centrale de Pékin/Beihang University, 37 Xueyuan Road, 100191 Beijing, China
e-mail: maqiao909@gmail.com

A. Briassouli et al. (eds.), *Health Monitoring and Personalized Feedback using Multimedia Data*, DOI 10.1007/978-3-319-17963-6_10

sensors at the waist, chest, and sides of a person body for the detection of daily living activities, where data fusion was carried out by classification methods (e.g., Naïve Bayes, C4.5). Although wearable inertial sensors provide a rich representation of body dynamics, they are subjected to problems such as motion noise, inter sensor-calibration, and in case of large scale research studies, the need of placing sensors in a relatively similar body position among monitored people, what may introduce noise in experimental data. Fleury et al. [7] have presented a multi-modal event monitoring system using actimeters, microphones, PIR (Passive Infrared) presence sensor, and door contact sensors. Data fusion is performed using a SVM classifier. Medjahed and Boudy [14] have proposed a smart-home setting which performs event detection relying only on ambient sensors like infrared, and physiological sensors; all fused by a Fuzzy classifier.

Computer Vision approaches for event detection may be summarized in three categories (adapted from [12]), classification methods, probabilistic graphical models (PGM), and semantic models; which rely on at least one of the following data abstractions: pixel-level, feature-level, or event-level. Probabilistic Graphical Models refer to techniques such as Conditional Random Fields, Dynamic Bayesian Networks, and Hidden Markov Models. Kitani et al. [11] have proposed a Hidden Variable Markov Model approach for event forecasting based on people trajectories and scene features. Examples of classification methods are Artificial Neural Networks, Support-Vector Machines (SVM), and Nearest Neighbor. In this context, Le et al. [13] have presented an extension of the Independent Subspace Analysis algorithm applied for learning invariant spatio-temporal features from unlabeled video data for event detection. Wang et al. [23] have proposed new descriptors for dense trajectory estimation in action representation as input for non-linear support vector machines. Although PGMs and classification methods have considerably increased the event detection performance in benchmark data sets, as they focus on pixel-and feature-based representations, they have limitations at describing the scene semantics, the temporal dynamics and hierarchical structure of complex events. Moreover, these approaches only focus on video data, ignoring other modalities which could provide additional information in the presence of ambiguous data.

In the recent domain of video search in internet videos, multimodal event analysis have investigated event representations consisting of different image cues, like motion and appearance, combined with other modalities such as audio and text, and exploring fusion in different data abstraction levels. Jhuo et al. [10] introduced a feature-level representation which combines audio and video data by mapping the joint patterns among these two modalities. Myers et al. [15] have learned a set of base classifiers, each from a single data type/source (low-level vision, motion, audio, high-level visual concepts, or automatic speech recognition), and evaluated their fusion using different methods at event level (late fusion scheme). They report average output was one of the most effective fusion schemes. Similarly, Oh et al. [17] have presented a multimodal (audio and video) system, where base classifiers are learned from different subsets of features, and score fusion are used to combine their output into complex events. Mid-level features, such as object detectors, were employed to enrich event model semantics. Even though multimedia event analysis approaches have demonstrated significant advances by seeking to capture the hierarchical nature of events and incorporating auxiliary sources of information, most methods rely on learning steps involving large amounts of training data.

Semantic (or Description-based) models make use of a description language and logical operators to build event representations incorporating knowledge of domain experts. These languages allow to explicitly model the semantic information and hierarchical structure of events, besides to not require as much data as PGMs and classification methods. For instance, Zaidenberg et al. [24] have presented a generic model-based framework for group behavior detection on surveillance applications such as airport, subway, and shopping center.

Cao et al. [3] proposed a multimodal event detection where two context models are defined: the human and the environment contexts. The human context (e.g., body posture) is obtained from data of a set of cameras, while the environment context (semantic information about the scene) is based on accelerometer devices attached to objects of daily living which once manipulated trigger an event, (e.g., TV remote control or doors use). A rule-based reasoning engine is used for combining both context types at event detection level. Although semantic models ability to easily incorporate scene semantics, they are sensitive to noise of underlying process, like image segmentation and people tracking in vision systems. To overcome such limitations, probabilistic frameworks may be adopted to handle data uncertainty as in [25] and [14]. For example, Zouba et al. [25] have evaluated a multimodal monitoring system at the identification of activities of daily living of older people in a model apartment. Video-camera data was used to track people over the scene and environmental sensorS to obtain complementary data on object interaction. Dempster–Shafer theory was employed for reasoning under imprecise data.

This paper presents a hierarchical model-based framework to multiple sensor context. We extend the generic ontology proposed by Vu et al. [22] to describe event models in terms of elementary (low-level) events coming from different sensors, as a basis to infer Multimodal Complex Events. Event level fusion is chosen as it provides a flexible way to deal with sensor heterogeneity, and has been reported to present a higher performance than early fusion schemes based on pixel- and feature-level representations [15, 21]. A Dempster–Shafer-based probabilistic approach is presented to handle event conflict using an adapted combination rule. The framework is evaluated on real multisensor recordings of participants of a clinical protocol for Alzheimer disease study.

2 Hierarchical Model-Based Framework

The proposed framework is composed of two main components: an event ontology and a temporal event detection algorithm [22]. The temporal algorithm is responsible for event inference based on the event models defined by domain expert and available input data. The video event ontology proposed in [22] is extended

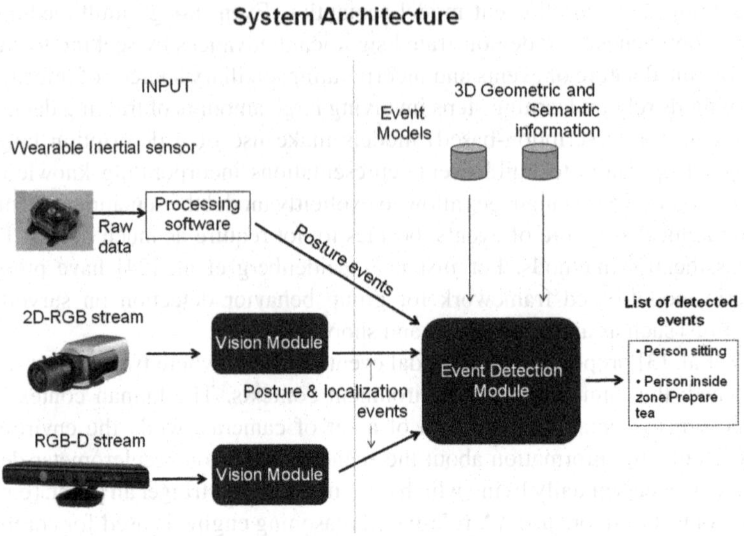

Fig. 1 Overall architecture of the video monitoring system (adapted from [5])

for multiple sensor scenario (then referred as Event Ontology), and the temporal algorithm to deal with mutually exclusive conflicting events of different sensors during people monitoring.

Figure 1 presents the architecture of the extended event detection framework, where a wearable inertial sensor and two video-cameras are given as examples of sensors. Sensor data is individually processed and their resulting output is taken as input for the multisensor framework (Event Detection Module). For instance, inertial sensor data would consist of a set of attribute-based events (e.g., for posture: person bending, person lying down), while for video camera data it would be a set of people detected in the scene and/or elementary events provided by vision module. All sensors are assumed to be time-synchronized.

2.1 Event Ontology

The event models are described using a constraint-based ontology language based on natural terminology to allow domain experts to easily add and change them. An event model is composed of up to six parts [22]:

- **Physical Objects** refer to real objects involved in the recognition of the modeled event. Examples of physical object types are: mobile objects (e.g. person herein, or vehicle in another application), contextual objects (equipment) and contextual zones (chair zone);
- **Components** refer to sub-events that the model is composed of;

- **Forbidden Components** refer to events that should not occur in case of the event model is recognized;
- **Constraints** are conditions that the physical objects and/or the components should hold. These constraints could be logical, spatial and temporal;
- **Alert** describes the importance of a detection of the scenario model for a given specific treatment; and
- **Action** in association with the Alert type describes a specific action which will be performed when an event of the described model is detected (e.g. send a SMS to a caregiver responsible to check a patient over a possible falling down).

Three types of Physical Objects are defined: Mobile, Person, and Contextual Objects. Mobile class defines a set of attributes which are common to any mobile object (e.g., height, width, position, speed). Person class extends Mobile by adding person-related attributes like body posture, appearance, etc. Contextual Objects refer to *a priori* knowledge of the scene. *A priori* knowledge refers to a decomposition of a 3D projection of the scene floor plan into a set of spatial zones (e.g., TV zone, Armchair Zone), and equipment, (e.g., home appliances and furniture such as TV, armchair, and Coffee machine), which hold semantic information to the modeled events. Constraints define conditions that physical object property(-ies) and/or components should satisfy. They can be non-temporal, such as spatial and appearance constraints; or they could be temporal and specify that a specific time ordering of two event model instances should generate a third event, for example, *Person crossing from Zone*1 *to Zone*2 is defined as *Person in zone*1 before *Person in zone*2. Temporal constraints are expressed using Allen's interval algebra (e.g., BEFORE, MEET, and AND) [2].

The ontology hierarchically categorizes models according to their complexity on (in ascending order):

- **Primitive State** models an instantaneous value of a property of a physical object (e.g., Person posture, or Person inside a semantic zone).
- **Composite State** refers to a composition of two or more primitive states.
- **Primitive Event** models a change in a value of a physical object's property (e.g., Person changes from Sitting to Standing posture).
- **Composite Event** refers to the composition of two event models which should hold a temporal relationship (e.g., Person changes from Sitting to standing posture before Person in Corridor Zone).

Figure 2 presents the description of Primitive State called Person sitting, which checks whether the attribute *posture* of a *person* object assumes the value *sitting*. Figure 3 presents an example of Composite Event, called Person sitting and using Office Desk, which defines a constraint between two sub-events (components). First

Fig. 2 Person_sitting

```
PrimitiveState (Person_sitting,
  PhysicalObjects ( ( p1 : Person ) )
  Constraints   ( ( p1->Posture = sitting) )
)
```

```
CompositeEvent(Person_sitting_and_using_OfficeDesk,
  PhysicalObjects( (p1:Person), (z1:Zone) )
  Components(
    (c1:CompositeEvent  P_insideOfficeDeskZone(p1,z1))
    (c2:PrimitiveState  P_sitting (p1)))
  Constraints( (c1->Interval AND c2->Interval) )
)
```

Fig. 3 Person_sitting_and_using_OfficeDesk

Fig. 4 Class person

```
class Person:Mobile
{
  String PostureV;
  String PostureWI;
}
```

Fig. 5 Person_sitting_WI

```
PrimitiveState( Person_sitting_WI,
  PhysicalObjects ((p1 : Person))
  Constraints(
              (p1->PostureWI = Sitting)
             )
)
```

component checks whether the person position lies inside of *a priori* defined zone relative to an office desk, while the second component verifies whether the person posture is sitting (using Fig. 2 model). The constraint defines that the model will be satisfied when both of its components happen at the same time (c1 AND c2).

2.2 Modeling Events from Different Sensors

Previous section has described how the event ontology categorizes and models events. We have chosen to model events generated by different sensor data using Primitive States, since they are the most basic building block of the event ontology. Handling sensor input at an early stage of the event hierarchy avoids the propagation of noise to high-level events, and also abstracts the derived models from the sensor data they are conditioned on.

Figure 4 describes the class *Person* where an attribute is created for each posture estimation, e.g., *PostureWI* for the estimation from wearable inertial sensor, and *PostureV* for the estimation of the video-based algorithm.

Figure 5 illustrates an example of Primitive state using the posture estimation from a inertial sensor. If one aims to increase system precision over recall, a Composite Event may be devised to combine (be composed of) both posture estimation (primitive states) and to restrict the targeted posture detection to when all sensor estimations agree, see Fig. 6 for an example.

```
CompositeEvent( Person_Sitting_MS,
  PhysicalObjects(
    (p1:Person), (z1:Zone), (eq1:Equipment))
  Components(
    (c1: PrimitiveState      Person_sitting_V (p1))
    (c2: PrimitiveState      Person_sitting_WI(p1)))
  Constraints( (c1->Interval AND c2->Interval) )
)
```

Fig. 6 Person_Sitting_MS

```
CompositeEvent( Person_sitting_and_using_OfficeDesk,
   PhysicalObjects(
    (p1:Person), (z1:Zone), (eq1:Equipment))
   Components(
    (c1: CompositeEvent P_inside_OfficeDeskZone(p1, z1))
    (c2: CompositeEvent Person_sitting_MS(p1))
   )
   Constraints(  (c1->Interval AND c2->Interval)   )
)
```

Fig. 7 Person_sitting_and_using_OfficeDesk

Figure 7 presents the event model "Person sitting and using Office Desk" which relies on a multisensor event for the detection of posture sitting. Using an ontology language for event modeling on multisensor scenarios allows to decompose event complexity and provides a flexible way to add or change sensor-based events.

The presented model examples described how to combine estimations of multiple sensors over the same attribute of Person class. But, there is no restriction on how event models from different sensors are combined. A complex event model may have a person posture estimated from an inertial sensor (*PostureWI*) while his/her localization is provided by a vision system.

2.3 Event Conflict Handling

To address conflicting evidence among (mutually exclusive) events generated by different sensors, a probabilistic framework is proposed to assess event reliability for event fusion. The conflict handling framework works as follows: firstly, event instantaneous likelihood is computed; secondly, event temporal reliability is computed from the current and close past event instantaneous likelihood (see [18]); finally, a variant of Dempster–Shafer rule of combination is used to decide upon event reliability which of the events is being performed.

The event conflict handling framework is performed at primitive state level to reduce the propagation of noise from low-level components to hierarchically higher event models, abstract high-level events from the sensor estimated events, and derive high-level events only from consolidated information.

2.3.1 Instantaneous Likelihood of a Primitive State

The instantaneous likelihood of Primitive States is computed based on the feature(s) the event constraints are conditioned on. Assuming the Primitive state feature (e.g., height) follows a Gaussian distribution, a learning step is performed *a priori* to obtain the expected feature distribution parameters (mean, μ, and variance, σ^2) given a primitive state and a sensor. The learning procedure is performed for each mutually exclusive event model affected by the analyzed feature.

Learned distribution parameters are then used during event inference (detection) to compute the instantaneous likelihood of an event given the feature value and the sensor providing it using Eq. 1.

$$P^{inst}_{\Omega,k,i} = \frac{\exp(-(Height_{\Omega,k,i} - \mu^2_{\Omega,i}))}{2\sigma^2_{\Omega,i}} \tag{1}$$

where,

k: video frame number (current instant), Ω : event model, i: sensor identifier

2.3.2 Temporal Reliability of a Primitive State

The instantaneous likelihood of the Primitive State considers the probability of a given primitive state (e.g., sitting, standing) be recognized at the current frame. But, noise from underlying vision algorithms can compromise the feature value which a primitive state is based on for a short interval of time, (e.g., problems at image segmentation can harm the height estimation of a person). To cope with instantaneous deviations of primitive state probabilities we compute the event temporal reliability, which considers the instantaneous likelihood of an event and its previous values for a given time interval (time window). Equations 2 and 3 present an adapted computation of temporal reliability using a time window of fixed size [18]. A cooling function is used to reinforce the information of near frames and lessen the influence of farther ones. The window size parameter used in these equations was set to match the minimum expected duration of the modeled primitive states.

$$P^{temp}_{\Omega,k,i} = \frac{P^{inst}_{\Omega,k,i} + M}{\sum_{t=k-w}^{t=k-1} exp(-(k-t))} \tag{2}$$

$$M = \sum_{t=k-w}^{t=k-1} [\exp(-(k-t))(P^{temp}_{\Omega,k,i} - P^{inst}_{\Omega,k,i})] \tag{3}$$

where,

k: video frame number (current instant), Ω: event model, i: sensor identifier, w: temporal window size

Primitive State Temporal Reliability is then considered as a belief level value on "how strongly it is believed that the event generated by the sensor i is true at the evaluated time instant". From here on Primitive State Temporal Reliability will be referred as Primitive State Reliability.

2.3.3 Primitive State Conflict Handling

Once the reliabilities of all mutually exclusive Primitive States are computed it is then necessary to decide which events are being actually performed. To perform such task we have adopted Dempster–Shafer Theory (DS). DS theory was proposed by Dempster [6] and then improved by Shafer [20]. It extends the Bayesian inference by allowing uncertainty reasoning based on incomplete information. The major components of evidence theory are the frame of discernment (Θ, Eq. 4), and the basic probability assignment (BPA). The frame of discernment contains all possible mutually exclusive hypotheses.

$$\Theta = \{Sitting, Standing, \ldots\} \tag{4}$$

The BPA is a function m: $2^{\Theta} \rightarrow [0, 1]$ related to a proposition satisfying conditions (5) and (6) [1]:

$$m(\emptyset) = 0 \tag{5}$$

$$\sum_{A \in \Theta} m(A) = 1 \tag{6}$$

where, A is any subset of the frame of discernment, and EMPTYSET refers to the empty set. For any $A \in 2^{\Theta}$, m(A) is considered as the subjective confidence level on the event A. Accordingly, the whole body of evidence of one sensor is the set of all the BPAs greater than 0 (zero) under one frame of discernment. The combination of multiple evidences defined on the same frame of discernment is the combination of the confidence level values based on BPAs (e.g., pre-defined by experts). Given two sensors (1 and 2), where each sensor has its body of evidence (m_{s1} and m_{s2}), these are the corresponding BPA functions of the frame of discernment. The combination rule of the classical DS theory can be implemented to fuse data from two sensors, but it can lead to illogical results in the presence of highly conflicting evidence [1]. We herein adapt the combination rule proposed by Ali et al. [1], as it has been demonstrated to provide more realistic results than the standard DS rule when combining conflicting evidence from multiple sources. Equations 7 and 8 present the mass function for computing Sitting (Sit.) and Standing (Sta.) primitive states, respectively:

$$(m_s 1 \bigotimes m_s 2)(Sit.) = \frac{(1 - (1 - m_s 1(Sit.))(1 - m_s 2(Sit.)))}{(1 + (1 - m_s 1(Sit.))(1 - m_s 2(Sit.)))} \tag{7}$$

$$(m_s1 \bigotimes m_s2)(Sta.) = \frac{(1-(1-m_s1(Sta.))(1-m_s2(Sta.)))}{(1+(1-m_s1(Sta.))(1-m_s2(Sta.)))} \tag{8}$$

Among a set of mutually exclusive events the framework chooses the event with the highest probability (mass function). The combination rule can be used on an iterative fashion to combine more than two bodies of evidence.

3 Evaluation

To evaluate the proposed framework we have used multisensor recordings from real participants of a clinical protocol for Alzheimer disease study. This data set is chosen due to the growing applicability of monitoring systems for older people care, assisted living, and frailty diagnosis.

The event detection performance is evaluated in two scenarios: firstly, we compare the crisp multisensor approach using data from an 2D-RGB camera and a wearable inertial sensor to a mono-sensor (camera) approach. Inertial sensor raw data is pre-processed using its (proprietary) software to generate the list of postures performed by the participant during the experimentation. Multi-sensor event models use wearable inertial sensor data for posture-based events and video-based data for person localization in the scene. Second scenario evaluates the proposed probabilistic approach for event conflict handling on events generated by two vision modules (the 2D-RGB camera vision system and a variant of it using a RGBD sensor). For this scenario, posture data is obtained per vision module and then propagated for fusion in the form of events.

All sensors are assumed to be time synchronized, but no spatial correspondence is computed among the cameras in the second scenario. Briefly, we assume the multi-sensor system does not know the transformation function amongst the coordinate-systems of the cameras.

3.1 Performance Evaluation

Event detection performance is measured using the indexes of sensitivity, precision, and F-Score described in Eqs. 9, 10, and 11, respectively. System event detection is compared to event annotation performed by domain experts.

$$Sensitivity = \frac{TP}{TP + FN} \tag{9}$$

$$Precision = \frac{TP}{TP + FP} \tag{10}$$

where, TP: True Positive rate, FP: False Positive rate, FN: False Negative rate.

$$F - Score = \frac{2 * (Sensitivity * Precision)}{Sensitivity + Precision} \tag{11}$$

3.2 Vision Module

The Vision Module used to test the proposed framework is an evaluation platform locally developed that allows the testing of different algorithms for each step of the computer vision chain (e.g., video acquisition, image segmentation, physical objects detection, physical objects tracking, actor identification, and actor events detection). Image segmentation is performed by an extension of the Gaussian Mixture Model algorithm for background subtraction proposed by [16]. People tracking is performed by an implementation of the multi-feature tracking algorithm proposed in [4], using the following features: 2D size, 3D displacement, color histogram, and dominant color. The vision component is responsible for detecting and tracking mobile objects on the scene. These objects (so-called physical objects) are classified according to a set of *a priori* defined classes, e.g., a person, a vehicle. The detected physical objects are then passed to the event detection module which assesses whether the actions/activities of these actors match the event models defined by the domain experts.

3.3 Data Set

Participants aged more than 65 years are recruited by the Memory Center (MC) of Nice Hospital. Inclusion criteria of the Alzheimer Disease (AD) group are: diagnosis of AD according to NINCDS-ADRDA criteria and a Mini-Mental State Exam (MMSE) [8] score above 15. AD participants which have significant motor disturbances (per the Unified Parkinson's Disease Rating Scale) are excluded. Control participants are healthy in the sense of behavioral and cognitive disturbances. The clinical protocol asks the participants to undertake a set of physical tasks and Instrumental Activities of Daily Living (IADL) in a Hospital observation room furnished with home appliances. Experimental recordings use a RGB video camera (AXIS®, Model P1346, 8 frames per second), a RGB-D camera (Kinect®sensor), and a wearable inertial sensor (MotionPod®).

The set of monitored IADLs is composed as follows:

1. Watch TV,
2. Prepare tea/coffee,
3. Write the shopping list of the lunch ingredients,
4. Write a check to pay the electricity bill,

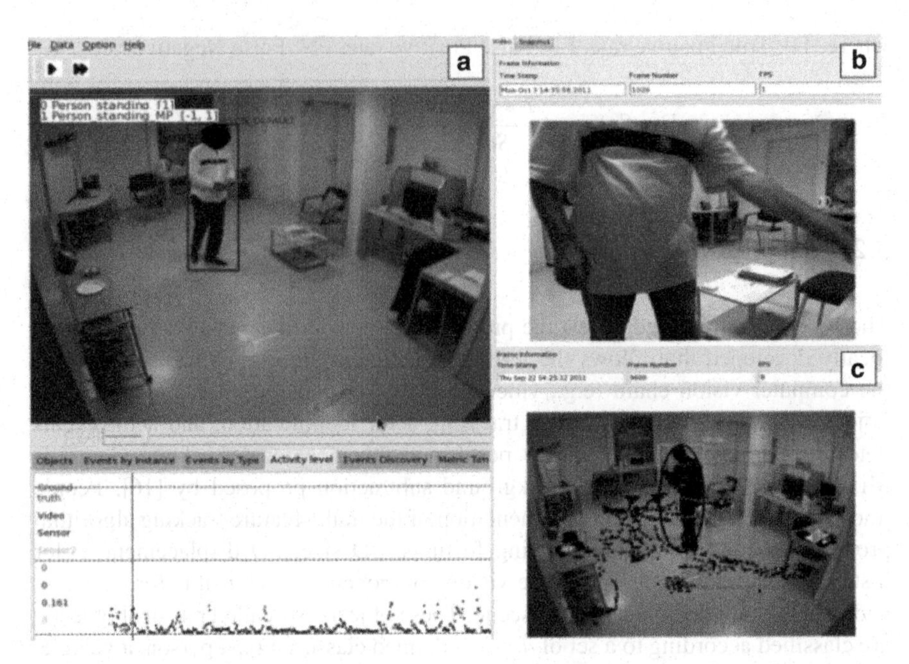

Fig. 8 Participant' activities by the point of view of different sensors: (**a**) RGB camera view and actimetry provided the inertial sensor (the *bottom* of image **a**); (**b**) RGB-D camera view of participant, which shows the inertial sensor worn by the participant; and (**c**) Drawn points on the ground represent the trajectory information of the participant during the experimentation

5. Answer/Call someone on the Phone,
6. Read newspaper/magazine,
7. Water the plant
8. Organize the prescribed drugs inside the drug box according to the weekly intake schedule.

Figure 8 shows the recording viewpoint of the 2D-RGB and RGB-D cameras in Fig. 8a,b, and WI sensor at Fig. 8b.

3.4 Event Modeling

Each one of the eight focused IADL is modeled using two composite models and three primitive states. First composite model is composed of two of the primitive states: one for the recognition of the person position inside a contextual zone (*a priori* defined), and another for his/her proximity to a static object (equipment) located into the respective zone (also *a priori* defined, e.g., phone table, coffee machine). Second composite model is composed of the first composite model to include the recognition a given IADL, and a primitive state model related to the posture of the person. The primitive states for posture recognition used data from the inertial sensor only. The activities "writing a check" and "writing a shopping list" are not differentiated and are referred instead as "Person using Office Desk" due to the lack of information about the object been manipulated by the patient. The name of the activity "Organize the prescribed drugs" is shortened as "Person using pharmacy basket".

4 Results and Discussion

Table 1 presents the performance of the framework at recognizing the IADLs a person is undertaking and his/her posture. Results are presented for mono- and multisensor approaches (2D-RGB camera and wearable inertial sensor). Average performance is presented for IADLs with and without posture sub-events. The row "Average of IADL without Posture" refers to event models based only on the person localization in the scene provided by the video-camera, therefore no difference is expected between Mono- and multisensor approach in this case.

The deterministic (or crisp) modeling of multisensor events has improved by ~ 19 % the precision index of IADLs involving sitting posture by the replacement of the vision system by an inertial sensor for posture estimation. However, the multisensor event models had a slightly lower performance on the detection of IADL involving standing posture than the mono-sensor approach. These results point that none of the two employed sensors can completely replace the other, and limiting the detection performance to the quality of the individual sensors output.

The difference in performance between IADL detection with and without posture component shows that by reducing the number of model constraints a higher detection performance can be achieved at the expenses of less information about how the event was performed. To tackle this problem, a probabilistic approach should be used to combine both posture estimations and also make models more robust to noise.

Table 2 presents the results of the proposed framework for conflict handling on the recognition of the Person posture using events from two different video-cameras (2D-RGB and RGB-D). The individual performance of the hierarchical model-based framework per camera is presented for comparative purposes.

The results in Table 2 showed the conflict handling framework improves the detection of posture-related primitive states on both posture categories.

Table 1 Comparison of mono and multisensor approaches

F-SCORE	Mono-	Multisensor
IADLs + Sitting posture	52.00	71.00
IADLs + Standing posture	73.15	71.00
Average of IADL with posture	68.00	71.00
Average of IADL without posture	81.22	81.22

N: 9; 15 min. each; Total: 64,800 frames (135 min)

Table 2 Postures recognition in physical tasks

Posture	Sitting		Standing	
Sensor	Precision	Sensitivity	Precision	Sensitivity
RGB	84.29	69.41	79.82	91.58
RGB-D	100.00	36.47	86.92	97.89
Fusion	82.35	91.30	91.04	95.31

N: 10. A window of 5 s is used for temporal probability

The precision achieved at standing recognition is higher than the one achieved by each video camera individually, demonstrating the suitability of the conflict handling framework for the assessing of event reliability and the combination of multiple sensor events for a more accurate detection.

5 Conclusions

We highlight as contributions of this paper a hierarchical model based framework for multisensor combination and a probabilistic approach for event conflict handling and fusion. The hierarchical model-based framework following a crisp combination of events from different sensors improves the detection of people seated while undertaking IADLs, and presents similar results to the mono-sensors approach in the other cases. Therefore in the crisp modeling case the detection performance is limited to the quality of the output of individual sensors. However, with the event conflict handling approach we showed it is possible to obtain better results than the ones individually achieved by the combined sensors by measuring the event reliability before the fusion process. Moreover, the probabilistic approach would also reduce the influence of errors from low-level sensor in the inference of high-level events.

The hierarchical model based framework (event ontology + event conflicting handling) is a hybrid approach between the hand-crafted semantic-models and the completely learned parameters of Probabilistic Graphical Models, but requiring a much smaller amount of training data. Future work will extend the evaluation of the framework for a larger variety of sensors (heterogeneous and homogeneous) and types of primitive states, and verify possible alternatives to remove the learning step.

References

1. Ali, T., Dutta, P., & Boruah H. (2012). A new combination rule for conflict problem of Dempster-Shafer evidence theory. *International Journal of Energy, Information and Communications 3*(1), 35-40.
2. Allen J. F. (1983). Maintaining knowledge about temporal intervals. *Communications of the ACM, 26*(11), 832–843.

3. Cao, Y., Tao, L., & Xu, G. (2009). An event-driven context model in elderly health monitoring. In *Proceedings of the Symposia and Workshops on Ubiquitous, Autonomic and Trusted Computing* (pp. 120–124).
4. Chau, D. P., Bremond, F., & Thonnat, M. (2011). A multi-feature tracking algorithm enabling adaptation to context variations. In *Proceedings of International Conference on Imaging for Crime Detection and Prevention.*
5. Crispim-Junior, C., Joumier, V., & Bremond, F. (2012). A multi-sensor approach for activity recognition in older patients. In *Proceedings of the Second International Conference on Ambient Computing, Applications, Services and Technologies*, AMBIENT 2012. Barcelona, September 23–28.
6. Dempster, A. P. (1968). Generalization of Bayesian inference. *Journal of the Royal Statistical Society, 30*, 205–247.
7. Fleury, A., Noury, N., & Vacher, M. (2010). Introducing knowledge in the process of supervised classification of activities of daily living in health smart homes. In *Proceedings of 12th IEEE International Conference on e-Health Networking Applications and Services* (pp. 322–329).
8. Folstein, M. F., Robins, L. M., & Helzer, J. E. (1983). The mini-mental state examination. *Archives of General Psychiatry 40*, 812.
9. Gao, L., Bourke, A.K, & Nelson, J. (2011). A system for activity recognition using multi-sensor fusion. In *Proceedings of Annual International Conference of the IEEE Engineering in Medicine and Biology Society* (pp. 7869–7872).
10. Jhuo, I.-H., Ye, G., Gao, S., Liu, D., Jiang, Y.-G., Lee, D. T., et al. (2014). Discovering joint audio visual codewords for video event detection. *Machine Vision and Applications, 25*(1), 33–47.
11. Kitani, K. M., Ziebart, B. D., (Drew) Bagnell, J. A., & Hebert, M. (2012). Activity forecasting. In *European Conference on Computer Vision.*
12. Lavee, G., Rivlin, E., & Rudzsky, M. (2009). Understanding video events: A survey of methods for automatic interpretation of semantic occurrences in video. *IEEE Transactions on Systems, Man, and Cybernetics, Part C: Applications and Reviews, 39*(5), 489–504.
13. Le, Q. V., Zou, W. Y, Yeung, S. Y., & Ng, A. Y. (2011). Learning hierarchical invariant spatio-temporal features for action recognition with independent subspace analysis. In *Proceedings of IEEE Conference on Computer Vision and Pattern Recognition* (pp. 3361–3368).
14. Medjahed, H., Istrate, D., Boudy, J., Baldinger, J.-L., & Dorizzi, B. (2011). A pervasive multi-sensor data fusion for smart home healthcare monitoring. In *Proceedings of IEEE International Conference on Fuzzy Systems* (pp. 1466–1473).
15. Myers, G. K., Nallapati, R., van Hout, J., Pancoast, S., Nevatia, R., Sun, C., et al. (2014). Evaluating multimedia features and fusion for example-based event detection. *Machine Vision and Applications, 25*(1), 17–32.
16. Nghiem, A. T., Bremond, F., & Thonnat, M. (2009). Controlling background subtraction algorithms for robust object detection. In *Proceedings of 3rd International Conference on Imaging for Crime Detection and Prevention*, London, UK, 1–6 December 2009.
17. Oh, S., McCloskey, S., Kim, I., Vahdat, A., Cannons, K. J., Hajimirsadeghi, H., et al. (2014). Multimedia event detection with multimodal feature fusion and temporal concept localization, *Machine Vision and Applications, 25*(1), 49–69.
18. Romdhane, R., Bremond, F., & Thonnat, M. (2010). Complex event recognition with uncertainty handling. In *Proceedings of the 7th IEEE International Conference on Advanced Video and Signal-Based Surveillance*, Boston, USA.
19. Rong, L., & Ming, L. (2010). Recognizing human activities based on multi-sensors fusion. In *Proceedings of 4th International Conference on Bioinformatics and Biomedical Engineering*, June 1–4.
20. Shafer, G. (1976). *A mathematical theory of evidence*. Princeton, NJ: Princeton University Press.
21. Snoek, C. G. M., Worring, M., & Smeulders, A. W. M. (2005). Early versus late fusion in semantic video analysis. In *Proceedings of the 13th Annual ACM International Conference on Multimedia (MULTIMEDIA 2005)* (pp. 399–402).

22. Vu, T., Bremond, F., & Thonnat, M. (2003). Automatic video interpretation: A novel algorithm for temporal scenario recognition. In *Proceedings of the Eighteenth International Joint Conference on Artificial Intelligence*, Acapulco, Mexico, August 9–15.
23. Wang, H., Klaser, A., Schmid, C., & Liu, C. (2011). Action recognition by dense trajectories. In *Proceedings of IEEE Conference on Computer Vision and Pattern Recognition* (pp. 3169–3176).
24. Zaidenberg, S., Boulay, B., Bremond, F., & Thonnat, M. (2012). A generic framework for video understanding applied to group behavior recognition. In *Proceedings of the IEEE International Conference on Advanced Video and Signal-based Surveillance* (pp. 136–142).
25. Zouba, N., Bremond, F., & Thonnat, M. (2010). An activity monitoring system for real elderly at home: Validation study. In *Proceedings of the 7th IEEE International Conference on Advanced Video and Signal-Based Surveillance*, Boston, USA.

Part III
Multimedia-Based Personalized Health Feedback Solutions

The Use of Visual Feedback Techniques in Balance Rehabilitation

Vassilia Hatzitaki

1 How Our Brain Uses Vision to Learn Novel Tasks

Vision supports achievement of a variety of goals of action, including transportation of the whole body to a new location, as in locomotion, and movement of a body part or limb to a new position, as in reaching and aiming. Visual feedback based correction of such goal-directed actions is a slow process that results in intermittencies because the processing of visual information is faster than the accompanying motor output corrections [1, 2]. To overcome this limitation of the on-line, closed loop processing of visual information, humans can also plan movements on the basis of visual cues. Thus, through short-term visuo-motor practice the motor system can learn novel visuo-motor transformations and apply them in order to predict future actions. This is because humans can pre-program movements based on predictable visual cues. Experience in a particular visuo-motor coordination task leads to the storage of appropriate control parameters which are used in programming subsequent movements, via a short-term motor memory [3]. This is the main principle visual feedback training is based on.

The coordination between vision and posture constitutes a familiar element in our daily life activity repertoire. The precise and incessant adaptation of posture and locomotion to visual cues is a complex task that requires the coordination of the body's multiple and often redundant degrees of freedom. The multiple degrees of freedom that need to be controlled can be a burden in accomplishing any complex visuo-motor transformation. In addition, the high mechanical resonance of the

V. Hatzitaki (✉)
Faculty of Physical Education and Sport Science, Motor Control and Learning Laboratory, Aristotle University of Thessaloniki, Thessaloniki, Greece
e-mail: vaso1@phed.auth.gr

A. Briassouli et al. (eds.), *Health Monitoring and Personalized Feedback using Multimedia Data*, DOI 10.1007/978-3-319-17963-6_11

effectors involved in postural control imposes additional constraints in visuo-motor adaptation. Yet, visually driven postural control is highly relevant to the performance of daily actions that preclude the risk of falling.

2 The Design of Visual Feedback Systems

On-line visual feedback of the body's Centre of Gravity (CoG) and Centre of Pressure (CoP) position has powerful control mechanisms over static and dynamic postural sway. Based on the capacity of the postural control system to make use of the available visual information, assistive and training devices have been developed which provide augmented feedback during performance of voluntarily controlled postural sway actions [4]. A typical example of such a system is shown in Fig. 1. The individual performing the exercise is standing on a force platform which records the three components of the ground reaction force or the CoP position at a sampling rate of at least 100 Hz. This performance signal is visually represented and fed back to the standing subject with negligible time delay. In Fig. 1 the instantaneous CoP displacement is indicated by the continuous black line while the subject is instructed to voluntary shift his/her CoP over the pre-specified stationary visual targets, indicated here by the squares. Another example of a visual feedback practice

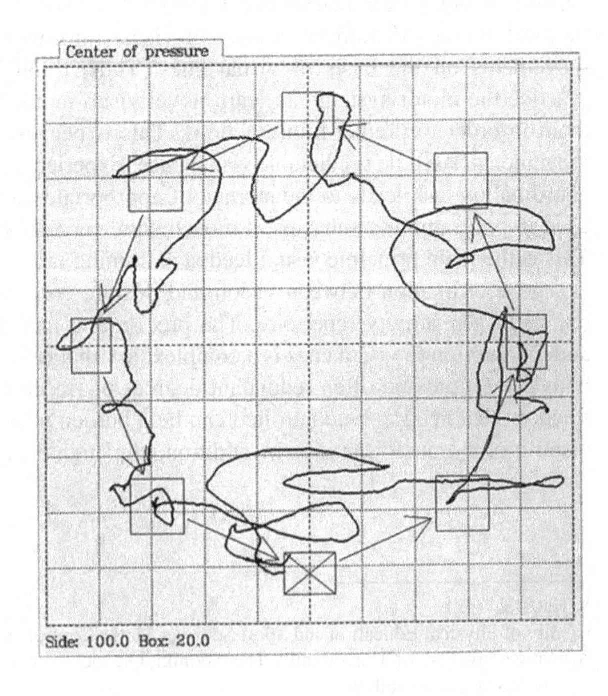

Fig. 1 A visual display of a visual feedback exercise protocol. The individual performing the exercise is standing on a force platform which records the CoP position. This performance signal is visually represented on a computer display with negligible time delay. The instantaneous CoP position is indicated by the continuous *black line* while the subject is instructed to voluntary shift his/her CoP over the pre-specified stationary visual targets, indicated here by the *squares*

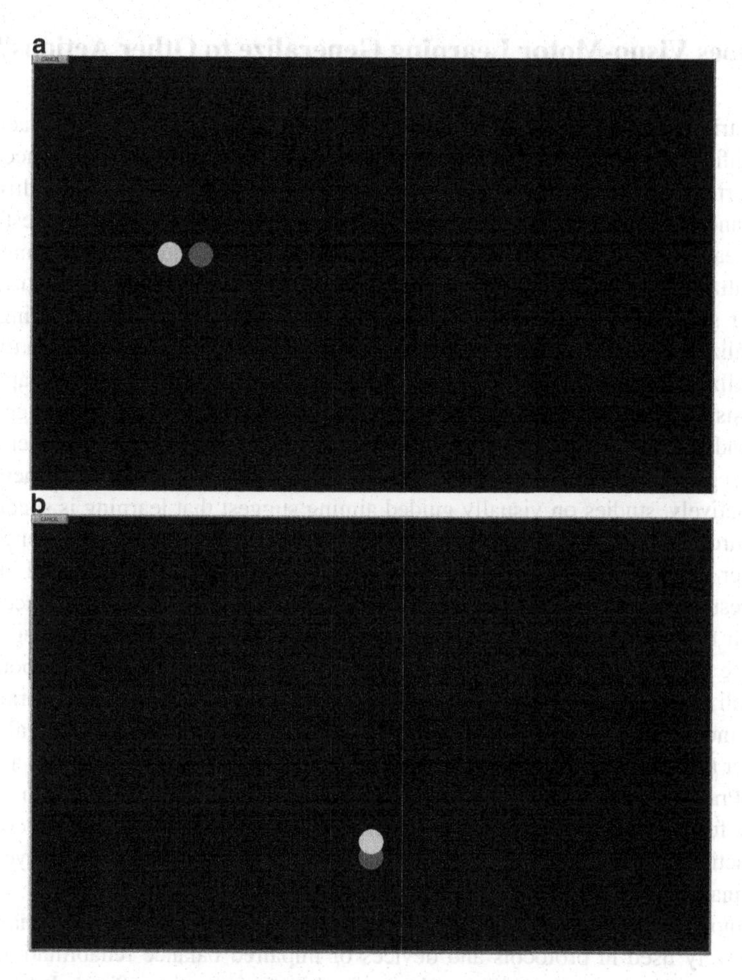

Fig. 2 Another example of a visual feedback practice protocol. The subject is instructed to track with his body the *red ball* as precisely as possible which moves in an oscillating pattern in either the horizontal (**a**) or vertical (**b**) direction. The *yellow ball* represents the instantaneous body weight (force) distribution between two force platforms while the subject is swaying between the two platforms (Color figure online)

protocol is illustrated in Fig. 2. Here, the subject is instructed to track with his body the red ball as precisely as possible which moves in an oscillating pattern in either the horizontal (Fig. 2a) or vertical (Fig. 2b) direction. The yellow ball represents the instantaneous body weight (force) distribution between two force platforms while the subject is swaying between the two platforms. Several commercially available systems such as the "Nintendo Wii" balance board are also based on the same training principle and have been extensively used in balance rehabilitation [5].

3 Does Visuo-Motor Learning Generalize to Other Actions?

One intriguing question is how well a learned transformation acquired by practicing a specific visuo-motor task generalizes to another task. It is generally accepted that during visuo-motor practice, an internal representation appropriate for different tasks and environments is acquainted. This notion however has been questioned by research evidence showing that adaptation to prisms during a walking task generalizes to an arm pointing or reaching task but the opposite is not true [6]. A later study indicated that a prism adaptation acquired in an arm pointing task generalized to a lower limb pointing task but a lower limb prism adaptation did not generalize to the arm [7]. This line of evidence from research in prism adaptation suggests that practicing visually guided motor behavior invokes a more general, limb-independent visuo-motor remapping, involving recalibration of higher-order brain regions in which the cerebellum has a critical role in the brain network. Alternatively, studies on visually guided aiming suggest that learning is specific to the sources of afferent information available during practice and does not generalize to other tasks [8]. Withdrawing visual feedback after practice of a manual aiming task results in a severe decrease in aiming accuracy. This decrease in accuracy is such that participants are often less accurate than controls who are beginning practice of the task without visual feedback. These opposing views about the generalization of visuo-motor practice could be due to the nature of the practiced visuo-motor tasks. Specifically, visually guided aiming does not generalize to another task whereas prism adaptation during walking can be transferred to another task. Practically, this means that practicing whole body movements with use of visual feedback results in a more generalized visuo-motor learning in contrast to practicing visual tracking of specific limb movements (i.e. aiming, eye-hand coordination).

Although the idea of practicing visually guided whole body movements has been extensively used in protocols and devices of impaired balance rehabilitation, one raising concern is the durability and transferability of practice effects. It has been shown for example that learning a novel visually-driven ankle-hip coordination may lead to improved control over certain task imposed ankle-hip relationships but can also destabilize the spontaneous, pre-existing coordination patterns [9]. In another study examining the efficacy of providing visual feedback of the centre of gravity motion during exercise [10], it was shown that this type of practice improved visual tracking performance but did not improve static balance control. Several researchers have questioned the use of learning specific visuo-motor coordination tasks through visual feedback practice because the consequences of learning do not generalize across different types of tasks, even when similar coordination solutions are involved. More evidence about the efficacy of visual feedback training in balance rehabilitation is reviewed in a subsequent section where age-related interventions are discussed.

4 Visual Stimuli and Task Parameters That Optimize Visuo-Motor Learning

The type of visual feedback cues provided during visuo-motor practice greatly influences the extent to which an internal representation of the visuo-motor task would be developed and generalized to other tasks. Specifically, continuous feedback cues allow for accurate movement production when employing on-line visual closed-loop control without access to an internal model. This type of feedback however does not allow the buildup of an internal visuo-motor transformation because continuous cues increase the dependency on feedback as closed-loop corrections permit automatic recalibration of the visuomotor mapping without access to an internal model [11]. On the other hand, an internal model of the practiced visuo-motor transformation is acquired only when feedback is provided as end-point information that is only when visual feedback about the end-point movement positions is provided.

In our laboratory, we extended and explored this hypothesis comparing the use of continuous and end-point visual cues in postural learning [12]. Specifically, we examined the impact of two types of visual cues, continuous and terminal, that were provided either as target or performance feedback information (Fig. 3). We sought to determine whether the acquisition of visually-guided postural sway can be extended to auditory driven sway and the extent to which such a transfer is dependent on the type of visual cues provided during practice. The results of this study revealed that practicing with continuous visual target cues (Fig. 3b) increase the load of on-line closed loop corrections resulting in greater accuracy at the target peak positions of the postural sway task but at the cost of increasing movement intermittency and variability. When visual feedback is continuously available, an extrinsic visuo-motor transformation can generally be mastered without the need to acquire an internal model. End-point cues on the other hand, enforce the buildup of an internal representation of the visuo-motor transformation resulting in less variable performance, improved coupling but serious target overshooting. The study concluded that the amount of visual information provided to guide voluntary postural sway has a strong influence on the relative contribution of feed forward and feedback control processes during task performance. Therefore, practicing postural tracking of periodic visual cues of repetitive nature enforces use of feed-forward control and the development of an internal model. The use of periodic visual cues in visual feedback training is critically challenged in the last section of this chapter where directions for future research and development of visual feedback tools are discussed. Another important finding of our study was that the differential adaptations acquired through visuo-motor practice did not generalize in the audio-driven postural workspace. This is because visual cues are better suited for the perception of space whereas auditory cues are more relevant for the perception of time [13]. It seems that differential practice does not impact the generalization of postural learning in the audio-motor workspace.

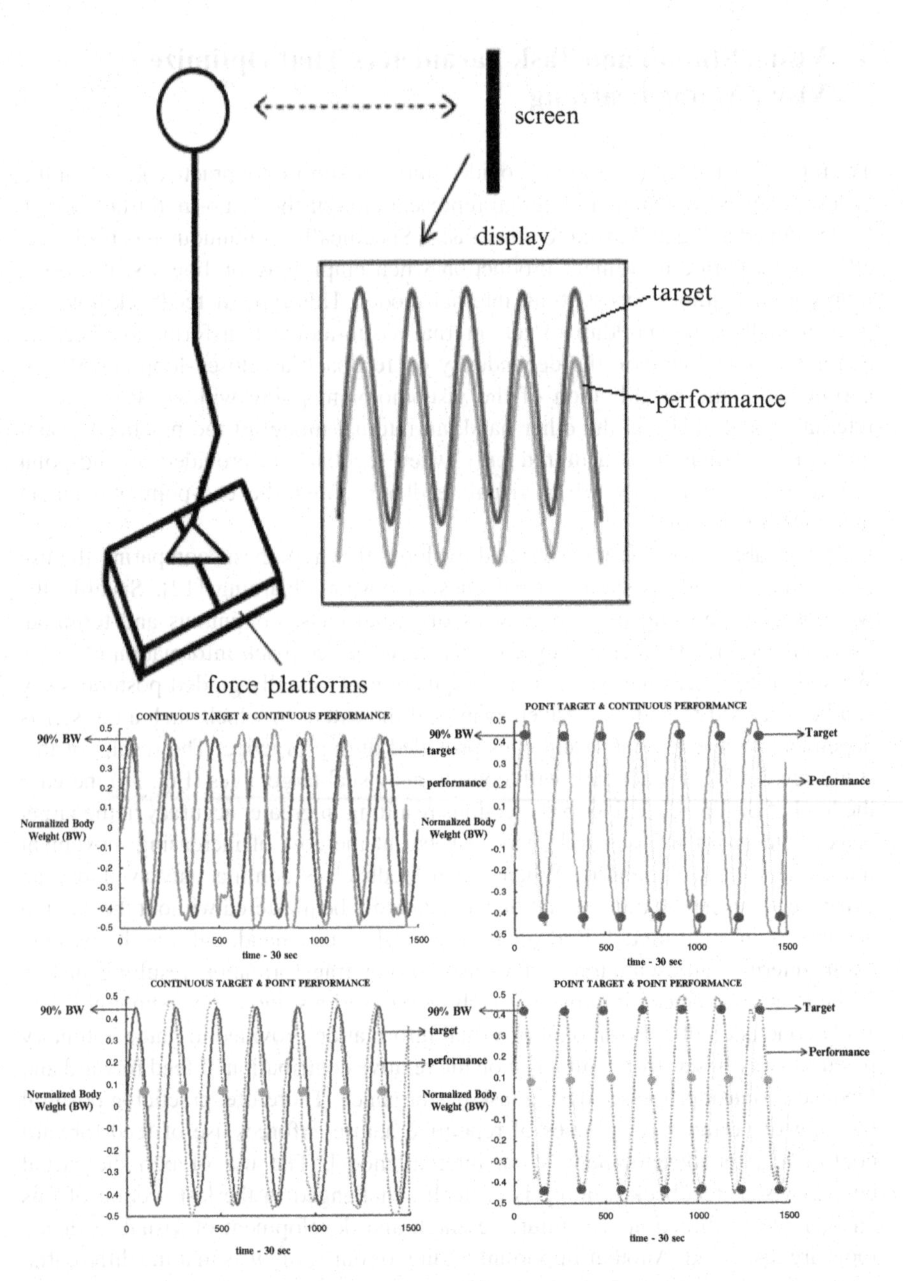

Fig. 3 (**a**) A graphical representation of the experimental protocol showing the dual force platform and the visual display (adopted from [12]). (**b**) Performance (in *green*) and target (in *red*) traces for continuous target and continuous performance feedback, continuous target and point (*dotted*) performance feedback, point target and continuous performance feedback, point target and point performance feedback (Color figure online)

Several other issues need to be considered when employing visually-driven postural learning in balance rehabilitation. Research evidence for example suggest that the color of the feedback cues may also play a role in visuo-motor adaptation and the acquisition of a novel visuo-motor transformation [14]. Moreover, the capacity to use visual feedback information to control dynamic postural sway depends on the amplitude and frequency constraints of the sway task. Specifically, a "natural" sway frequency around 0.6 Hz permits the optimal use of visual feedback information [15] whereas feedback is less effectively used by the central nervous system when sway amplitude or frequency increase.

Another important question is whether visual feedback should be provided concurrently and continuously during task performance or only at a particular time after the end of the trial. The concurrent provision of visual feedback during the trial evokes an automatic recalibration of the visuo-motor mapping which allows for more accurate performance which is lost however when feedback is not available any longer [16]. Individuals therefore may become so reliant on visual feedback that their performance seriously deteriorates in the absence of it. Post trial feedback on the other hand allows for the buildup of a "cognitive strategy" or internal model that can be optimally used to perform without feedback or to adapt to a novel environment.

The additional value of providing feedback information about sway when individuals already use visual target information to control their posture is also arguable [17]. Individuals using visual target information were no better in controlling their leaning posture when CoP feedback was provided in addition to visual target information. This is because performance feedback is usually contaminated with more stochastic variability in the signal whereas performers prefer to rely on less variable and simpler structure information.

In summary, there are several stimuli properties and task parameters that can influence visuo-motor learning in visual feedback training systems. There is no doubt that more research along these lines is required to unravel the important mechanisms underlying visuo-motor learning. Moreover, a critical issue that is addressed in the following sections is whether and how visuo-motor learning is constrained by aging. This is important particularly when considering that the target end user group of visual feedback technology is the aging population and particularly those older adults with balance impairments.

5 How Aging Affects the Ability to Learn Novel Visuo-Motor Tasks

With increasing age, individuals prioritize the use of vision and rely heavily on visual inputs for controlling their posture since somatosensory inputs become less reliable due to peripheral neuromuscular degeneration [38]. Nevertheless, the ability to use visual feedback information for controlling static or dynamic balance also diminishes with age. Age-induced delays in the sampling and processing of visual information required for the on-line control of posture and locomotion might be a contributing factor to impaired balance function and associated increased incidence of falling [18]. These are reported in precision stepping [19] and obstacle avoidance [20] during locomotion and have been attributed to neural degeneration of the visual and visuo-motor pathways in the brain. In a study investigating the use of visual feedback about CoP position to control static postural sway, only young participants were able to decrease the amplitude and increase the frequency of their sway based on visual feedback information [21]. Moreover, removal of the visual feedback resulted in a 'destabilizing' of standing in elderly adults whereas young individuals were able to maintain the speed and precision of dynamic sway when visual feedback was withdrawn.

In our laboratory we have investigated the capacity of older women (>65 years of age) to learn a novel visuo-postural coordination task [22]. Elderly women stood on a dual force platform and were asked to shift their body weight between the two platforms while keeping the vertical force produced by each foot within a ±30% force boundary that was visually specified by a target sine-wave signal (Fig. 4a, white circles inside the target sine curve indicate the % of bodyweight applied by each foot). Practice consisted of three blocks of five trials performed in 1 day followed by a block of five trials performed 24 h later. Older women made longer weight-shifting cycles and had lower response gain and higher within-trial variability compared to younger women suggesting a weaker coupling between the visual stimulus and the response force (Fig. 4b). Regardless of age however, visuo-postural coupling improved with practice suggesting that learning of the particular visuo-motor coordination task was apparent in both groups. Older women however employed a different functional coordination solution in order to learn the task that was imposed by age-specific constraints in the physiological systems supporting postural control.

In summary, with aging humans rely more heavily on visual information for controlling their balance despite the age induced delays in the sampling and processing of visual information for the online control of posture and locomotion. Visuo-motor processing delays however might be compensated by reinforcing the natural process of sensory substitution with the provision of augmented visual feedback.

6 Improving Balance Control in Aging Using Visual Feedback Training

Based on the idea that older people maintain their ability to learn novel visuo-motor transformations, visual feedback protocols and devices have been developed promising to improve control of balance and locomotion and prevent falling due to age degeneration or a specific pathology. Visually guided postural training with

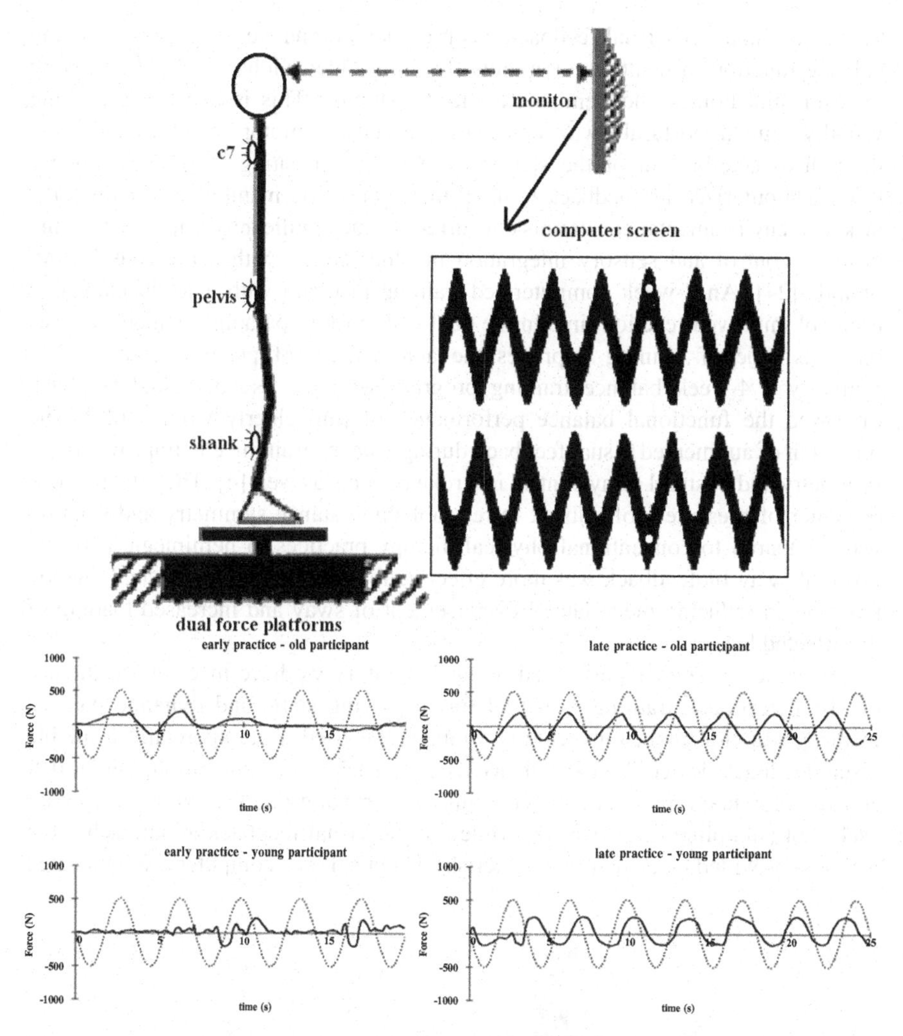

Fig. 4 (**a**) Schematic representation of the visual feedback task used by Hatzitaki and Konstadakos [22]. The subject stands on a dual force platform and is asked to shift body weight between the two platforms while keeping the vertical force produced by each foot within a ±30 % force boundary that was visually specified by a target sine-wave signal (*white circles* inside the target sine curve indicate the % of bodyweight applied by each foot). (**b**) Time series of the ground reaction force (*solid line*) produced during one weight shifting trial superimposed on the stimulus force (*dotted line*) for an old (*top*) and young (*bottom*) participant during early (1st trial, *left panel*) and late (15th trial, *right panel*) practice. Target force is scaled to individual body weight

the use of enhanced visual feedback has provided alternative means of improving balance function in healthy community-dwelling older adults, frail elderly living in residential homes and hemiparetic stroke patients. This is because practicing visually guided postural sway enhances the sensory-motor integration process through a recalibration of the sensory systems contributing to postural control [23]. Computerized biofeedback training that focuses on manipulating individual, task, and environmental constraints concurrently can significantly improve dynamic postural control and sensory integration in older adults with a previous history of falls [24]. An 8-week computerized training program with a feedback fading protocol improved reaction time in the dual-task posture paradigm which suggests that this type of training improves the automaticity of postural control [25]. Similarly a 4-week balance training program with the use of visual feedback improved the functional balance performance of frail elderly women [26]. The provision of augmented visual feedback during balance training can improve stance symmetry and postural sway control in stroke patients as well [4]. The effectiveness of center of pressure biofeedback in reestablishing stance symmetry and stability was compared to conventional physical therapy practices in hemiplegic patients. Postural sway biofeedback was more effective than conventional physical therapy practices in reducing mean lateral displacement of sway and increased loading of the affected leg.

In a series of studies performed in our laboratory we have investigated the use of visual feedback training as a tool for improving static and dynamic balance in the population of healthy elderly women. We used a commercially available visual feedback device (ERBE Balance System, Elektromedizin Gmbh, Fig. 5) that consists of a dual force platform recording the vertical ground reaction force under each foot (sampling rate: 100 Hz) while on-line visual feedback about each force vector is provided by a cursor (i.e. asterisk) displayed on a computer screen located

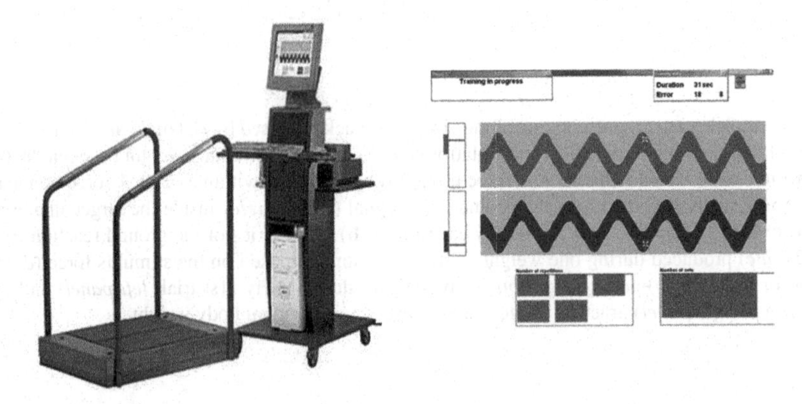

Fig. 5 A commercially available biofeedback device consisting of a dual force platform recording the vertical ground reaction force under each foot while on-line visual feedback about each force vector is provided by a cursor (i.e. *asterisk*) displayed on a computer screen located in front of the subject

in front of the subject (1.5 m ahead, eye-level). The training task requires shifting body weight between the right and left foot (for Medio-Lateral, M/L training) or from toes to heel (for Anterior–Posterior, A/P training), while maintaining each platform's force vector within a visually specified sine waveform constraints. The bandwidth of each sine waveform curve can be individually adjusted to $\pm$ % of bodyweight. The duration of the training trial is specified by the number of required weight shifting cycles. Each time either force vector exceeds the $\pm$ % limit in either direction, the movement of the cursor on the screen stops and a spatial error is recorded. Our participants practiced with this type of visual feedback for 4 weeks (3 hourly sessions per week). Visual feedback training improved static postural control only in the group practicing in the A/P direction whereas no adaptations were observed in the group practicing sway in the M/L direction [27]. In addition, the same visually guided training increased the limits of stability and enhanced the use of the ankle strategy for maintaining balance during leaning and dynamic swaying tasks [28]. More importantly, practicing visually guided weight shifting in the M/L direction improved the postural adjustments associated with a moving object avoidance task [29]. Specifically, as a result of practice, postural response onset shifted closer to the time of collision with the obstacle and sway amplitude during the avoidance decreased substantially. These observations suggest that visually guided practice enhances older adults' ability for on-line visuo-motor processing when avoiding collision eliminating reliance on anticipatory-predictive mechanisms. Contrary to static balance improvements that were apparent in the group practicing in the A/P direction adaptations in obstacle avoidance skill were more evident to the group practicing in the M/L direction. These findings suggest that specifying the direction of visually guided sway seems to be critical for optimizing the transfer of training adaptations.

In conclusion, the impact of visual feedback training on balance performance is an issue surrounded by continuing controversy. On one hand, training-induced improvements of both static [27] and dynamic [29] balance have been reported in healthy and frail older adults [26]. On the other, several investigations have failed to reveal long-term benefits on static and/or dynamic balance [30, 31] although acute improvements of weight bearing symmetry are reported in stroke patients. More importantly, the observed improvements, noted only in those dynamic balance tasks that are specific to the training, give rise to the possibility of motor learning effects. Controversies between different study results may be attributed to the diversity of tasks employed in visual feedback training that vary in several directions (anterior, posterior, lateral, diagonal), movement forms (weight shifting, leaning, stepping), stance positions and support surfaces configurations as well as in the type of target and performance cues used in the visual representations. It appears that the benefits of visual feedback training are mainly limited to those tasks that are specific to the training. This could be due to the direct link between visual information and specific motor response loci developed through visually guided actions that does not permit the generalization between vision and action found in one posture to other postures [32]. Thus, task specificity during training could be implicated for the absence of training effects on functional measures of static and dynamic balance performance.

Transfer of skill learning is therefore an important issue that needs to be addressed when considering the efficacy of visual feedback training for improving balance function.

7 Future Research and Technology Development Directions

The regularity of visual target motions used in visual feedback training enhances the use of predictive mechanisms in motor learning [3]. The problem though with practicing of stereotypic and highly predictable visual cues is that this type of visual feedback training does not enable the visuo-motor system to use its online sensory-motor calibration and perception-action processes. As a result, practice leads to the development of an internal motor plan of a particular visuo-motor task that does not generalize to other types of tasks. This lack of generalization seriously limits the number of functional coordination solutions the motor system uses in order to adapt to a novel task or environmental condition.

A quite promising solution to the problem of task specificity is to introduce a more variable approach in the presentation of visual cues adopting target cues that are non periodic and therefore less predictable. The problem though with using more variable and less predictable visual cues in visuo-motor practice is that the human central nervous system perceives the stationary, low frequency visual target cues whereas it is less reliant on the performance feedback signal that has higher frequency and less stationary properties. Nevertheless, recent pilot research evidence from our laboratory suggest that adults can couple their sway and visual (gaze) motions much better to a chaotic (i.e. Lorenz attractor) visual target motion (Fig. 6a) than a periodic target motion (Fig. 6b) [39]. Although this is a surprising finding it may be explained based on the fact that the evolution of healthy movement variability can be described in terms of deterministic processes as well as being complex [33]. It is this optimal combination of determinism and complexity that enables us to adapt in our environment in a stable but flexible manner. Optimal movement variability degrades with aging and pathology (e.g. Parkinson's Disease) and has also been associated with a risk of falling. Since the goal of balance rehabilitation is to restore optimal sway variability i.e. the balance between regularity and complexity, we believe that the use of more variable and less predictable visual cues to guide postural sway practice holds the potential of optimizing the effects of visual feedback training. More research is required however for indentifying the stimuli parameters that will optimize the learning process.

A relatively recent development of visual feedback based systems is **exergaming**. Exergaming refers to technologically advanced virtual reality systems which combine exercise with video games. These systems offer an attractive and motivating balance training tool that has been widely used in community dwelling older adults [34, 35]. A computer game-play system that employs user interface visual feedback loops, so as to physically but also mentally involve the user. Researchers and practitioners in the field of aging have introduced elderly audiences to cognitive training and stimulation games which are now enjoyed within the context of computer and/or game consoles. Numerous recent studies have investigated how exergaming can be specifically applied to the needs of seniors. The most common platform used for balance training is the "Nintendo Wii". Most studies that have tested exergaming as a balance training tool in healthy elderly report positive results with respect to improvements in balance ability after a training period, at least as much as the conventional/traditional exercise. The greater attainment of exergames, is the greater extend, to witch, individuals, seem to enjoy the exercise by interacting in a virtual reality environment. The use of virtual reality in balance rehabilitation

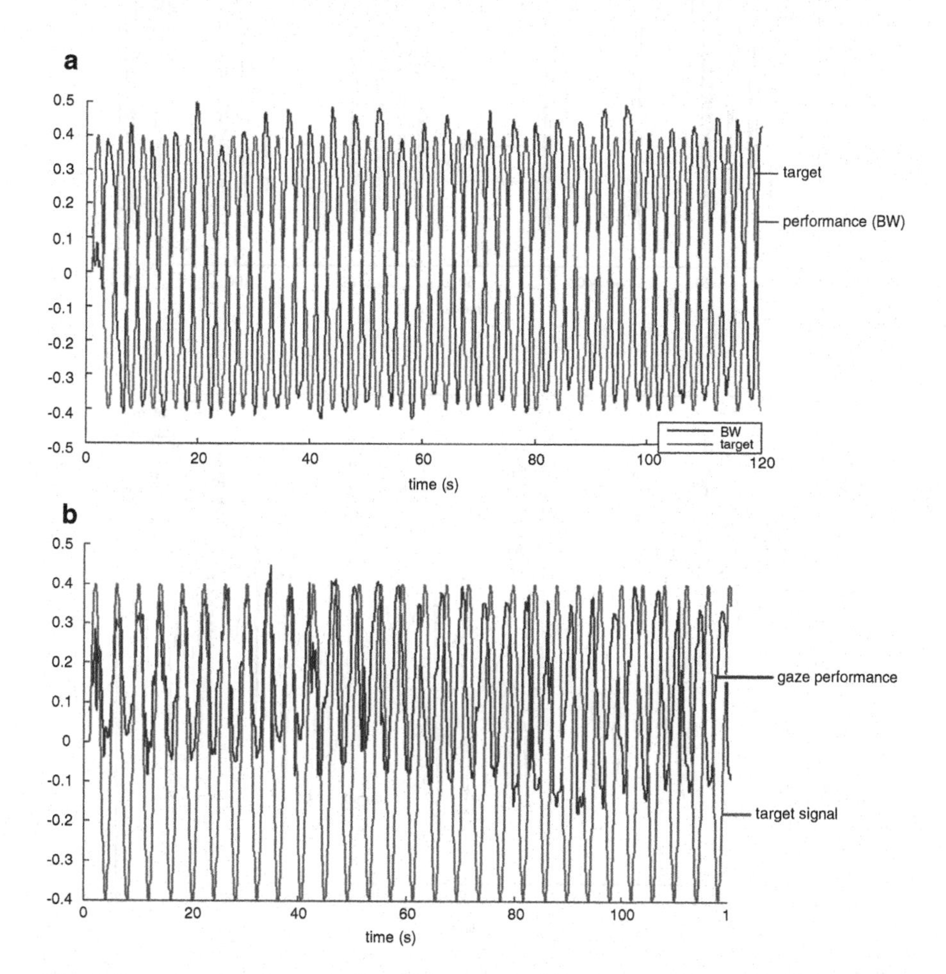

Fig. 6 Examples of performance (*blue line*)–target (*red line*) curves during postural tracking of different complexity stimuli motions: (**a**) Body Weight (BW)–target (sine), (**b**) gaze–target (sine), (**c**) Body Weight (BW)–target (Lorenz attractor) and (**d**) gaze–target (Lorenz attractor). Data published in [39] (Color figure online)

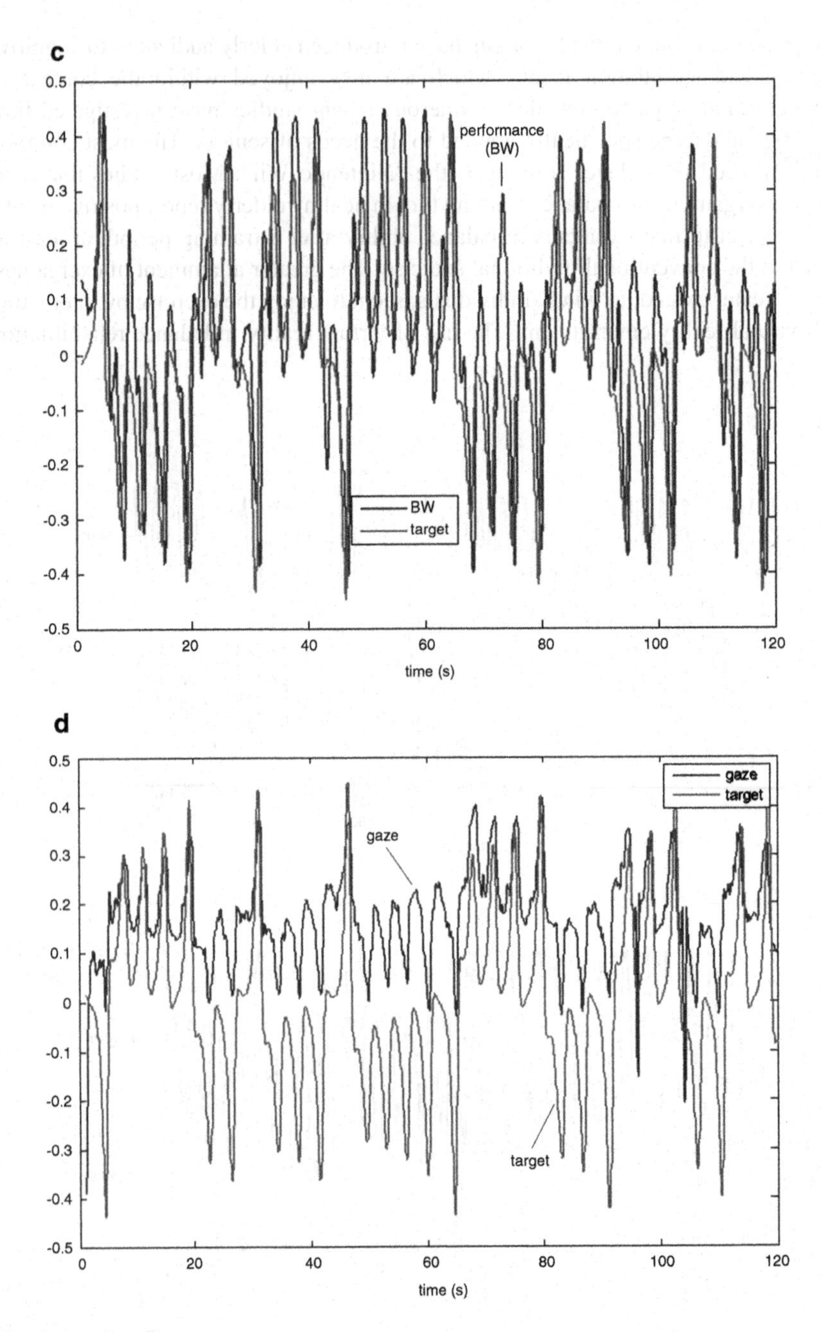

Fig. 6 (continued)

has the potential to offer experiences which are engaging and rewarding and more importantly simulate real life balance challenging conditions [36]. This line of research is certainly still growing and has been the subject of recent systematic review [37].

References

1. Miall, R. C., Weir, D. J., & Stein, J. F. (1993). Intermittency in human manual tracking tasks. *Journal of Motor Behavior, 25*(1), 53–63.
2. Slifkin, A. B., Vaillancourt, D. E., & Newell, K. M. (2000). Intermittency in the control of continuous force production. *Journal of Neurophysiology, 84*(4), 1708–1718.
3. Miall, R. C., Haggard, P. N., & Cole, J. D. (1995). Evidence of a limited visuo-motor memory used in programming wrist movements. *Experimental Brain Research, 107*(2), 267–280.
4. Shumway-Cook, A., Anson, D., & Haller, S. (1988). Postural sway biofeedback: Its effect on reestablishing stance stability in hemiplegic patients. *Archives of Physical Medicine and Rehabilitation, 69*(6), 395–400.
5. Gil-Gomez, J. A., Llorens, R., Alcaniz, M., & Colomer, C. (2011). Effectiveness of a Wii balance board-based system (eBaViR) for balance rehabilitation: A pilot randomized clinical trial in patients with acquired brain injury. *Journal of Neuroengineering and Rehabilitation, 8*, 30.
6. Morton, S. M., & Bastian, A. J. (2004). Prism adaptation during walking generalizes to reaching and requires the cerebellum. *Journal of Neurophysiology, 92*(4), 2497–2509.
7. Savin, D. N., & Morton, S. M. (2008). Asymmetric generalization between the arm and leg following prism-induced visuomotor adaptation. *Experimental Brain Research, 186*(1), 175–182.
8. Mackrous, I., & Proteau, L. (2007). Specificity of practice results from differences in movement planning strategies. *Experimental Brain Research, 183*(2), 181–193.
9. Faugloire, E., Bardy, B. G., & Stoffregen, T. A. (2009). (De)stabilization of required and spontaneous postural dynamics with learning. *Journal of Experimental Psychology: Human Perception and Performance, 35*(1), 170–187.
10. Hamman, R. G., Mekjavic, I., Mallinson, A. I., & Longridge, N. S. (1992). Training effects during repeated therapy sessions of balance training using visual feedback. *Archives of Physical Medicine and Rehabilitation, 73*(8), 738–744.
11. Heuer, H., & Hegele, M. (2008). Constraints on visuo-motor adaptation depend on the type of visual feedback during practice. *Experimental Brain Research, 185*(1), 101–110.
12. Radhakrishnan, S. M., Hatzitaki, V., Vogiannou, A., & Tzovaras, D. (2010). The role of visual cues in the acquisition and transfer of a voluntary postural sway task. *Gait & Posture, 32*(4), 650–655.
13. Bausenhart, K. M., de la Rosa, M. D., & Ulrich, R. (2013). Multimodal integration of time. *Experimental Psychology*, 1–13.
14. Hinder, M. R., Woolley, D. G., Tresilian, J. R., Riek, S., & Carson, R. G. (2008). The efficacy of colour cues in facilitating adaptation to opposing visuomotor rotations. *Experimental Brain Research, 191*(2), 143–155.
15. Danion, F., Duarte, M., & Grosjean, M. (2006). Variability of reciprocal aiming movements during standing: The effect of amplitude and frequency. *Gait & Posture, 23*(2), 173–179.
16. Hinder, M. R., Tresilian, J. R., Riek, S., & Carson, R. G. (2008). The contribution of visual feedback to visuomotor adaptation: How much and when? *Brain Research, 1197*, 123–134.
17. Duarte, M., & Zatsiorsky, V. M. (2002). Effects of body lean and visual information on the equilibrium maintenance during stance. *Experimental Brain Research, 146*(1), 60–69.

18. Klein, B. E., Klein, R., Lee, K. E., & Cruickshanks, K. J. (1998). Performance-based and self-assessed measures of visual function as related to history of falls, hip fractures, and measured gait time. The Beaver Dam Eye Study. *Ophthalmology, 105*(1), 160–164.
19. Chapman, G. J., & Hollands, M. A. (2006). Evidence for a link between changes to gaze behaviour and risk of falling in older adults during adaptive locomotion. *Gait & Posture, 24*(3), 288–294.
20. Schillings, A. M., Mulder, T., & Duysens, J. (2005). Stumbling over obstacles in older adults compared to young adults. *Journal of Neurophysiology, 94*(2), 1158–1168.
21. Dault, M. C., de Haart, M., Geurts, A. C., Arts, I. M., & Nienhuis, B. (2003). Effects of visual center of pressure feedback on postural control in young and elderly healthy adults and in stroke patients. *Human Movement Science, 22*(3), 221–236.
22. Hatzitaki, V., & Konstadakos, S. (2007). Visuo-postural adaptation during the acquisition of a visually guided weight-shifting task: Age-related differences in global and local dynamics. *Experimental Brain Research, 182*(4), 525–535.
23. Hu, M. H., & Woollacott, M. H. (1994). Multisensory training of standing balance in older adults: I. Postural stability and one-leg stance balance. *Journal of Gerontology, 49*(2), M52–M61.
24. Rose, D. J., & Clark, S. (2000). Can the control of bodily orientation be significantly improved in a group of older adults with a history of falls? *Journal of American Geriatrics Society, 48*(3), 275–282.
25. Lajoie, Y. (2004). Effect of computerized feedback postural training on posture and attentional demands in older adults. *Aging Clinical and Experimental Research, 16*(5), 363–368.
26. Sihvonen, S. E., Sipila, S., & Era, P. A. (2004). Changes in postural balance in frail elderly women during a 4-week visual feedback training: A randomized controlled trial. *Gerontology, 50*(2), 87–95.
27. Hatzitaki, V., Amiridis, I. G., Nikodelis, T., & Spiliopoulou, S. (2009). Direction-induced effects of visually guided weight-shifting training on standing balance in the elderly. *Gerontology, 55*(2), 145–152.
28. Gouglidis, V., Nikodelis, T., Hatzitaki, V., & Amiridis, I. G. (2011). Changes in the limits of stability induced by weight-shifting training in elderly women. *Experimental Aging Research, 37*(1), 46–62.
29. Hatzitaki, V., Voudouris, D., Nikodelis, T., & Amiridis, I. G. (2009). Visual feedback training improves postural adjustments associated with moving obstacle avoidance in elderly women. *Gait & Posture, 29*(2), 296–299.
30. Geiger, R. A., Allen, J. B., O'Keefe, J., & Hicks, R. R. (2001). Balance and mobility following stroke: Effects of physical therapy interventions with and without biofeedback/forceplate training. *Physical Therapy, 81*(4), 995–1005.
31. Walker, C., Brouwer, B. J., & Culham, E. G. (2000). Use of visual feedback in retraining balance following acute stroke. *Physical Therapy, 80*(9), 886–895.
32. Goodale, M. A., & Milner, A. D. (1992). Separate visual pathways for perception and action. *Trends in Neurosciences, 15*(1), 20–25.
33. Harbourne, R. T., & Stergiou, N. (2009). Movement variability and the use of nonlinear tools: Principles to guide physical therapist practice. *Physical Therapy, 89*(3), 267–282.
34. Heiden, E., & Lajoie, Y. (2010). Games-based biofeedback training and the attentional demands of balance in older adults. *Aging Clinical and Experimental Research, 22*(5–6), 367–373.
35. Lamoth, C. J., Caljouw, S. R., & Postema, K. (2011). Active video gaming to improve balance in the elderly. *Studies in Health Technology and Informatics, 167*, 159–164.
36. Sveistrup, H., Thornton, M., Bryanton, C., McComas, J., Marshall, S., Finestone, H., et al. (2004). Outcomes of intervention programs using flatscreen virtual reality. *Conference Proceedings: IEEE Engineering in Medicine and Biology Society, 7*, 4856–4858.
37. Bamidis, P. D., Vivas, A. B., Styliadis, C., Frantzidis, C., Klados, M., Schlee, W., et al. (2014). A review of physical and cognitive interventions in aging. *Neuroscience and Biobehavioral Reviews, 44*, 206–220.

38. Redfern, M. S., Furman, J. M., & Jacob, R. G. (2007). Visually induced postural sway in anxiety disorders. *J Anxiety Disord, 21*(5), 704–716.
39. Hatzitaki, V., Stergiou, N., Sofianidis, G., & Kyvelidou, A. (2015). Postural sway and gaze can track the complex motion of a visual target. *PLoS One, 10*(3), e0119828.

Recommending Video Content for Use in Group-Based Reminiscence Therapy

Adam Bermingham, Niamh Caprani, Ronán Collins, Cathal Gurrin, Kate Irving, Julia O'Rourke, Alan F. Smeaton, and Yang Yang

1 Introduction

One of the unfortunate consequences of the current trend of our ageing population is the increase in age-related diseases and human conditions and among those one of the most worrysome is the expected increase in prevalence of dementia. With an ageing population, it is inevitable that there will be an increase of dementia incidence, making it one of our the biggest, global public health challenges. Today, there are an estimated 35.6 million people living with dementia worldwide and it is estimated that this number will increase to 65.7 million by 2030 and 115.4 million by 2050 [51].

Apart from the personal implications, dementia is a costly condition as it draws on a variety of public and private, formal and informal resources to provide appropriate care [40]. It is estimated that the total worldwide costs of dementia

A. Bermingham • N. Caprani • C. Gurrin • A.F. Smeaton (✉) • Y. Yang
Insight Centre for Data Analytics, Dublin City University, Glasnevin, Dublin 9, Ireland
e-mail: adambermingham@gmail.com; niamhcaprani@gmail.com; cathal.gurrin@dcu.ie; alan.smeaton@dcu.ie; alan.smeaton@dcu.ie

R. Collins
Age-Related Health Care, Adelaide and Meath Hospital Tallaght, Dublin, Ireland
e-mail: ronan.collins2@amnch.ie

K. Irving
School of Nursing and Human Sciences, Dublin City University, Glasnevin, Dublin 9, Ireland
e-mail: kate.irving@dcu.ie

J. O'Rourke
Department of Speech and Language Therapy, Adelaide and Meath Hospital Tallaght, Dublin, Ireland
e-mail: Julia.ORourke@amnch.ie

A. Briassouli et al. (eds.), *Health Monitoring and Personalized Feedback using Multimedia Data*, DOI 10.1007/978-3-319-17963-6_12

for the year 2010 were US$604 billion [51] and the cost of care increases with the progression of the disease as a person in the late stages usually requires admission into long-term care [25].

Dementia is an umbrella term for many different disease processes, all of which cause some memory, behavioural andor cognitive problems. The disease is progressive and currently incurable. There has been some small progress in the pharmaceutical industry in finding treatments for dementia but psychosocial interventions have been relatively under researched [4, 15]. Psychosocial interventions are based on the premise of person-centred care which promotes actively engaging the will and preference of people with dementia and promoting choice and rights [13, 26]. In the context of communication difficulties, common in dementia, getting to know the person, their will and preference and their valued identity is problematic. Putting all these together, the result is that caring for people with dementia is often complex, and costly, and we often have the situation that people living in a nursing home or care facility, especially those who are reported to have a dementia, have a poor quality of life.

Reminiscence therapy serves two purposes; it allows the care-giver a structured approach to communicating with people with dementia in order to engage with them in a meaningful manner and have insight into will, preference and identity, and it gives the person with dementia a legitimate and safe place to express those valued identities thus maintaining or curating them.

Reminiscence therapy (RT) has seen success in recent years as a method of therapy for people with Alzheimer's and other dementias. RT refers to the guided recollection of previous life experiences or subjects of interest either in a group or individual context. RT has been proven to have a positive effect in terms of increased life satisfaction, decreased depression, and increased communication skills and patient-caregiver interactions [7].

In a typical RT session, a facilitator (for example a clinician or activity co-ordinator) uses *cues* to stimulate recall of memories. These cues may be objects from a person's past or old family photographs, for example. More recently, digital cues have been used in the form of multimedia content.

Identifying relevant content to use in reminiscence therapy can be a time-consuming and resource-intensive task. Traditionally, therapy facilitators have kept either paper or mental records of a person's life history and interests so that they may make an informed decision about which content would likely be beneficial for use in an RT session. RT participants are also encouraged to maintain scrapbooks of their own past, known as *lifebooks* and these generally depend on loved ones gathering such material. These methods have significant drawbacks in terms of scalability because of the resources necessary to produce them, if such information is available at all, and the challenges inherent in trying to re-purpose materials or identify generalisable materials for use in a group setting.

Other factors which make identification of reminiscence materials a challenging task include generational and cultural barriers between the facilitator and person with dementia, acquired communication difficulties in dementia and a lack of a collateral history to inform patient biography where such difficulties exist.

During group or even individual RT sessions, the facilitator often needs to make decisions quickly and monitor participants' reactions, which limits the time they can devote to finding new materials, say from a digital library, during the RT sessions. A common approach therefore is to plan sessions beforehand. However, apart from the extra time required, such a rigid approach limits flexibility in terms of adapting when a pre-planned stimulus has proven ineffective during a session or one that is proving upsetting, or to follow a topic thread of material which proves to be of interest, and so sessions need to be dynamic and reactive to the circumstances of how it is unfolding.

Thus the requirements of a system to support group RT are that it should be efficient, accurate, personalised and provide a high degree of utility to the facilitator, ultimately leading to successful group RT outcomes. It should also not distract from other tasks and should seek to be relevant to *all* group members. Our system, REMPAD (**R**eminiscence Therapy **E**nhanced **M**aterial **P**rofiling in **A**lzheimers and other **D**ementias), addresses these requirements using a novel group-recommender approach to multimedia RT.

Many of the existing digital solutions to RT have focused on personal content to support people with dementia. For example, Yashuda et al. [53] proposed a system to use personalized content with predefined themes. Sarne-Flaischmann et al. [44] concentrated on patients' life stories as reminiscent content while Hallberg et al. [19] developed a reminiscence support system to use lifelog entities to assist a person with mild dementia.

The REMPAD system uses generic content rather than personal material, and operates as shown in the following series of screenshots which are taken from a tablet interface. Figure 1a shows the screen used to start a new session where the facilitator can add or edit an existing participant or group.

This leads to a recommendation of two pieces of multimedia video content, a selection of photographs from Dublin in the 1980s and a clip from a Hurling final from 1975, shown in Fig. 1b. The facilitator can play either of these, add

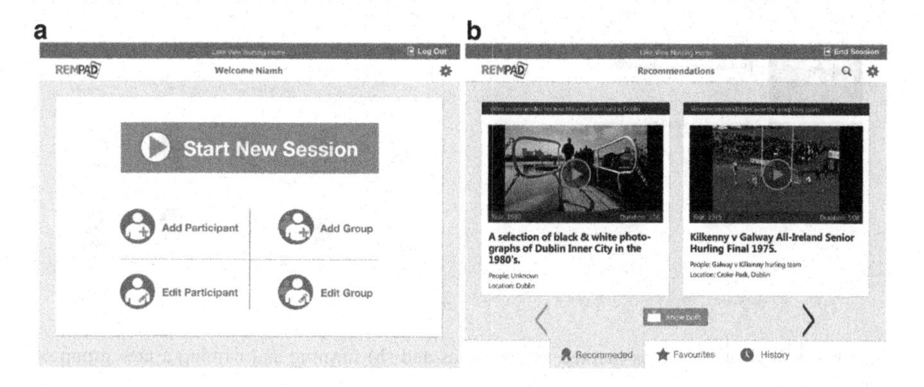

Fig. 1 (**a**) Start screen and (**b**) video recommendation

to favourites or can request the next two recommendations from the system. Our feedback from facilitators has been that they always like to make an A-B choice among content.

Once a video selection has been made and played on an accompanying wall-mounted big screen, the facilitator is then asked to give feedback on the chosen video in order to improve recommendations for the benefit of the present RT session as well as improving the accuracy of participants' profiles. Feedback is done based on a group, and optionally on a per-participant basis and Fig. 2 shows two feedback screens, Fig. 2a indicating the content as appropriate and useful and Fig. 2b indicating the opposite.

Any given video can be marked as a favourite or "like/starred" and this can be based on a per-group as well as a per-participant basis, as shown in Fig. 3a. If an RT group session is starting and the group composition is new, i.e. not a regular

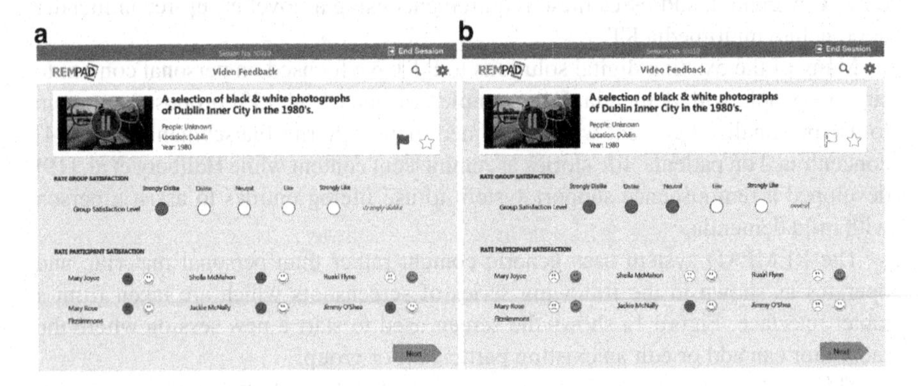

Fig. 2 Video feedback screen (**a**) indicated as not useful or inappropriate and (**b**) indicated as useful

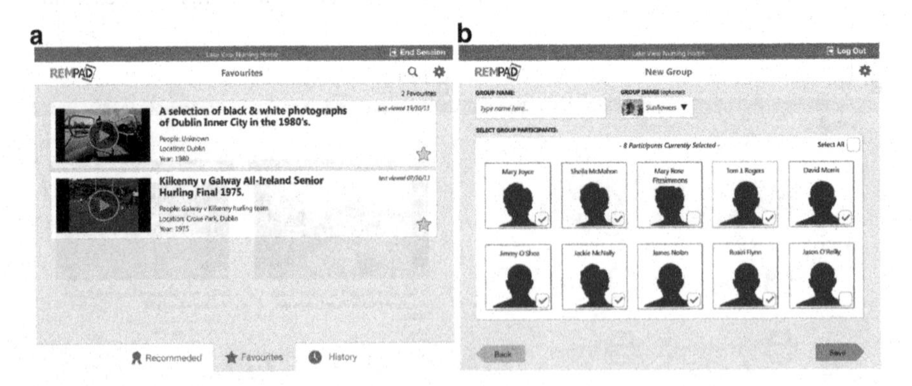

Fig. 3 (**a**) Reviewing videos marked as favourites and (**b**) forming and naming a new group of participants

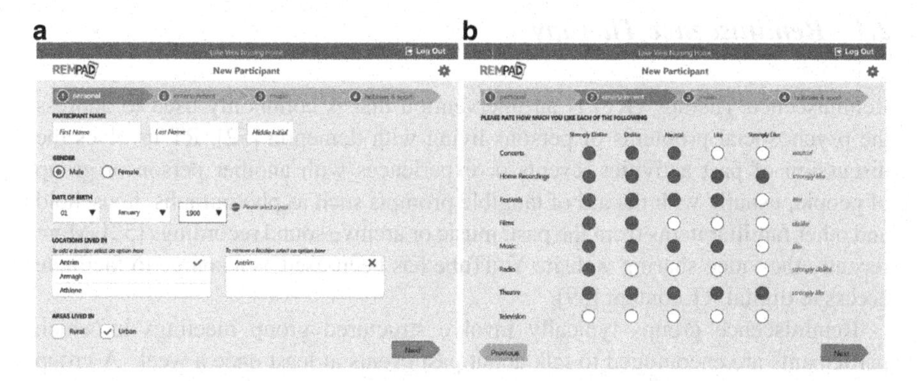

Fig. 4 Registering a new participant, screens (a) and (b)

group arrangement then a new group can be formed and named as shown in Fig. 3b, selecting participants from those registered and already known to REMPAD, in this case Mary, Sheila, Tom, etc. (names anonymised).

For participants new to the system, we need to learn something about them in order to bootstrap the recommendation process and Fig. 4a,b shows two screens used to enter details including name, age, places where the participant has lived in the past, a 1.5 rating for the kind of entertainment he/she likes, music preferences, interests in different sports, etc. Thus selection of profile aspects come from various interactions with nursing homes and facilitators and indicate the kinds of topics which make a good basis for collaborative and group RT.

The rest of this chapter is structured as follows. In the next section we provide some background and description of related work so that we can position our research in the context of the fields of Reminiscence Therapy and Recommender Systems. We then elucidate the challenges around building a system which combines these, and we then describe our approach to the design and development of the REMPAD system in Sect. 4, followed by Experiments and Results and Discussion in Sects. 5 and 6. We conclude in Sect. 7.

2 Background and Related Work

We now discuss related work in the field of recommender systems. First however, we look more closely at reminiscence therapy, and in particular how it is suited to a recommender systems approach.

2.1 Reminiscence Therapy

Reminiscence Therapy (RT) is an intervention that is commonly used to address the psychosocial problems of persons living with dementia [52]. RT involves the discussion of past activities, events or experiences with another person or group of people, usually with the aid of tangible prompts such as photographs, household and other familiar items from the past, music or archive sound recordings [52]. More recently the video sharing website YouTube has been used as a source to facilitate access to digital RT content [39].

Reminiscence groups typically involve structured group meetings in which participants are encouraged to talk about past events at least once a week. A group leader or facilitator assists and guides the group members to recall previous life experiences and facilitates the group's affirmation of the value of these experiences [9]. This activity aims to improve mood, well-being, communication and to stimulate memory and strengthen a sense of personal identity [8, 52]. This treatment is based on the assumption that autobiographical memory remains intact until the later stages of dementia and may be used as a form of communication with the person with dementia [37].

There is evidence to suggest that RT is effective in improving mood in older people without dementia and its effects on mood, cognition and well-being in dementia are present, but less well understood [52]. Improvements in autobiographical memory selectively in RT groups for mild-to-moderate degree dementia have also been described [18, 36]. Despite the limited empirical study of reminiscence undertaken, the most results indicate the positive effects of reminiscence [18, 52].

Autobiographical memory is characterized by multiple *types of knowledge*, and refers to a memory system consisting of episodes recollected from an individual's life. This is based on a combination of episodic memories (personal experiences and specific objects, people and events experienced at a particular time and place) and semantic memories (general knowledge and facts about the world) [17, 50]. Flashbulb memories are a particular type of autobiographical memory of vivid mile-marker events with associated personal and meaningful experiences [45]. They rely on elements of personal importance, consequentiality, emotion, and surprise [16]. They may include collectively shared public events marked by their uniqueness and emotional impact. Autobiographical memories may be accessed more easily and with greater frequency in old age, precisely because they are more robust and less likely to dissipate than memories of everyday commonplace experiences such as what you had for dinner last week. Autobiographical memories include multi-sensory information about the experiential context, including sights, sounds and other sensory and perceptual information. A song, a scent, or simply a word can evoke a cascade of autobiographical memories.

RT can also be conducted on a one-to-one level but is distinct from life review therapy (LRT). LRT typically involves individual sessions in which the person is guided chronologically through life experiences, encouraged to evaluate them, and may produce a life story book as a result. Although the procedures are different, both RT and LRT involve the recollection of past experiences (events, emotions and relationships).

Facilitated reminiscence exploits the relatively well-preserved autobiographical memories to enhance communication opportunities for older adults who may differ in abilities, cultural background, or life experiences [22]. In facilitating a traditional RT group the facilitator manages the selection of topics, scheduling, group composition and communicative interactions between and among group participants. An understanding of the participants' shared historical experiences is the starting place for topic selection [22]. This is achieved by firstly, considering the personal interests, likes and dislikes of individuals. Secondly, the flashbulb memories shared by particular age cohorts and thirdly, universally experienced developmental life events such as childhood, schooldays, adulthood, marriage, work life, and retirement.

Creative therapeutic approaches are required to facilitate the socialization needs of residents and to appeal to an increasingly culturally and linguistically diverse population. Facilitated RT programmes therefore need to be simultaneously engaging, relevant, cost effective and culturally sensitive [21]. Mismatches can arise in age, life experience and culture between the majority of culture clinicians and older adults from non-mainstream populations [41] or vice versa. Hence the need for detailed group member profiling is important to enable positive and successful reminiscence facilitation. REMPAD builds on our previous research in which the use of video and other digital multisensory content to stimulate conversation and social interaction was found to be a feasible in group reminiscence therapy sessions [39].

A comprehensive approach towards the person with dementia that takes into account their life history is essential. The *person-centred approach* to dementia situates the person with dementia at the centre of all aspects of caregiving [12, 27]. The focus is on identifying and meeting the needs of the person, in contrast to the medical model which focuses on identifying and treating symptoms. The person-centred approach aims to enhance well-being by improving relationships and communication between people with dementia, their families and professional caregivers. This is achieved by taking into account the life experiences and the likes and dislikes of each person with dementia in order to develop a greater understanding of the individual. This in turn allows for care tailored specifically to the individual to take place. *Person-centredness* can be achieved when carers and family members focus more on the individual than on the illness. This is enabled by knowing the person which is challenging in the context of communications difficulties. RT allows a structured communication strategy to enable care givers to engage with the person with dementia, to get to know them and to acknowledge their unique identity.

To address the needs of the residential care population and their associated activity coordinators REMPAD proposes a solution to enhance facilitator knowledge and provide access to personalized reminiscence material for the benefit of aiding conversation and memory recollection amongst nursing home participant users in a group context.

2.2 Recommender Systems

In this section we provide background and related work in the area of recommender systems. There are broadly three categories of recommender system: those based on user matching (collaborative), those based on learning content preferences (content-based) and those that use a knowledge base approach (case-based reasoning). We describe each of these in turn and how they relate to the REMPAD system.

Much work in recent years in the area of recommender systems has focussed on user-item rating prediction through inference over large datasets. A common approach is to make predictions of user-item ratings based on the previous ratings for that item of similar users, known as *collaborative recommendation*. Perhaps the most salient example is the Netflix prize [5] which pushed forward the state of the art in large-scale collaborative recommendations systems. A characteristic of collaborative recommender systems is that they rely on the availability of large amounts of data. Also, a collaborative approach relies solely on user-item rating information, rather than any information about the items themselves. These user ratings may not be able to model certain aspects of the recommendation task.

A second popular category of recommendation is *content-based recommendation*. In content-based recommender systems, a user's preferences are stored based on their previous interactions or ratings of items. The system then learns from these preferences so that they may identify new items to recommend. Content-based recommenders rely on the system being able to explicitly model properties of objects. The advantages of content-based systems include transparency in the recommendation decisions and the ability to recommend new items never seen before by the system, provided the necessary features can be extracted. A drawback is the uncertainty when a new user uses the system and the limitations in terms of how items can be modelled, sometimes referred to as the *semantic gap*. Pazzani and Billsus provide an overview of content-based recommenders [42] and Lops et al. provide a recent review of the state of the art [30].

A third approach to recommender systems is based on *case-based reasoning* (CBR). CBR approaches are those which rely on a knowledge base representation of known items and item context. Although CBR approaches can vary, in particular to the extent that they implement the full CBR process, the most common CBR Recommenders use a stated preference from a user and a similarity function to match the parameters in that preference to the item descriptors in the knowledge base [31]. This process could also involve other information relevant to the recommendation task such as user profile, preference refinement and previous uses of the system. A limitation of CBR systems is the need to create and maintain a knowledge base of items and as with content-based recommenders, the semantic gap. An advantage is the intuitiveness with which a user can express their preference and if necessary, refine their requirements, in many ways similar to interactive search systems. CBR systems are of particular use in e-commerce where users are looking for products to purchase. Overviews of the adaptation of the CBR process to recommendation tasks are provided by Bridge et al. [11] and by Smyth [47].

For group-based RT, a collaborative approach is not feasible due to the small size of the user group. Our system uses a hybrid approach consisting of a CBR and a content-based recommender, supported by a traditional search feature for query refinement, and a novelty multiplier. To mitigate the limitations of these approaches we use the CBR approach to bootstrap the content-based approach. In order to create our knowledge base of users and items we adapt traditional methods for profiling RT participants and use an efficient curation and annotation process to produce low-cost item descriptors.

Some recent works have examined the more complex task of recommending content for groups of individuals. In groups with disparate sets of preferences, it is not clear how to optimally recommend content for a given group context. Popular approaches seek to minimize misery, maximize individual utility or use an aggregated measure of group satisfaction.

McCarthy et al.'s work has tackled the group problem from a case-based perspective using iterative interactive critiquing of cases among group members to reach an optimal solution [33–35]. An early review of group recommenders is provided by Jameson and Smyth, outlining the significant challenges in moving from individual to group recommendation [24]. Another early work from O'Connor et al. uses collaborative filtering to produce lists of movie recommendations for groups to watch [38]. They introduce a *minimum misery* strategy i.e. the overall satisfaction in a group is directly related to the satisfaction of the least happy group member. Later we will see this is a principle we employ in REMPAD. Recently, Masthoff has compared group recommender systems from the literature, noting the different strategies used for aggregating individual profiles [32]. Although many systems use relatively straightforward strategies to simulate group recommender systems using *individual* recommenders, more complex approaches have been tried to explicitly model *group* preferences [14]. However, perhaps due to the typical dearth of group-level ratings, or the complexity of the task, most approaches use an array of individual recommenders.

Although we are aware of some recent works which investigated using digital systems for RT [3, 20, 29], to the best of our knowledge the REMPAD system is the first system to implement an algorithm to recommend content in the context of group RT. We take inspiration from the aforementioned CBR and content-based approaches, but design our system with specific considerations for the RT application domain such as minimizing interactivity and task complexity, and maintaining tight constraints on preventing dissatisfaction among group members and recommendation dead-ends.

Later in this chapter we describe the design, development, deployment, testing and evaluation of the REMPAD system but first we highlight the underlying challenges that the situation with using recommender systems in a reminiscence therapy application, presents.

3 REMPAD Challenges

There are a number of benefits of performing reminiscence therapy in a group context. In particular, therapy sessions enjoy a social component as participants can share experiences and discussion. Whereas it can be challenging to identify suitable content in a one-on-one context, identifying suitable content for a group of individuals is a much more difficult task for facilitators. The facilitators must identify content which optimally benefits the group, while minimising any negative effect. For example, a video which some group members find engaging might be undesirable overall if this induces a negative effect in other members.

Thus there are motivations and challenges for the application of a recommendation and search approach to supporting group digital reminiscence therapy. Due to the nature of RT, there are a number of task-specific requirements and constraints which make it different from other group recommender systems which have been traditionally focused on tasks in areas such as e-commerce and entertainment. The approach we take addresses specific challenges related to RT as an application area.

Public or more generalised content are now being recognised as valuable reminiscence prompts, from which individuals obtain personal meaning. The benefit of this type of content for RT is that different people have their own memories associated with the same public event, which can stimulate conversation about shared experiences and interests, as well as personal reminiscence. André and colleagues [1] explored the concept of workplace reminiscence by creating personally evocative collections of content from publicly accessible media. Other studies examined the use of interactive systems, displaying generalized content to support people with dementia in clinical settings, such as hospitals or nursing homes. For example, Wallace et al. [49] designed an art piece for people with dementia and hospital staff to interact with. This consisted of a cabinet containing themed globes, which when placed in a holder initiated videos displayed on a TV sçreen, which were based on the associated theme, for example nature, holiday, or football. CIRCA, an interactive computer system designed to facilitate conversation between people with dementia and care staff, used a multimedia database of generic photographs, music and video clips to support reminiscence [2]. Astell et al. maintain that generic content is more beneficial than personal content as it promotes a failure-free activity for people with dementia as there are no right or wrong memories in response to the stimuli.

However, what all these systems have in common is that their content is static and requires uploading and selection by either system developers or reminiscence facilitators. Multimedia websites potentially hold a wide range of subject matter that can be easily accessed. The question naturally arises: can we leverage the extensive range of online multimedia content, so that the reminiscence experience is maximized ? We postulate that video sharing websites, in particular YouTube which is what we use in our work, are a valuable tool in promoting interaction and social engagement during group RT [6, 39].

In our work we have developed a multimedia system for modelling group preferences and recommendation algorithms and integrating them into an RT system based on video content from YouTube. Our approach uses a combination of case-based reasoning recommendation, content-based recommendation and search to address RT facilitators' content needs in real time. The focus of our evaluation is to assess the efficacy of the recommendation algorithm. Our results are based on a user trial we conducted in residential care homes with seven user groups. We examine the accuracy and utility of content suggested by the REMPAD system through analysis of system usage logs as well as explicit ratings from users, comparing a number of system configurations. We also report on usability interviews with RT facilitators who participated in the trial.

4 Approach

In our approach we model a system for use in a care setting with a group of people with mild-moderate dementia and an activity co-ordinator. In this section we describe the design of the system, the data curation process, user profiling and our approach to recommendation. There are two types of users in our system: the activity co-ordinator, or clinician, who facilitates the session, and the therapy participants themselves. We use *item* to refer to a video indexed by our system; *user* to refer to a therapy participant; *group* to refer to a therapy group, consisting of a set of users; and *facilitator* to refer to the clinician or therapist who runs the session and physically interacts with the system.

4.1 Design of the REMPAD System

As mentioned earlier, REMPAD is a cloud-based service which is accessed through a mobile device such as a tablet. This interface controls the application flow, interpreting participant requirements, selecting content to display on a second larger screen and providing online feedback to the system.

A typical session involves creating a new session and registering participants, examining the recommended videos and selecting one to play on the shared viewing screen and then feeding back to the system indicators of perceived user and group satisfaction on which are based subsequent recommendations. The system is designed to support sessions with minimal intrusion on the facilitator who must also monitor and engage with the group participants during the sessions. Typically a session lasts about 45 min and a group will watch several videos in a session.

Healthcare systems are characterized by complex user requirements and information-intensive applications. Usability research has shown that a number of potential errors can be reduced by incorporating users perspectives into the development life cycle [23]. Thus, employing a user-centred design (UCD) approach throughout the development cycle, may lead to high quality intelligent healthcare systems. In order to conduct a UCD research study, we need to define user characteristics, tasks, and workflows so that we can understand different stakeholder needs.

4.1.1 Participant Sample

The primary stakeholders of the REMPAD system are the facilitators who lead group RT sessions and interact directly with the system. For this study we focused on how the system supports these users to conduct RT sessions. The participant sample consisted of 14 health professionals, including 7 speech and language therapists (SLTs). All participants currently run RT sessions in hospitals, day care centres or residential nursing homes.

The secondary stakeholders of the system are the therapy participants—people with dementia who attend RT sessions. Although these participants do not directly interact with the tablet PC, information is displayed to the group through the TV monitor and information is also relayed through the facilitator. Current practice requires the facilitator to make subjective judgments after a session regarding the success of the material used in RT sessions to support inter-group interaction and their communication, mood and well-being. This was the method we used to gauge secondary stakeholder satisfaction in our field trials study.

The study was designed in three parts: (1) exploratory interview, (2) low-fidelity prototype test, and (3) field evaluation. We implemented findings from each stage into the system design which we then re-examined. We now discuss these methods.

4.1.2 Study 1: Exploratory Interviews

The purpose of the exploratory interviews was to understand current RT practices, the types of technology used in these sessions if any, and the challenges that facilitators experience during these sessions. The types of questions that the facilitators were asked included: what types of technology do you use during a RT session? Do you prepare material before a group session? What are the challenges you experience? The findings were divided into four categories: current practices; technical skills; session challenges; and technical challenges.

Facilitators spoke about their RT practices using physical and digital prompts. Each facilitator may work with several groups, in several different locations. It was most common for them to use paper-based objects in these sessions, such as photos, newspaper clippings, or printed images. Physical objects were selected for their texture and smell to stimulate memories, for example polish or lavender. The most common method used throughout the RT sessions was to begin with general or current themes. The conversation would then develop from these topics. After the session, the facilitator would write up a report on what material or topics worked well.

We learned that facilitators had different levels of technological expertise, from novice (n = 1), average (n = 5), to above average (n = 1) skills, and some (43 %) had little or no exposure to using tablet PCs. This poses a need for clear and intuitive interfaces with easy-to-use interaction modalities. Facilitators reported experiencing several challenges when using technology in the RT sessions. For example, internet connectivity might be very good in some sections of a nursing home or care facility but poor in others, while some locations also have blocked access to certain websites, including YouTube. Facilitators told us that most of their working time is spent preparing for sessions, searching for appropriate material based on previous discussions or group preferences. On the one hand, this meant that the facilitators were confident that the material would stimulate conversation, but it also meant that topics were fixed and did not allow for spontaneous deviation. Five of the seven facilitators had used video websites (such as YouTube) during their sessions to support spontaneous deviation. They reported difficulties finding content about a topic before the conversation drifted onto another topic. Currently, the practice is to prepare a number of video clips prior to the RT session to ensure that they are of good visual and sound quality.

The facilitators also commented on the challenge of preparing for a group RT session when they do not know the participants or groups preferences. These challenges revolve around learning about an individual's interests if they are unable to suggest topics or interaction, and in a group setting it can be unhelpful to direct attention to them by putting them on the spot. We know that the best way to present the technology behind the proposal is through a worked example. Based on the functional requirements provided, we created initial wireframe prototypes of the REMPAD system, consisting of a series of 12 use cases including Start a new session; Edit an existing group; Browse video clips; and Enter feedback (see Fig. 5). A use case walkthrough was undertaken to familiarize participants (7 SLTs from Study 1) with the proposed task flow and interaction paradigm of the prototype system.

Participants expressed high enthusiasm and positive response towards the initial prototype design believing it to be simple and straightforward, and that users with low technology experience would feel comfortable interacting with it.

One of the crucial elements of an intelligent reminiscence system is to offer customizable content to users. Diversity exists inside a group in areas of individual backgrounds, interests and preferences. As one of the facilitators mentioned, "the biggest challenge is finding relevant videos". It is believed that automatic recommendation would save facilitators a significant amount of time, which is currently used planning RT sessions and would allow them to interact with the group rather than searching for appropriate material.

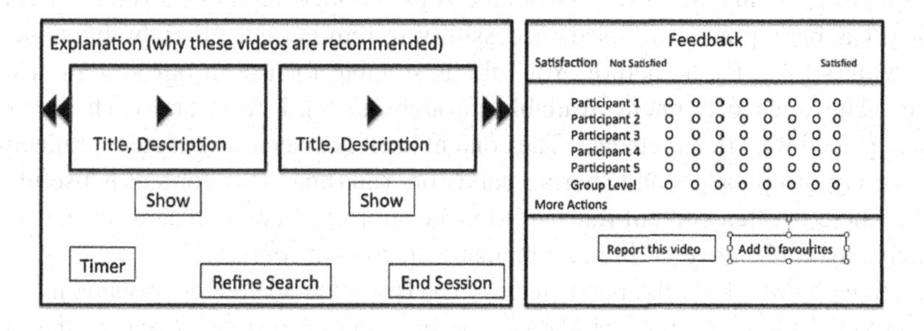

Fig. 5 Example wireframe screens used in use case walkthrough method

The presentation of the videos was also discussed with the facilitators. It was decided that an option of two videos at a time was preferable as the facilitator could then relay this choice to the group without overloading them. Information about the video is also necessary so that the facilitator can have some knowledge about the subject being discussed. Finally, facilitators emphasised the importance of having control over topics. Maintaining the current practice of beginning a session with general topics and moving into more specific topics, facilitators said that they would use recommended videos for the most part, but would like to have the option to search for a video based on how a discussion develops.

Design alternatives were displayed to participants to search for a topic, or refine by category. We decided that the most appropriate design would be to include a search bar, which the user could refine according to a different year or decade. The ability to save successful video clips into a "favourites" section for future sessions was also requested by participants.

Another challenge highlighted in building an intelligent reminiscence system is to ensure content is of high quality. In order to maximize group reminiscence experience, it was proposed that the recommendation engine should monitor patients' engagement levels, and adapt based on real-time user feedback. We designed the feedback screen layout as showed in Fig. 5. After each video, the group facilitator enters individual patient and group reactions to the presented video, so the selection of videos is improved in future sessions. However, we were unsure whether this function would add too much burden on the facilitator. During the discussion stage, participants unanimously confirmed that this level of feedback was achievable and understood and valued the benefit it would bring. The facilitators reported that they currently use pictures and icons to rate group satisfaction and topics discussed etc., in order to keep track of group progress. It was suggested that an end-of-session feedback report also be included in the system for the facilitator's records. This feedback was used to improve user interface design and justify design decisions, which were then implemented into a fully functional REMPAD system.

4.2 Data Curation and Annotation

The data we use in our system is from the popular video sharing website, YouTube, which has been previously used successfully in reminiscence therapy by some of the authors [39]. By its nature, YouTube is suitable for use in our system. There is an abundance of content available through standard APIs and each video is accompanied with rich metadata. The content itself is diverse and esoteric, reflecting the variety of uploaders and sharing habits on YouTube. This content is useful for RT as there is often content relevant to niche subjects, people, places, events, etc., which may not be covered in more mainstream content sources.

Although we had intended using YouTube metadata for organising and presenting videos in the REMPAD system, initial testing revealed that the quality and consistency of metadata were not of sufficient standard to support the system requirements. To address this we used a curation and annotation process.

A project team consisting of research assistants, clinicians, and postdoctoral researchers, curated content using a custom curation interface. This interface offers a search functionality which uses the YouTube search API to find videos relevant to areas of interest, times and locations which are suggested to the curator. We provided curators with subject matter targets reflecting a broad range of media types and content. The curator then previews videos and if happy with the content can queue the videos for annotation.

The index used in the system contains a wide range of video content. Examples include documentary excerpts, home videos, music recordings, interviews and sports. Curators were advised to search for videos ideally less than 5 min and no more than 10 min in duration so that they were appropriate for use in RT.

An important concept in RT is orientation towards people, places and times. To offer a personalised experience, we also wish to model a user's preferences and interests. The metadata produced by the annotation process for a video includes title, description, location(s), date, people, seasons/holidays as well as vectors describing relevance to a variety of genres, media, music, interests and sports.

Initially the authors annotated 343 videos. This can be a time-intensive task, taking approximately 3 or 4 min per video. To reduce the cost of indexing content, we obtained a further 258 video annotations using the crowdsourcing service, CrowdFlower.[1]

The crowdsourced annotations were added to the video index at approximately the halfway stage in the trials to prevent staleness of content. In order for the system to perform effectively, it needs to provide usable recommendations amongst the top results (ideally top 2) or otherwise risk slowing the facilitator and disrupting the momentum of the RT session. Even though the index we use in these trials is relatively small, it is still a significant challenge to produce useful recommendations at the very top of the results list. We have designed the system and processes to be scalable as significantly expanding the user base and index is a goal of future work.

4.3 User Profiling

The user profiles are gathered through short interviews with users before their first use of the system. This is inspired by existing practices in care settings where a record is often made of people's life history and interests. Similar to video metadata, the metadata we collect for users includes date of birth, locations lived in, and interest vectors related to genre, media, music, interests, sports, similar to the video vectors. A key difference with users is we allow them to also express dislike using a 5-point Likert scale whereas the equivalent for video was either categorical or on a 3-point relevance scale: not relevant, relevant, highly relevant. In the following section we refer to the concatenation of the genre, medium, music, interests, sports vectors as simply the *feature vector* for users and items.

[1] http://www.crowdflower.com.

4.4 A Recommender Model for RT

Our recommender algorithm consists of a scoring function which is used to proactively rank items for a given recommendation context consisting of a group of users, their previous item ratings and interactions, and optionally a search query.

We model a user u as having three features: a location, a date of birth and a feature vector whose values are normalised to between -1 and 1.

$$u =< u_l, u_d, u_f > \tag{1}$$

Similarly, we model an item i as having four features: a location i_l, a date i_d, an interest vector with values normalised between 0 and 1 i_f, and a textual description i_t.

$$i =< i_l, i_d, i_f, i_t > \tag{2}$$

A search query q is given by two optional fields: a text query q_t and a decade q_d.

$$q =< q_t, q_d > \tag{3}$$

The scoring function for an item i, given a group of users, G, and an optional search query, q is:

$$S(i, G, q) = \left(\frac{w_1 S_{CBR}(i, G) + w_2 S_C(i, G) + w_3 S_{Rel}(i, q)}{\sum_{j=1,2,3} w_j} \right) * N \tag{4}$$

where S_{CBR} is the CBR scoring function; S_C is the content-based scoring function; S_{Rel} is the relevance function; and N is a novelty multiplier. In our system, we present two options for S_{CBR}. In the first, $S_{CBRlate}$, we aim to aggregate individual preferences at a late stage using a minimum misery approach.

$$S_{CBRlate}(i, G) = \min_{u \in G}(S_{CBRlate}(i, u)) \tag{5}$$

For each individual user the function uses a linear combination of three similarity functions:

$$S_{CBRlate}(i, u) = Sim_{date}(i_d, u_d) + Sim_{loc}(i_l, u_l) + Sim_{feat}(i_f, u_f) \tag{6}$$

In line with the priorities of good reminiscence content, the date similarity function upweights items related to recent events or to events that occurred when the user was aged below 30. We also provide a small bonus to items from before the user was born which may be of historical or cultural interest.

$$Sim_{date}(i_d, u_d) = \begin{cases} 1 & when\ i_d - u_d < 30\ yrs \\ 0.75 & when\ now - i_d < 10\ yrs \\ 0.25 & when\ i_d < u_d \\ 0 & otherwise. \end{cases} \tag{7}$$

Similarly, the location similarity function upweights the best specific matches between user and item:

$$Sim_{loc}(i_l, u_l) = \begin{cases} 1 & when\ regions\ match \\ 0.5 & when\ countries\ match \\ 0.1 & when\ continents\ match \\ 0 & otherwise. \end{cases} \tag{8}$$

The similarity between feature vectors is given by the Cosine Similarity between the feature vectors:

$$Sim_{feat}(i_f, u_f) = Cosine\ Similarity(\mathbf{i_f}, \mathbf{u_f}) \tag{9}$$

In the second of our CBR scoring functions, we aggregate preferences into a single meta profile for the group from the outset. $S_{CBRearly}$, consists of a linear combination of similarity functions, but this time interpreted at a group level:

$$S_{CBR_{early}}(i, G) = Sim_{date}(i_d, G_d) + Sim_{loc}(i_l, G_l) + Sim_{feat}(i_f, G_f) \tag{10}$$

$$where\ G_x = \{u_x : u \in G\} \tag{11}$$

This can be seen as treating the group as a *meta-user*. For date, we simply model the date for the group as the mean point in time, given the range of dates of birth:

$$Sim_{date}(i_d, G_d) = Sim_{date}(i_d, \overline{u_d}) \tag{12}$$

$$where\ \overline{u_d} = \frac{1}{|G_d|} \sum_{u_d \in G_d} u_d \tag{13}$$

For locations we use the best match for a common location in the group:

$$Sim_{loc}(i_l, G_l) = \max_{u_l \in G_l}(i_l, u_l) \tag{14}$$

where $u_l \in G_l$ and u_l is common to 2 or more members of group G.

To compare features at a group level we consider positive features and common negative features:

$$Sim_{feat}(i_f, G_f) = Common_{pos}(i_f, G_f) - Common_{neg}(i_f, G_f) \tag{15}$$

In order to identify the common positive features we rank the features according to the number of users in the group who have declared each feature as an interest or strong interest and take the top m features, F_{pos}. Similarly, in order to identify the common negative features we rank the features according to the aggregate score from the users in the group, and take the top n features, F_{neg}. For the negative ranking we assign 1 to a *dislike* and 2 to a *strong dislike*, thus emphasising extreme negative preferences. We also create a set of relevant features for each item, F_{rel}. The commonality scores are then given by:

$$Common_{pos}(i_f, G_f) = \frac{|F_{pos} \cap F_{rel}|}{m} \tag{16}$$

$$Common_{neg}(i_f, G_f) = \frac{|F_{neg} \cap F_{rel}|}{n} \tag{17}$$

In our experiments we set m to 40 and n to 20.

$S_C(i, G)$ is given by the output classification probability of the positive class from a multinomial naive Bayes classifier trained on positive and negative examples for the group G. An item i is a positive example for group G if it satisfies the following criteria:

- There has been no negative item ratings from group G for item i.
- There has been no negative item ratings for user u for item i, $u \in G$.
- There has previously been a positive item rating from group G, or from $u \in G$, for item i.

There is just a single criterion for an item to become a negative example:

- There has been negative group-level or individual feedback from Group G for item i.

If the number of examples in the positive set is below a threshold r, we bootstrap the process by adding the r top-ranked examples by S_{CBR} to the positive set. Similarly if the size of the negative set is less than r, we add r lowest-ranked examples by S_{CBR} to the negative set. In our experiments we set r to 5. The features used for classification are the item feature vector i_f.

In a case where a user has chosen to enter a search query, the search query-item relevance is given by:

$$S_{Rel}(i, q) = \frac{w_4 Rel_{text}(i_t, q_t) + w_5 Rel_{date}(i_d, q_d)}{\sum_{j=4,5} w_j} \tag{18}$$

where $Rel_{text}(i_t, q_t)$ is the score given by a search over an index item text fields (title, description, people), i_t, using the search platform SOLR.[2] We reward queries if they are from the same decade or a neighbouring decade as a candidate items:

[2]http://lucene.apache.org/solr/.

$$Rel_{date}(i_d, u_d) = \begin{cases} 1 & \textit{when from same decade} \\ 0.5 & \textit{when from neighbouring decades} \\ 0 & \textit{otherwise.} \end{cases} \quad (19)$$

We set $w_4 = 2$ and $w_5 = 1$, emphasising the specificity of a text query, particularly as a common search task is known-item search, where the facilitator is trying to find an item they are aware is in the index.

Novelty often has an important role in recommender systems [48]. In order to prevent the results list becoming predictable and familiar, we penalise results if they have been recently browsed or played. This novelty function has a decay so as to allow familiar videos to move back up the results list as the time since they were last browsed or played increases. In REMPAD there is both a requirement to show novel results in the list *and* to ensure that known familiar and useful content is re-discoverable.[3]

Let $n_b(i, G)$ be the number of queries since item i was last browsed in a results list for group G. Let $n_p(i, G)$ be the number of queries since item i was last played in group G. We define the novelty multiplier N then to be:

$$N(i, G) = \frac{w_6 \log(\min(n_p(i, G), h)) + w_7 \log(\min(n_p(i, G), k))}{\sum_{j=6,7} w_j} \quad (20)$$

We set $h = 5$ and $k = 10$ in our experiments, and upweight the importance of playing an item over browsing, with $w_6 = 2$ and $w_7 = 1$.

5 Experiments

We trialled our system over a period of several weeks involving over 50 users in 7 therapy groups across 6 locations, each being a residential care home. See Table 1 for details of groups and sessions for those groups.

Our evaluation has two focuses. First we wish to ascertain the degree to which the recommender has supported the reminiscence therapy sessions for the groups in our study. Secondly, we wish to investigate the comparative performances of different configurations of our algorithm.[4] The four configurations we use are (i) $S_{CBRearly}$

[3]For this reason we also provide *favourites* and *history* functions which are sometimes used.

[4]As this is not a controlled study, our ethical approval does not extend to using a control as one of our experimental conditions. This has precluded us from exposing people with dementia to potentially weak experimental conditions such as randomly selected content which might not suit their tastes.

Table 1 Session and video play counts for trial groups

Group	Members	Sessions completed	Videos played	Videos per session
A	7	4	21	5.25
B	8	9	59	6.56
C	8	9	61	6.78
D	7	6	55	9.17
E	8	11	72	6.55
F	7	5	26	5.2
G	11	10	68	6.8
Total	56	54	362	6.7

without S_C, (ii) $S_{CBRearly}$ with S_C, (iii) $S_{CBRlate}$ without S_C, (iv) $S_{CBRlate}$ with S_C. These configurations were assigned to sessions for groups in a latin squares arrangement.[5]

In both cases, our evaluation focuses on three aspects: (i) accuracy, (ii) utility and (iii) perceived usefulness. Unlike some recommenders, our multimedia system is based on ranked recommendation lists, akin to a search system. For accuracy we compare system-ranked lists to reference rank lists as rated using a given group of annotators, using Spearman's rank correlation coefficient, ρ. In this approach, we construct ideal lists for users and groups given knowledge of their item ratings. We then use these as references with which we correlate a given ranking produced by the system [46].

For utility we use *R-Score*. R is appropriate in scenarios like ours, where the user can only use a small set of items and the user is unlikely to be exposed to the majority of the items in the ranked list. R incorporates a half-life, α, which is equivalent to approximately the rank at which the user has a 0.5 chance of browsing the item, thus incorporating likelihood of observation of a given recommendation [10, 46]. In our experiments we set α to 5. For calculating both ρ and R we use user-item ratings and group-item items and present them as mean values over a given set of ranked lists returned by the system.

Recently there has been an emphasis on the importance of user experience and the perceived usefulness in evaluation of recommender systems [28, 43]. To reflect this in our evaluation we also use end-of-session group and user ratings. For reporting these scores, we conflated any Likert or other ordinal scales to a three-point scale: positive, neutral, negative. We then average these values assigning $+1$ to positive, -1 to negative, 0 to neutral, giving an average score, r, in the range $(-1,1)$ for a rating of feedback values.

It is worth noting that in our multimedia system, these ratings are important as they are the clinician's interpretation of the satisfaction of the individuals, and group, in the therapy sessions. This is a natural extension of the facilitator's role in terms of monitoring, interpreting reacting to therapy participants' reaction to stimulus.

[5]In practice this was difficult to maintain as users often created impromptu sessions for training and testing purposes which were later removed from the trial data.

6 Results and Discussion

The results overall from our trials are positive and show that the system is effectively supporting the content discovery task for the facilitator during a group RT session. Sixty-nine percent of queries successfully resulted in a played video. Typically, unsuccessful queries resulted in the facilitator either refreshing to obtain a new list of recommendations, refining the query using the search function, or playing a previously viewed video from favourites or history. Inspection of our logs reveals that search was only used in a minority of cases. The search query terms suggest that the most common search need was to find a result either viewed previously or previously browsed in a results list, a pattern sometimes called *known-item search*.

In 43 % of successful queries, the video chosen was on the first screen (top two recommendations), 73 % in the first three screens (top six recommendations) and 86 % in the first five screens (top ten recommendations, see Fig. 6). Facilitators appear to be comfortable choosing from near the top of the results list, consistent with a high level of satisfaction and trust in the recommendations.

Looking closer at the explicit online ratings that the facilitators provide to the system, we see they are overall very positive (see Table 2). Sixty-two percent of user-item ratings were positive, with just 1 % negative. Similarly, 49 % of group-item ratings were positive, with just 3 % negative. The reason the group-item ratings were not as positive as the user-item ratings likely reflects the comparative difficulty in recommending items for a group rather than an individual. Looking at end-of-session feedback, we observe the same pattern.

For six of the seven groups, the user session-ratings were more positive than item-ratings. This pattern also holds for group ratings for five of the seven groups. This is interesting as it agrees with the intuition that the probability of overall satisfaction

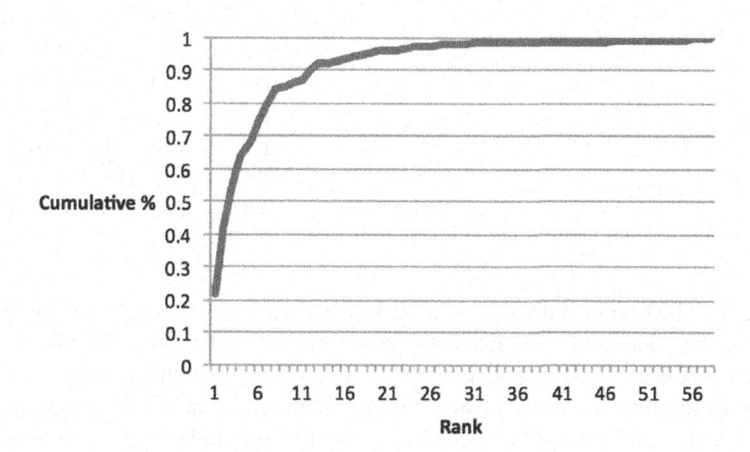

Fig. 6 Cumulative rank of selected item for successful queries

Table 2 Ratings for each group (A to G) and for total

	User-session					Group-session				
	n	+1	0	−1	*r*	n	+1	0	−1	*r*
A	25	0.36	0.64	0.00	0.36	4	0.25	0.75	0.00	0.25
B	56	0.64	0.36	0.00	0.64	9	0.33	0.67	0.00	0.33
C	51	0.71	0.22	0.08	0.63	9	0.56	0.44	0.00	0.56
D	29	0.83	0.17	0.00	0.83	6	0.67	0.17	0.17	0.50
E	63	0.86	0.11	0.03	0.83	11	0.91	0.09	0.00	0.91
F	26	0.73	0.27	0.00	0.73	5	0.80	0.20	0.00	0.80
G	70	0.61	0.39	0.00	0.61	10	0.60	0.40	0.00	0.60
All	320	0.69	0.29	0.02	0.67	54	0.61	0.37	0.02	0.59

	User-item					Group-item				
	n	+1	0	−1	*r*	n	+1	0	−1	*r*
A	132	0.35	0.64	0.02	0.33	59	0.34	0.66	0.00	0.34
B	348	0.57	0.41	0.02	0.55	72	0.63	0.38	0.00	0.63
C	317	0.61	0.35	0.04	0.56	55	0.55	0.45	0.00	0.55
D	255	0.70	0.29	0.01	0.69	21	0.24	0.76	0.00	0.24
E	417	0.71	0.28	0.00	0.71	61	0.33	0.52	0.15	0.18
F	128	0.77	0.23	0.00	0.77	26	0.77	0.23	0.00	0.77
G	478	0.56	0.43	0.01	0.56	68	0.57	0.41	0.01	0.56
All	2075	0.62	0.37	0.01	0.60	362	0.49	0.48	0.03	0.47

Table 3 Utility (*R* score) and accuracy (Spearman's ρ) scores for groups and total

	R		Mean ρ	
Group	User	Group	User	Group
A	17.51	28.13	0.11	0.10
B	11.22	9.34	0.19	0.13
C	3.51	4.89	0.20	0.23
D	6.49	0.62	0.06	0.04
E	11.47	12.33	0.09	0.09
F	3.40	3.64	0.16	0.20
G	13.30	12.87	0.08	0.10
All	9.62	9.42	0.13	0.13

is higher if the user or individual is evaluating over a series of recommendations, as they may be tolerant of some inaccuracies i.e. a user may be satisfied with a session without necessarily giving a positive rating for each video in that session.

Unlike the ratings, R and ρ show no significant difference between groups and users when it comes to either accuracy or utility (see Table 3). The R-Score for groups does vary in some cases, showing much higher group utility than user utility for group A and a lower group utility than user utility for group D. In the former case, group A has by far the lowest proportion of non-neutral item ratings, so

Table 4 Utility (R Score) and accuracy (Spearman's ρ) scores for four system configurations

		Mean R		ρ	
S_{CBR}	S_C	User	Group	User	Group
Late	No	11.42	10.20	0.08	0.09
Early	No	8.37	8.20	0.19	0.19
Late	Yes	9.80	10.53	0.08	0.08
Early	Yes	9.03	8.78	0.12	0.13

perhaps this has an effect, although how is unclear. For accuracy, we see that each of the rank correlations are positive, although relatively weak. It should be noted that novelty has had negative effect on both accuracy and utility as we report it here. The novelty multiplier deliberately pushes recently seen videos far down the results list. As we have seen, the majority of these will have had a positive rating, and R and ρ will be negatively affected as a result (Table 4).

With the recommender configurations, we wish to compare the two forms of S_{CBR} and to look at the impact of including S_C. Thus, two important questions in our experiments are (a) does altering the method of computing S_{CBR} have an effect? and (b) does integrating S_C into the scoring function have an effect? In Table 5 we examine the difference in four system configuration comparisons: comparing early aggregation CBR with late aggregation CBR (i) and (ii); and comparing CBR with and without content-based recommendation (iii) and (iv). See Table 6 for user and group ratings according to system configuration and Table 4 for ρ and R.

For the base case (i) we find $S_{CBRearly}$ performs better than $S_{CBRlate}$ for ρ, but lower on all other measures. Thus our method for combining profiles into a meta-profile before employing CBR similarity functions does not perform as well as a minimum misery late aggregation approach in terms of utility or ratings, but has a higher accuracy. Integrating S_C (ii) appears to reduce the disparity between the CBR approaches. In this case, ρ is still significantly higher for $S_{CBRearly}$ than $S_{CBRlate}$, but the difference is smaller. We also see the gap lessen for R and ratings, particularly user-session ratings, where $S_{CBRearly}$ performs significantly better, producing the highest ratings for user-session, group-session and group-item. It would appear that $S_{CBRearly}$ with S_C is somewhat of a sweet spot, balancing the individual and group preferences in S_C with the meta-profile used for $S_{CBRearly}$.

Adding S_C to $S_{CBRlate}$ (iii) appears to significantly hurt performance from a user perspective, but not for groups. This is the standout case in which we observed a difference in how users and groups respond to different experimental conditions.

Our results show that ρ is at odds with some of the other measures. For example, configurations using $S_{CBRlate}$ have a higher R but a lower ρ; adding S_C to $S_{CBRearly}$ dis-improves ρ but performance improves across other measures. This intriguing observation suggests it is possible that switching between late aggregation and early aggregation CBR, or indeed using a weighted combination, would enable us to tune the system by trading accuracy for utility.

After our trials, the facilitators participated in a semi-structured interview. Two aspects we focused on were ease of use and perceived usefulness of the

Table 5 Difference in recommender configurations (B-A) with statistical significance at $p < 0.05$ (*) and $p < 0.001$ (**)

	Rec. Config. (A,B)				ρ		R		$r_{session}$		r_{item}	
	S_{CBR}	S_C	S_{CBR}	S_C	User	Group	User	Group	User	Group	User	Group
i	Late	No	Early	No	0.10**	0.11**	−3.00**	−2.00	−0.21*	−0.20	−0.08*	−0.09
ii	Late	Yes	Early	Yes	0.04**	0.04**	−0.77	−1.75	0.23*	0.20	0.05	0.07
iii	Late	No	Late	Yes	0.00	0.00	−1.62	0.32	−0.19*	−0.06	−0.07*	0.04
iv	Early	No	Early	Yes	−0.07**	−0.07**	0.61	0.58	0.24*	0.33	0.07*	0.20

Table 6 Ratings for four system configurations, altering method for computing S_{CBR}, and optionally including S_C

		User-session				Group-session				User-item				Group-item			
S_{CBR}	S_C	+1	0	−1	r	+1	0	−1	r	+1	0	−1	r	+1	0	−1	r
Late	No	0.78	0.22	0.00	0.78	0.64	0.36	0.00	0.64	0.67	0.31	0.02	0.64	0.90	0.00	0.10	0.79
Early	No	0.59	0.39	0.02	0.58	0.50	0.44	0.06	0.44	0.57	0.43	0.01	0.56	0.85	0.01	0.14	0.70
Late	Yes	0.63	0.34	0.04	0.59	0.57	0.43	0.00	0.57	0.59	0.39	0.02	0.58	0.90	0.04	0.06	0.83
Early	Yes	0.83	0.16	0.01	0.81	0.77	0.23	0.00	0.77	0.64	0.35	0.01	0.63	0.94	0.01	0.04	0.90

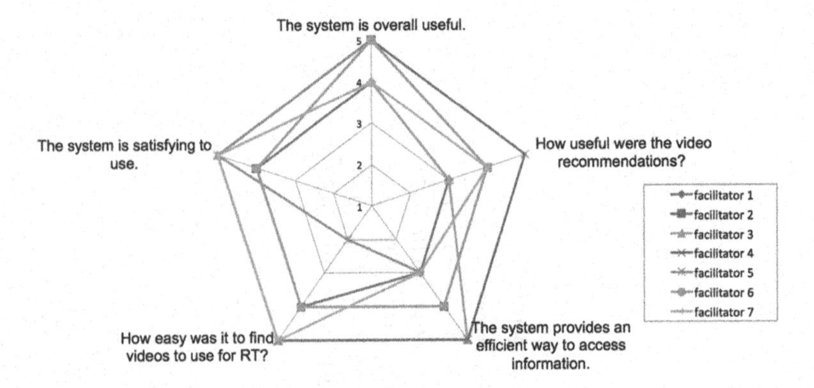

Fig. 7 Facilitators' responses to usability questions

recommendations (see Fig. 7). The responses were positive, with all facilitators agreeing that the system was satisfying and useful. They were also predominantly positive about the ease and efficiency with which they could find those items and the usefulness of those items. Some unstructured feedback emphasised the requirement that speed, efficiency, novelty and accuracy are important, and even the smallest delay or frustration with the system can have a negative effect, unlike other applications where users are perhaps more tolerant.

7 Conclusion

In this chapter we have contributed a novel approach to recommending multimedia content for use in group RT. We provided background and related work in the fields of RT and recommender systems, motivating the work and outlining the limitations of existing approaches. We introduced a series of design discussions and interactions with stakeholders in order to design the interface and functionality of REMPAD, the system we built and tested. The recommendation method in REMPAD is based on a hybrid system using CBR recommendation, content-based recommendation and search to satisfy user requirements. We developed and trialled this system over a period of several weeks in residential care homes and have reported on the efficacy of the proposed approach in terms of accuracy, utility and perceived usefulness.

We find, in general, a higher proportion of positive item ratings for individual users than groups, reflecting the greater difficulty in recommending for groups. We also find that session ratings are higher both for groups and users than individual item ratings. These observations suggest that although it is harder to recommend for groups than individuals, recommending a set or sequence of videos (as in our sessions) may have significant advantages over single recommendations. We see some variance for utility across groups and in general we find accuracy and utility to be consistent between group and user ratings.

Our best performing system configuration uses a combination of an early aggregation CBR and a learning-based content method, possibly reflecting the richest representation of user and group preferences. We also observe that a late aggregation CBR approach with minimum misery appears to favour utility, whereas an early aggregation CBR approach favours accuracy. This potentially gives scope to build a system which is tunable for accuracy versus utility.

Finally in interview feedback from users we learn of a unanimous satisfaction with the system and a reinforcement of the initial requirements for a responsive, accurate, efficient and easy-to-use system to support facilitation of RT sessions. This is perhaps a strong motivation to focus on a utility as evaluation measure for systems in this area.

For the work we have done to date, the features we used are quite specific to the RT application and somewhat heuristic and so for future work we intend to enrich our preference representations further, in particular using text features such as TF-IDF. We will also expand our content collection and investigate the possibility of introducing collaborative recommendation approaches. In order to enrich our content collection we also will explore using other sources of video and other forms of content. It would have been interesting to compare testing of REMPAD with existing systems currently adopted by practitioners in Reminiscence Therapy but there simply are none to compare against.

Overall we find recommending content for use in group RT challenging task and one that is naturally suited to a recommender systems approach. A discussion point that naturally falls out of our work is one of the relationship between modelling group and user preferences. There is evidently an interplay between the two, as, although the ultimate goal is individual therapy participant satisfaction and successful reminiscence, this may not be possible without achieving group satisfaction. Similarly, group satisfaction is likely not attainable without individual satisfaction. This is an important question to address and provides an interesting avenue for future research both for group RT systems and more generally in the area of group recommender systems.

Acknowledgements This work is supported by Science Foundation Ireland under grants 07/CE/I1147 and SFI/12/RC/2289 and by Enterprise Ireland under grant CF/2011/1318.

References

1. André, P., Sellen, A., Schraefel, M., & Wood, K. (2011). Making public media personal: Nostalgia and Reminiscence in the office. In *25th BCS Conference on Human-Computer Interaction* (pp. 351–360).
2. Astell, A., Alm, N., Gowans, G., Ellis, M., Dye, R., & Vaughan, P. (2009). Involving older people with dementia and their carers in designing computer based support systems: Some methodological considerations. *Universal Access in the Information Society, 8*(1), 49–58.
3. Astell, A. J., Alm, N., Gowans, G., Ellis, M., Dye, R., & Vaughan, P. (2006). Developing an engaging multimedia activity system for people with dementia. In *International Workshop on Cognitive Prostheses and Assisted Communication (CPAC 2006)* (p. 16).

4. Ayalon, L., Gum, A. M., Feliciano, L., & Areán, P. A. (2006). Effectiveness of nonpharmacological interventions for the management of neuropsychiatric symptoms in patients with dementia: A systematic review. *Archives of Internal Medicine, 166*(20), 2182–2188.
5. Bennett, J., & Lanning, S. (2007). The netflix prize. In *Proceedings of KDD Cup and Workshop* (Vol. 2007, p. 35).
6. Bermingham, A., O'Rourke, J., Gurrin, C., Collins, R., Irving, K., & Smeaton, A. F. (2013). Automatically recommending multimedia content for use in group reminiscence therapy. In *Proceedings of the 1st ACM International Workshop on Multimedia Indexing and Information Retrieval for Healthcare* (pp. 49–58). New York: ACM.
7. Birren, J. E., & Deutchman, D. E. (1991). *Guiding autobiography groups for older adults*. London: Johns Hopkins Press.
8. Bohlmeijer, E., Roemer, M., Cuijpers, P., & Smit, F. (2007). The effects of reminiscence on psychological well-being in older adults: A meta-analysis. *Aging and Mental Health, 11*(3), 291–300.
9. Bowlby, C.(1993). *Therapeutic activities with persons disabled by Alzheimer's disease and related disorders*. Baltimore: Aspen.
10. Breese, J. S., Heckerman, D., & Kadie, C. (1998). Empirical analysis of predictive algorithms for collaborative filtering. In *Proceedings of the Fourteenth conference on Uncertainty in artificial intelligence* (pp. 43–52). San Francisco: Morgan Kaufmann Publishers Inc.
11. Bridge, D., Göker, M. H., McGinty, L., & Smyth, B. (2005). Case-based recommender systems. *The Knowledge Engineering Review, 20*(03), 315–320.
12. Brooker, D. (2004). What is person-centred care in dementia? *Reviews in Clinical Gerontology, 13*(3), 215–222.
13. Brooker, D. (2006). *Person-centred dementia care: Making services better*. London: Jessica Kingsley Publishers.
14. Chen, Y.-L., Cheng, L.-C., & Chuang, C.-N. (2008). A group recommendation system with consideration of interactions among group members. *Expert Systems with Applications, 34*(3), 2082–2090.
15. Clare, L. (2007). *Neuropsychological rehabilitation and people with dementia*. Hove: Psychology Press.
16. Conway, M. A. (1995). *Flashbulb memories*. Hillsdale, NJ: Lawrence Erlbaum Associates, Inc.
17. Conway, M. A., Collins, A. F., Gathercole, S. E., & Anderson, S. J. (1996). Recollections of true and false autobiographical memories. *Journal of Experimental Psychology: General, 125*(1), 69.
18. Cotelli, M., Manenti, R., & Zanetti, O. (2012). Reminiscence therapy in dementia: A review. *Maturitas 73*(2), 203–205.
19. Hallberg, J., Kikhia, B., Bengtsson, J., Sävenstedt, S., & Synnes, K. (2009). Reminiscence processes using life-log entities for persons with mild dementia. In *1st International Workshop on Reminiscence Systems* (pp. 16–21). Cambridge, UK.
20. Harley, D., & Fitzpatrick, G. (2009). YouTube and intergenerational communication: The case of Geriatric1927. *Universal Access in the Information Society, 8*(1), 5–20.
21. Harris, J. L. (1997). Reminiscence: A culturally and developmentally appropriate language intervention for older adults. *American Journal of Speech-Language Pathology, 6*(3), 19.
22. Harris, J. L. (2012). *SIGnatures: Speaking up about memories*. The ASHA Leader, 20 October 2012.
23. Hesse, B., & Shneiderman, B. (2007). E-Health research from the users perspective. *The American Journal of Preventive Medicine, 32*, 97–103.
24. Jameson, A., & Smyth, B. (2007) Recommendation to groups. In *The Adaptive Web* (pp. 596–627). New York: Springer.
25. Jönsson, L., Jönhagen, M., Kilander, L., Soininen, H., Hallikainen, M., Waldemar, G., et al. (2006). Determinants of costs of care for patients with Alzheimer's disease. *International Journal of Geriatric Psychiatry, 21*(5), 449–459.
26. Kitwood, T. (1997). Dementia reconsidered: The person comes first. In *Adult Lives: A Life Course Perspective*. Buckingham: Open University Press.

27. Kitwood, T., Bredin, K., et al. (1992). Towards a theory of dementia care: Personhood and well-being. *Ageing and Society, 12*(3), 269–287.
28. Konstan, J. A., & Riedl, J. (2012). Recommender systems: From algorithms to user experience. *User Modeling and User-Adapted Interaction, 22*(1–2), 101–123.
29. Kuwahara, N., Kuwabara, K., & Abe, S. (2006). Networked reminiscence content authoring and delivery for elderly people with dementia. In *Proceedings of International Workshop on Cognitive Prostheses and Assisted Communication* (pp. 20–25).
30. Lops, P., de Gemmis, M., & Semeraro, G. (2011). Content-based recommender systems: State of the art and trends. In *Recommender Systems Handbook* (pp. 73–105). New York: Springer.
31. Lorenzi, F., & Ricci, F. (2005). Case-based recommender systems: A unifying view. In *Intelligent Techniques for Web Personalization* (pp. 89–113). New York: Springer.
32. Masthoff, J. (2011). Group recommender systems: Combining individual models. In *Recommender Systems Handbook* (pp. 677–702). New York: Springer.
33. McCarthy, K., McGinty, L., Smyth, B., & Salamó, M. (2006). The needs of the many: A case-based group recommender system. In *Advances in Case-Based Reasoning* (pp. 196–210). Heidelberg: Springer.
34. McCarthy, K., Salamó, M., Coyle, L., McGinty, L., Smyth, B., & Nixon, P. (2006). Cats: A synchronous approach to collaborative group recommendation. In *Proceedings of the Nineteenth International Florida Artificial Intelligence Research Society Conference*, Melbourne Beach, FL (pp. 86–91).
35. McCarthy, K., Salamó, M., Coyle, L., McGinty, L., Smyth, B., & Nixon, P. (2006). Group recommender systems: A critiquing based approach. In *Proceedings of the 11th International Conference on Intelligent User Interfaces* (pp. 267–269). New York: ACM.
36. Morgan, S. (2000). *The impact of a structured life review process on people with memory problems living in care homes*. Bangor: University of Wales.
37. Norris, A. (1986). *Reminiscence: With elderly people*. London: Winslow.
38. O'Connor, M., Cosley, D., Konstan, J. A., & Riedl, J. (2002). Polylens: A recommender system for groups of users. In *European Conference on Computer Supported Cooperative Work, ECSCW 2001* (pp. 199–218). Berlin: Springer.
39. O'Rourke, J., Tobin, F., O'Callaghan, S., Sowman, R., & Collins, D. (2011). 'Youtube': A useful tool for reminiscence therapy in dementia? *Age and Ageing, 40*(6), 742–744.
40. O'Shea, E., & O'Reilly, S. (2000). The economic and social cost of dementia in Ireland. *International Journal of Geriatric Psychiatry, 15*(3), 208–218.
41. Payne, J. C. (2011). Cultural competence in treatment of adults with cognitive and language disorders. The ASHA Leader.
42. Pazzani, M. J., & Billsus, D. (2007). Content-based recommendation systems. In *The Adaptive Web* (pp. 325–341). New York: Springer.
43. Pu, P., Chen, L., & Hu, R. (2011). A user-centric evaluation framework for recommender systems. In *Proceedings of the Fifth ACM Conference on Recommender Systems* (pp. 157–164). New York: ACM.
44. Sarne-Flaischmann, V., & Tractinsky, N. (2008). Development and evaluation of a personalized multimedia system for reminiscence therapy in alzheimers patients. *International Journal of Social and Humanistic Computing, 1*, 81–96.
45. Schacter, D. L. (1996). *Searching for memory: The brain, the mind, and the past*. New York: Basic Books.
46. Shani, G., & Gunawardana, A. (2011). Evaluating recommendation systems. In *Recommender systems handbook* (pp. 257–297). New York: Springer.
47. Smyth, B. (2007). Case-based recommendation. In *The Adaptive Web* (pp. 342–376). Berlin: Springer.
48. Vargas, S., & Castells, P. (2011). Rank and relevance in novelty and diversity metrics for recommender systems. In *Proceedings of the Fifth ACM Conference on Recommender Systems* (pp. 109–116). New York: ACM.

49. Wallace, J., Thieme, A., Wood, G., Schofield, G., & Olivier, P. (2012). Enabling self, intimacy and a sense of home in Dementia: An enquiry into design in a hospital setting. In *SIGCHI Conference on Human Factors in Computing Systems* (pp. 2629–2638). Austin, Texas: ACM.
50. Williams, H. L., Conway, M. A., & Cohen, G. (2008). Autobiographical memory. *Memory in the real world* (p. 21). London: Psychology Press.
51. Wimo, A., & Prince, M. (2010). *World alzheimer report 2010: The global economic impact of dementia*. London: Alzheimers Disease International (ADI).
52. Woods, B., Spector, A. E., Jones, C. A., Orrell, M., & Davies, S. P. (2005). Reminiscence therapy for dementia. *Cochrane Database of Systematic Reviews*, (2). doi: 10.1002/14651858.CD001120.pub2. Art. No: CD001120.
53. Yasuda, K., Kuwabara, K., Kuwahara, N., Abe, S., & Tetsutani, N. (2009). Effectiveness of personalized reminiscence photo videos for individuals with dementia. *Neuropsychological Rehabilitation, 19*(4), 603–619.

Using Ontologies for Managing User Profiles in Personalised Mobile Service Delivery

Kerry-Louise Skillen, Chris Nugent, Mark Donnelly, Liming Chen, and William Burns

1 Introduction

In recent years, user dependence and reliance on smart-phone services has significantly increased in relation to advancements in developing miniaturised technologies. Such technologies focus on the use of smart-phones, tablets or wireless sensors, which are embedded into our surroundings and used to realise the environments that we inhabit on a daily basis [1]. Ambient Intelligence (AmI) refers to the vision whereby people are surrounded by an environment that is capable of responding to the needs of the people within it, and without their knowledge of it occurring [2]. AmI can be described as a user-centric paradigm that works invisibly to aid a particular user. Research within AmI has expanded and aims to deliver reliable assistance in the form of adaptive environments. Within AmI, pervasive technology has gained significant influence over how we use smart services. Pervasive computing systems have the goal of providing assistive technology 'on-demand' which is embedded seamlessly into our surroundings. Such systems have been considered as key enablers within the field of pervasive environments and mobile service delivery. An environment can only be considered smart if the underlying systems are able to understand a user's activities and behaviours and act upon these [3]. The vision of 'invisible' technology has sparked the growth of personalised context-aware services, in the form of technology that can meet user needs at different times, based on their current surroundings [4].

K.-L. Skillen (✉) • C. Nugent
University of Ulster, London, UK
e-mail: k.skillen@ulster.ac.uk; cd.nugent@ulster.ac.uk

M. Donnelly • L. Chen • W. Burns
University of Ulster, Northern Ireland, UK
e-mail: mp.donnelly@ulster.ac.uk; liming.chen@dmu.ac.uk; Burns-WP@email.ulster.ac.uk

A. Briassouli et al. (eds.), *Health Monitoring and Personalized Feedback using Multimedia Data*, DOI 10.1007/978-3-319-17963-6_13

Recent research within this area has highlighted a key challenge in providing user personalisation through context-aware services. User personalisation refers to the approaches used to enable the delivery of customised information or assistance to a particular user, based on their unique preferences, needs or wants over time [5]. Providing such a service raises several difficulties. These include how to effectively manage and provide a high quality of service, how to accurately represent and characterise unique human traits, and how to enhance the delivery or feedback of personalised content to the user, via the advent of smart, mobile-based technologies [3]. One of the most important elements of any personalised service is its ability to provide the correct information to that particular user, and adapt itself according to the changes in human lifestyles and/or behaviours over time. While context-awareness is an important factor in creating applications that are adaptable, it is not sufficient in providing an accurate representation of a user's changing characteristics when used alone. Similarly, the increase in user dependence on mobile-based technologies has led to an increase in the need to provide adaptable services 'on-demand'. Consequently, this highlights the necessity for user personalisation, particularly for the delivery of enhanced mobile service delivery [6, 7].

2 Context, Context-Awareness and Context Reasoning

Context and context-awareness are two key concepts within the area of AmI. Both terms were originally introduced in 1994 by Schilit and Theimer [8]. Their work described context in terms of a person's location, the change in objects surrounding them and the appearance of nearby people. Context plays an essential role in the running and management of a successful smart environment, where contextual information is used to aid user's activities at the right time and the right place. A system is perceived to be fully context-aware when it can extract, use and interpret relevant contextual information, and then further adapt this information to the current context of use [9]. Environments equipped with context-aware abilities can supply a user with information concerning their surrounding environment and this information can be used to guide technologies to adapt to changing situations.

Contextual information is a fundamental element in any pervasive computing system, in particular within mobile-based applications. To provide the user with relevant information at the right time, the system must have a clear understanding of what the context is. A context-aware system can then decide when an action is to take place based on the information it has gathered. This could be the action of alerting a carer remotely via a mobile phone if their patient has fallen or the provision of personalised reminders (for example, to remind a user to make an important hospital appointment) while they are out of their home. The success of developing an accurate context-aware system relies heavily on its ability to reason about raw data and infer meaningful information from this data. The imperfect characteristics of context information highlight a need for some form of reasoning within a context-aware, mobile system. Human errors can decrease the accuracy of contextual information [10]. Context reasoning is a central component and can be used as a decision-making mechanism in any context-aware system. A key challenge exists within current research to enhance current approaches to reasoning for the purposes of providing a better representation of a user and their surrounding environment. Many existing research studies focus on accurately representing different users, based on where they are at any particular time, and delivering services to them that reflect their needs.

Mark Weiser [11] first introduced the notion of pervasive technology in the early 1990s and described the term as invisible technology and devices that are integrated into the daily lives of users, normally without the user being aware of it happening. The emergence and rapid development of mobile technologies and adaptable user interfaces has sparked research into enhancing user-based applications that are personalised to suit human behavioural changes. This increase in context-aware, mobile-based technologies has led to an increase in user dependence upon such technologies; which effectively fuels the requirement to develop new methods of user personalisation to cater for the growth in demand. Consequently, this has enabled an increase in both the demand and need for user modelling for service personalisation. With a surplus of research conducted within smart home environments, it is evident that there is significantly less work within the area of adaptable, context-aware assistance for users that move between different environments [3]. Related research emphasises the development and use of smart assistance primarily within closed-world setting (i.e. within a confined environment such as a home). People naturally lead active lives and are always moving around, therefore changing their surroundings. The technology should therefore move with each dynamic situation and subsequently adapt in different places. Issues do exist in providing an accurate representation of a user within changing situations and in handling multiple and varying user preferences. In addition to this, challenges still exist in modelling conflicting user needs or wants within smart environments [9]. As the contextual data surrounding a user will change rapidly from one environment to the next, the mobile-based services offered must adapt accordingly to the user's time, location and environment and provide personalised help when required.

3 Context-Aware Modelling and Management

A context-model is primarily used to define the context of a situation, store the relevant data and manage this data in a machine-readable form [12]. It can be described as a schema, which precisely defines the relationships between various entities, including any entity-relationships, instances, or properties that may exist. By modelling contextual data within a defined model, we can then perform context reasoning on this data to check for inaccuracies and illogical occurrences (e.g., a wine cannot be both a red wine and a white wine at the same time). The quality of a context model relies heavily on how it can represent the information that it receives, therefore by creating a richer model we can enable a more accurate service within a context-aware environment that can both be maintained and evolve with technological advances and increased knowledge. The development of flexible, powerful and reusable context models is still a challenge in the existing research [12].

To successfully adapt contextual information to a context-aware system, some form of context modelling is required. Existing approaches to context modelling differ in the method by which the context is captured, how expressive the model is and how well they can reason about data in order to extract some meaningful information for providing a context-aware service to a user [10]. Such approaches include the development of key-value context models, graphical-based models, logical-based models, object-oriented models and finally, ontological-based models [10]. While each model has the same goal of capturing relevant contextual information, they differ in the processes they use to capture, analyse and reason about this information to provide context-aware services.

A key-value context model can be described as the simplest way to model any form of context data, through the use of key-value pairs. Key-value models are commonly used within distributed service frameworks and use attributes to describe the capabilities of a service and these are then matched with algorithms. These models, while easy to implement, are unable to provide a sufficient data structure, which can be used to accurately retrieve context from an environment.

Graphical context models have commonly been used (using the Unified Modelling Language) as a generic structure that can be used to model context in environments. While these models can be used to successfully represent context, they are not good at handling ambiguous data sets and struggle with the incomplete data and partial validation. Object-oriented context modelling makes use of object-oriented techniques and has the benefits of supporting reusability and inheritance. These models are based on the use of objects and the relationships between them and use objects to represent various context types [13]. Logic-based models can also be described as a series of facts, expressions and rules that are used to create individual context models within an environment.

Recent advancements have introduced the concept of ontology-based context models [14], which make use of a series of concepts and relationships to represent contextual information inside a data structure. They are a popular choice due to their expressiveness and support for reasoning. Ontologies are able to successfully specify a context model's core concepts along with sub-concepts and facilitate both knowledge sharing and reuse within pervasive computing applications. The use of ontologies to model context has been increasing and solves previous problems from other models, such as dealing with inaccurate and inconsistent data, validating data, and can be applied to various context-aware applications and mobile-based services [15].

4 User Modelling and Representation

Within context-aware environments, users can differ in terms of their needs from that environment and also behavioural patterns can change over time. As a result, context modelling alone is insufficient for providing a personalised service for a user; this allows the notion of ontological modelling to be explored and used to enable adaptable service delivery and feedback.

User modelling can be described as the process of characterising users and their needs of a specific application domain and creating a data structure; that is, a user model, that can then capture these user characteristics and behaviours [16]. Within any mobile, context-aware system, the characteristics of a user model are the core components that facilitate a higher quality of personalised services being delivered to a user. A successful user modelling system must be able to handle the ability of making inferences (or assumptions) about a user, even if the information gathered is incomplete, inaccurate or mistaken. There have been a number of studies based on the requirements of an efficient user modelling system [17, 18]. Such requirements include accurate user profiles to enable precise knowledge sharing, domain independence to facilitate cross- system capabilities, expressiveness and extensibility of user models and enhanced privacy particularly when handling sensitive user information among different applications. Over recent years, two main approaches have emerged with the domain of user modelling; these can be categorised into data-driven and knowledge-driven [14]. While both approaches are similar in their attempt to define a user's profile, characteristics and preferences, and each have the goal of maximising user satisfaction, they differ in the process that they employ to fulfil these goals.

4.1 Data-Driven User Modelling

Data-driven approaches to user modelling typically focus on the use of statistical or probabilistic analysis to explore the various relationships between the data and user properties [19]. These approaches are often referred to as learning-based approaches as they use machine learning and data mining techniques to detect relationships between the environment surrounding a user and the sensors used to collect user and context data. Predictions from such models have been used to adapt the behaviour of context systems. Some well-known examples of data-driven models have included the use of Bayesian Networks or Markov Models [19]. These models make effective use of learning-based techniques to modelling by learning user needs and highlighting relevant correlations between sensor data and context. Predictive models such as Bayesian Networks or Decision Trees make predictions about a user's wants and these are then used to adapt the system behaviour to perform actions for the user or to provide user recommendations based on previous knowledge [20]. User profile information can be declared as either static or dynamic. Static information usually consists of a user's personal information (i.e. information that does not frequently change over time) and the dynamic information is information that will change over time (i.e. time, location, preferences). Static information has previously been modelled using Attribute-Value Pairs, first introduced in [21] where attributes are shown as facts that represent the user and the environment.

One of the key issues with modelling static information alone is its inability to provide a precise representation of a person, ultimately due to the fact that human behaviours and actions change over time. Data-driven approaches typically focus on the use of static information. Due to this limitation, they are therefore not suitable for the domain of user profile management as there is a dynamic component within user models that needs to be tracked. While data-driven modelling approaches are popular within the area of pervasive computing, the focus has shifted towards the use of knowledge-based modelling approaches. This has come as a result of the increase in the use of mobile (smart-phone) technologies. Subsequently, this increase has defined a need for knowledge engineering and modelling due to the complexity involved in learning and representing user needs. This is particularly apparent in dynamic environments where users are moving from one situation to the next and therefore their needs change accordingly.

4.2 *Knowledge-Driven User Modelling*

Knowledge-driven approaches develop models of users and dynamically match the user to the closest model available. They provide a representation of users via the use of logical rules and apply reasoning engines (used to assume new information based on previous knowledge) for the purpose of inferring personalised services from the surrounding situation, user profiles and sensor inputs [22]. These approaches are based on the use of logic or ontological reasoning and are commonly referred to as specification-based approaches to modelling. These approaches make use of expert knowledge and reason about that knowledge with data input by various surrounding sensors. One of the issues in knowledge-based user modelling is how to accurately represent knowledge (i.e. how to provide an accurate profile of a person). Knowledge representation languages have frequently been used within the area of context-aware computing and include languages such as XML, UserML, RDF and OWL [23]. XML has been a widely used choice of model used due to its simplicity and reusability. UserML can be defined as a common and powerful user model exchange language, which facilitates communication between different context-aware applications. Nevertheless, both UserML and XML are meta-languages that don't formally define knowledge semantics (i.e. meaning), which can lead to difficulties when reasoning about important user data [24]. Two more commonly used representation languages include RDF and OWL. These languages exhibit strength in their expressivity when representing user information.

Data-driven approaches remain a popular choice for modelling domain knowledge, as they are less labour intensive than hand-building knowledge bases. They provide an accurate representation of user data; however, they are usually user-specific (i.e. focus on one particular user at a time), which can cause problems when trying to accurately model multiple users (alongside multiple, changing preferences or behaviours) within pervasive environments. Furthermore, such approaches fail to provide a scalable solution to modelling, particularly when compared to more user-centric knowledge-based modelling approaches that have grown significantly in recent years. Existing research highlights a key issue involving how to effectively represent human behavioural changes via the use of knowledge-based user modelling techniques. Further to this, data-driven approaches to modelling tend to focus on providing a representation of the 'average' person, rather than identifying the unique characteristics of different people. Consequently, this can prove useless within the context of an adaptable, pervasive environment [25]. Due to their powerful ability to accurately represent knowledge and make inferences based on existing information, ontological-based approaches to modelling are steadily becoming a popular standard within the area of context-awareness. Advantages of this approach include the ability to reuse knowledge, share existing knowledge among other domains and reason about data to extract meaningful information [26]. These key elements make ontologies an ideal solution for use within mobile-based, personalised user services. In particular, ontologies are able to efficiently deal with both contextual information and user information by dividing themselves into upper-level and lower-level hierarchies, enabling them to be used across different application domain [3]. On the other hand, these approaches require expert knowledge. For example, to initially populate the ontology models with basic information concerning a user (i.e. how to create unique user profiles). While this is important, using ontological models to represent people could overcome one of the existing issues of interoperability among different context-aware applications. Interoperability is a common issue with context-aware computing applications due to the complexity involved in developing models that can be reused across different application platforms [15]. After a careful conceptualisation and engineering process, ontologies are effectively able to overcome such an issue. They can be used to enable both syntactic interoperability (i.e. to allow two or more applications to exchange user information) and semantic interoperability [i.e. to enable applications to exchange meaningful information) of data and information [27].

Significant effort is required in the development of any context-aware, adaptive system. A knowledge base consisting of the system users, domain knowledge and context must be managed and integrated seamlessly into the core of the system. While the user is the core aspect when considering the development of a more personalised system, the contextual information that is received must be managed, maintained and updated over time. Similar to the idea of a database management system, user profile or model management occurs within the management layer of any context-aware system, as shown in Fig. 1. It is here that user profiles and models are stored and maintained. Above this, you can expect to find an adaptation layer, which focuses on learning user behaviours or interests, and feeds new information

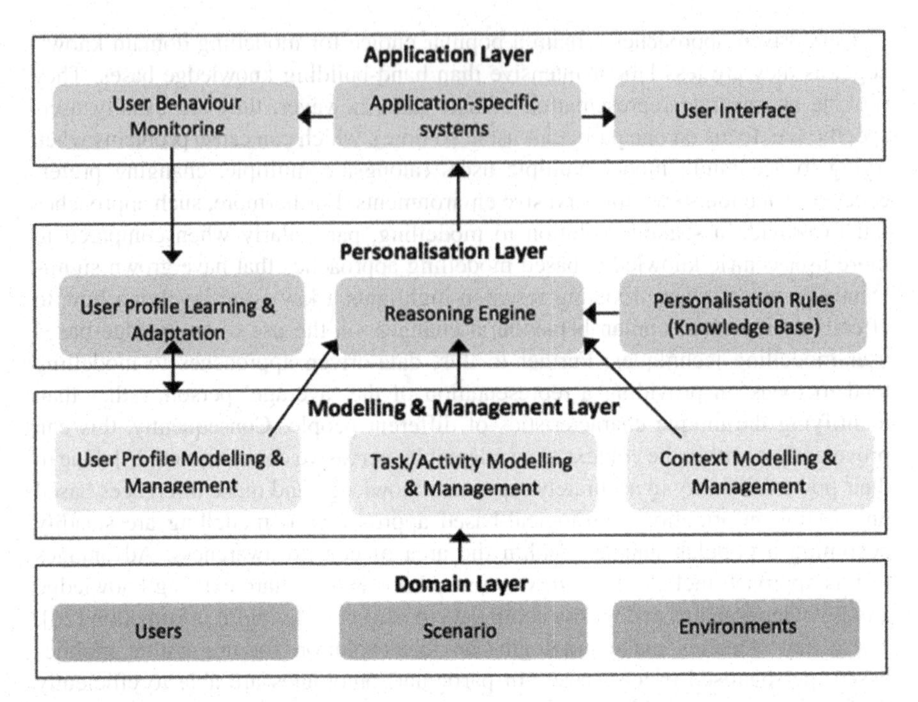

Fig. 1 A hierarchy of layers used to model users, activities and context within the context of adaptive applications. The key emphasis here is the use of user modelling, adaptation, reasoning to enable application personalisation

back to the management layer where existing profiles/models can be updated. By storing the user information within this layer it is possible to reason, query and infer new information in addition to reusing information for different application purposes. While a typical database management system primarily contains static information, dynamic user profile information must be maintained and updated frequently. This architecture was conceptualised and designed for use in context-aware applications, where user profiles are modelled and adapted over time. Further information on this work can be found in [28].

5 Ontological User Profiling for Personalisation

Ontology-based user modelling has grown in recent years and the process involves the construction of a number of concepts related to a particular domain (for example, the healthcare domain, structuring patient data), along with any number of properties or relationships that link these concepts.

Ontologies are seen as a 'representation vocabulary', where these user concepts are structured in a taxonomy based on various aspects. An ontology originates from the meaning of 'being' and has been used for several years by researchers to try and make sense of their environment and their surroundings. Typically, ontology models are structured into a hierarchical taxonomy, where the key domain concepts within that area are found. Ontological models can be used by logic reasoning mechanisms to deduce high-level information from raw data and have the ability to enable the reuse of system knowledge. This is of particular importance when modelling user aspects that can be remembered and reused for future services. For example, for an ontology concerning a person, some key concepts would include the person's personal information, their preferences or health conditions. Ontological modelling can therefore be referred to as the process of specifying these domain concepts, properties and the connections or relationships between these. As these concepts are organised into a hierarchical structure in terms of the properties they share, each concept (or class) may have one or more parent classes and thus a hierarchy of related data is formed [14]. Both data and object properties exist within each class, which describe features of that particular class and any restrictions or rules placed upon them [29].

Modelling via ontologies follows a specific process, which can be deconstructed into various key steps. This includes firstly (1) the analysis of user characteristics and needs in context-aware environments (i.e. gathering what the user wants, their personal attributes and behavioural characteristics), moving onto (2) the creation of various interrelationships among the ontology concepts, namely, users, environments and services. The next step (3) is to establish the key concepts that can model and represent these user aspects and also declare what properties these concepts may have. Once identified, the process would then involve (4) the classification of such concepts and properties into a hierarchy and finally (5) the use of ontology development tools (such as Protégé [30] to encode these user concepts and relationships, and represent them in a formal ontology language for use in adaptive applications.

Personalisation (particularly for smart-phone services) involves the use of several entities. These include the users themselves, the environment in which the user inhabits, the surrounding contextual information and the interactions/causal relationships between these entities [3]. User profile modelling can be viewed as a core component within personalisation for mobile-based services. A user model is a data structure that holds the characteristic attributes of users. This data structure is usually referred to as a model that serves as a template for creating multiple instances of user profiles for different people. These profiles are 'digital representations' of data associated with a particular person [31]. To achieve such a service that is both context-aware and user dependent, mobile-based services need to focus on the needs, wants and behaviours of the user and therefore consider user expectations from the system [31]. One of the challenges for a pervasive computing system is to decide the 'right' information and provide it for the 'right' user at the 'right' time in the 'right' way [16]. To realise such on-demand services, it is necessary to create user profiles within the system, with one user profile corresponding to one particular user. A user model serves as a template (or data structure) that maintains common user properties that are considered relevant for a user-adaptive application, including information such as user preferences, interests and behaviours [22]. For each individual, their user profile can be created by instantiating a user model with the individual's personal attributes, for example, their preferences, capabilities and interests. The generated user profiles can be saved and therefore form the user profile repositories. User models are commonly referred to as 'behaviour' models as they monitor user actions and try to discover various action patterns in order to learn user behaviours and use these behaviours to personalise context-aware applications. A well-known definition of the term 'user model' comes from Kobsa in [32] who defines a user model as: "A collection of information and assumptions about individual users which are needed in the adaptation process of systems to individual actions of users."

Nowadays, with the rapid development of user-adaptive, mobile systems and applications, there is a growing need to build user models that can be reused across different application domains. An example of this is presented in [22], which highlights an issue relating to how to improve future user model interoperability, where user adaptive systems within an environment could further co-operate and interact with each other. Within a typical context-aware system architecture, both user profiles and models are managed in the management layer. In the context of pervasive computing, user modelling is a requirement to enable the full potential of user models for mobile-based service delivery and adaptation.

5.1 Characterising Users Within Pervasive Environments

When personalising both the user's experience and the service that is delivered to them, modelling their personal profiles is a necessity. Personalisation can be presented in the form of adaptive user interfaces (particularly for mobile-based technologies), personalisation of information retrieval or service personalisation of content for context-aware applications.

As previously described, user profile modelling acts as a core component to enable the personalisation of content delivery in mobile-feedback applications. Nowadays people have busy, changing lifestyles, where they move between different dynamic environments seamlessly. People naturally move from their home environment to their workplace for example. The technology that exists must be able to cater for typical human behaviours, particularly in such a fast changing world. As people perform activities in different places, their behaviours, actions and needs also change accordingly. For example, people who are going to their workplace will normally be focused on planning the day's tasks, whereas people travelling on holiday will be concerned with directions or visiting tourist attractions.

Mobile-based technologies need to cater for the changes in variable human needs and wants, and also adapt to the changes over time. For pervasive applications that are user and context-aware, the technology needs to 'understand' the user as a whole, and delivery assistance that is tailored to both their environment and themselves [28]. Building specific user models (containing profiles) can be effectively used to facilitate the personalisation of adaptable applications. Human behaviours and concepts can be broken into various sub-levels of granularity. For example, a person may have a health condition, which can be split into a 'cognitive condition' or 'physical condition' which subsequently can be broken down further into the condition type or severity. The user's profile must model as much of that person as possible. The user profile should cover the person, their environmental and temporal contexts to build an extensible model of user attributes, which can be modified over time.

Core aspects that can be modelled in the ontology include the user's contexts, including their preferences, needs, wants, attributes and habits. Temporal contexts refer to user locations and time-specific activities. Modelling what activities a person performs at what times/locations can help tailor personalised mobile-services according to the overall context. Finally, environmental-based contexts refer to conditions surrounding the person. The core relationships between each of these contexts can also be modelled within the ontology, consequently building a comprehensive abstract view of different users.

A recent research study, described in [3] focused on the use of key focus groups and various questionnaire techniques to extract, analyse and identify a core set of user's attributes to enable the creation of an extensible, dynamic user profile ontology model. This work implemented a top-down design approach to deduce low-level user concepts from higher-level concepts. As shown in Fig. 2, the User Profile Ontology was designed to be extensible and interoperable among different application domains, with its particular focus on the personalised delivery of multimedia feedback within assistive applications (discussed further in [3]).

Figure 2 presents a fragment of the User Profile Ontology, where the key User-Profile class refers to the core concept within the model. It is within this class that all of the information concerning a particular user is held. Such information can include medical details, personal information, preferences, habits and characteristic traits. By building a relationship between these concepts, the ontology can be reasoned to infer new information, which can then be used as a basis for service personalisation.

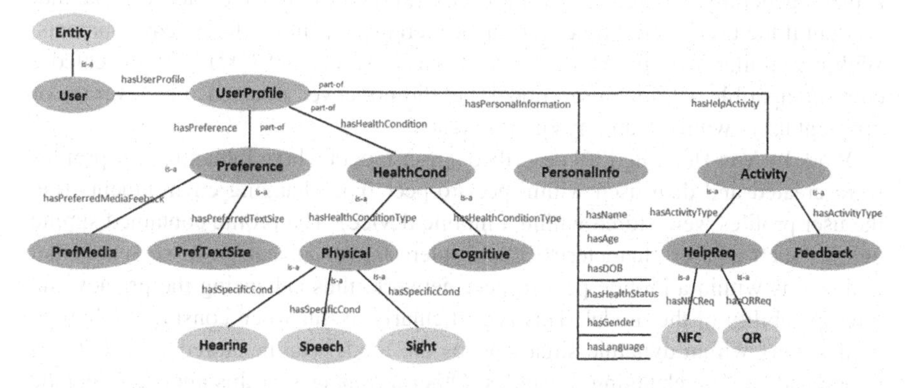

Fig. 2 An excerpt of the relationships classes listed in the User Profile Ontology developed in [3]

In this model, each class will contain several sub-classes, linked via properties. For example, a user will have a Preference class. This class may have various subclasses, which could include the PreferredMediaType and PreferredLanguage. To link these sub-classes, the PreferredLanguage class will be associated with the UserProfile class via an object property called hasPreferredLanguage (also linked to the User class). By adding different levels of granularity, mobile-based services can be personalised to suit a person's unique needs over time.

5.2 *Ontological User Models and Profiles*

Currently a very active research field, user modelling can be viewed as one of the most challenging and fundamental components to configure when creating adaptive, context-aware applications [28]. Ontology-based modelling is increasing in popularity, particularly within the area of mobile service personalisation. Two of the main advantages of using such an approach include their interoperability among different applications and ability to enable knowledge sharing and reuse [33].

Within existing literature, there are a wide variety of ontology-based user models and approaches available used primarily for context-aware personalisation. Some examples of the ontological models include work in [27], which presented a generic ontology-based user modelling architecture named OntobUM and GUMO, which was introduced by Heckmann [34]. The idea of dynamic user profiles was researched in [35] with the development of the User Profile Ontology with Situation-Dependent Preferences Support (UPOS). Defined within OWL, the ontology consisted of a series of sub-profiles, which categorised aspects of the user such as their characteristics, preferences and interests and stored them. These sub-profiles also enable meaningful conditions to be attached to them along with an extensible user attribute vocabulary. According to each condition, a corresponding sub-profile is added that contains all personalisation services or actions (e.g. put the mobile-phone profile onto silent when in a work meeting). One of the key issues with UPOS is that it generally only refers to a single context dimension, which doesn't take into account if the user has multiple, varying preferences within different environments. Within a similar area, the Unified User Context Model (UUCM) [27] presented a cross-compatible ontology user model, but did not cover an extensible coverage of different users within changing environments.

Work by Von Hessling [36] described a user model where semantic user profiles were created and then used within peer-to-peer type changing environments (i.e. the user profiles were stored within a mobile device). The profile contained simple entities such as user characteristics and interests. An advantage of such a model is that it is within a mobile peer-to-peer network, thus enhancing the privacy and interoperability of the model. This is particularly useful when considering how to model users within dynamic situations, as the model can be stored onto a smart-phone and capture changing dynamics. Other advantages of this approach include its ability to produce a scalable solution to user modelling as the management of the user profile is conducted within their own mobile devices and so no central store is needed. One of the key challenges highlighted with this model is that user profile information must be manually organised. Within a context-aware setting, this should not be the case, hence a key challenge that should be considered is how to automatically manage user profile information after it is captured and how to routinely reason upon this to provide personalised services to a user 'on-demand.'

6 Challenges in Mobile Service Personalisation

There are several components that are required to work together to form a context-aware system. Firstly, the system needs to know what is happening around it so relevant contextual information from the environment is gathered, stored and processed. As previously discussed, context alone is not enough when trying to meet the needs of a user within a smart environment hence some form of user modelling and representation is required. Nevertheless, the main aim of any context-aware system is to be adaptable and successfully modify its content to suit user needs, via service personalisation.

There are currently several challenges that still have to be addressed when aiming to personalise a service for different users, particularly when using mobile-based technologies. People tend to travel from different pervasive environments, and thus the technology must be able to adapt to these changes and present personalised services in relation to these. In order to personalise a service, service presentation and functionality are two core aspects that need to be adaptable to user behaviours. Existing problems include how to enhance the methods used to analyse user information in pervasive environments (i.e. modelling the travelling user) and how to infer (or assume) what the user needs at any time. Challenges also exist in maintaining an accurately personalised delivery of content in real time, and how to make changes to content upon learning differences in user behaviours over time [3].

In the next Section, we discuss the use of a 'Help-on-Demand' application and associated approaches to user modelling, with the aim of overcoming some of the challenges associated with current work within this field.

7 'Help-on-Demand' Personalisation

The work within this Section presents research that has previously been undertaken by the authors of this paper. This includes the development of a user modelling and personalisation approach for the personalisation of assistive services to users within pervasive environments.

When developing a system that is personalised 'on-demand', there are several core components that enable an accurate and efficient service to be delivered to the user at any time. Previous work from [3, 28, 37] has discussed in detail the use of ontologies as a means to provide adaptable, user-based services. In particular, the work of the authors has focused on the development of an architecture and mobile-based service called 'Help-on-Demand' (HoD). Working in collaboration with the EU-funded research project MobileSage [38] , this work has developed a novel user ontology and personalisation mechanism that can be integrated into various context-aware applications, for the provision of personalised media feedback via mobile technologies. MobileSage aims to provide users (older users in particular), with personalised assistance through the advent of mobile services. The ultimate goal is to enhance the quality of living through the use of personalised navigation, reminders and adaptable interfaces. The project aims to facilitate the personalisation of context-aware services as people travel between environments.

As presented in Fig. 3, the core components that enable this level of personalised media delivery include the User Ontology Models (including User Profiles) for characterising different users, the Personalisation mechanisms, which contain rules and reasoning services and the Adaptation service that can modify content and user interface design over time. Other components include the front-end mobile application, where the user interacts with the system and the back-end content management system, which is responsible for storing and retrieving the correct content to be delivered to the user.

One of the key features of this work is its interoperability among several domains. Ontology-based models incorporate a number of generic user concepts, resulting in

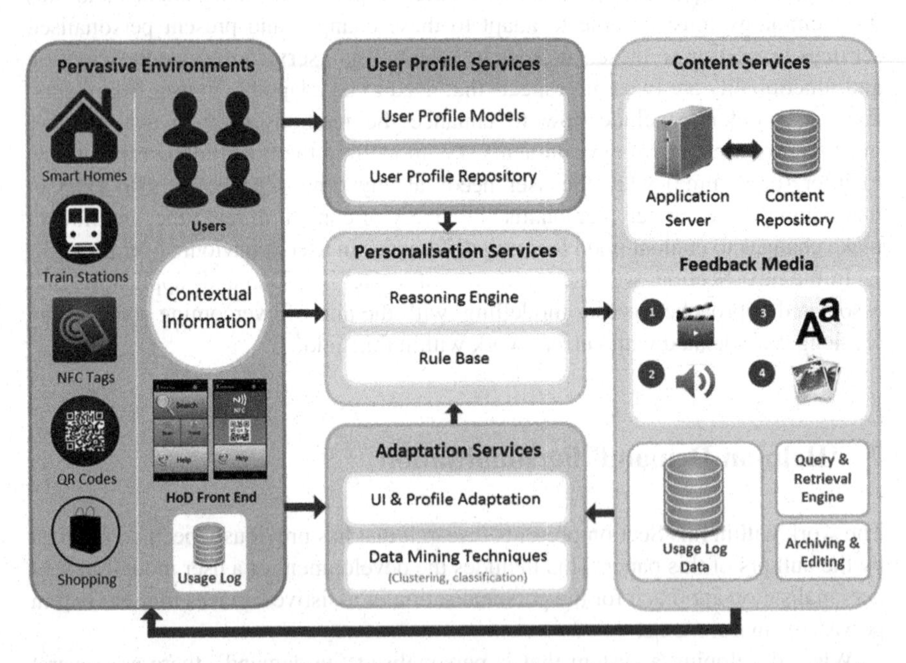

Fig. 3 The service-oriented system architecture, highlighting the core components required for the personalisation of mobile-based services in context-aware environments

an extensible, cross-compatible structure that can be used across many application areas. These user models can be used to personalise the healthcare information provided to patients according to their medical records, activities and previous diagnosis, or be tailored to provide personalised travel assistance to users as they move between different environments, enhancing their activities of daily living. The User Profile Ontology from [3], along with other existing models, formally defines user concepts using semantic (meaningful) information. Another key component from this work is the use of a personalisation mechanism via rules and reasoning.

7.1 Rule-Based Personalisation and Reasoning

From existing literature, personalisation and reasoning can be divided into several sub-categories. Such categories include the use of user profiling for personalisation, content filtering to include collaborative and rule-based approaches, and content modelling, as reviewed in [39]. Figure 4 illustrates some of the key categories associated with the personalisation of user-based services in mobile technologies. In particular, the popularity of rule-based personalisation has grown in recent years, and works efficiently alongside the use of ontological user modelling and profiling. Rules make use of domain level knowledge to enable the development of causal relationships between user profiles and service delivery (for example, the

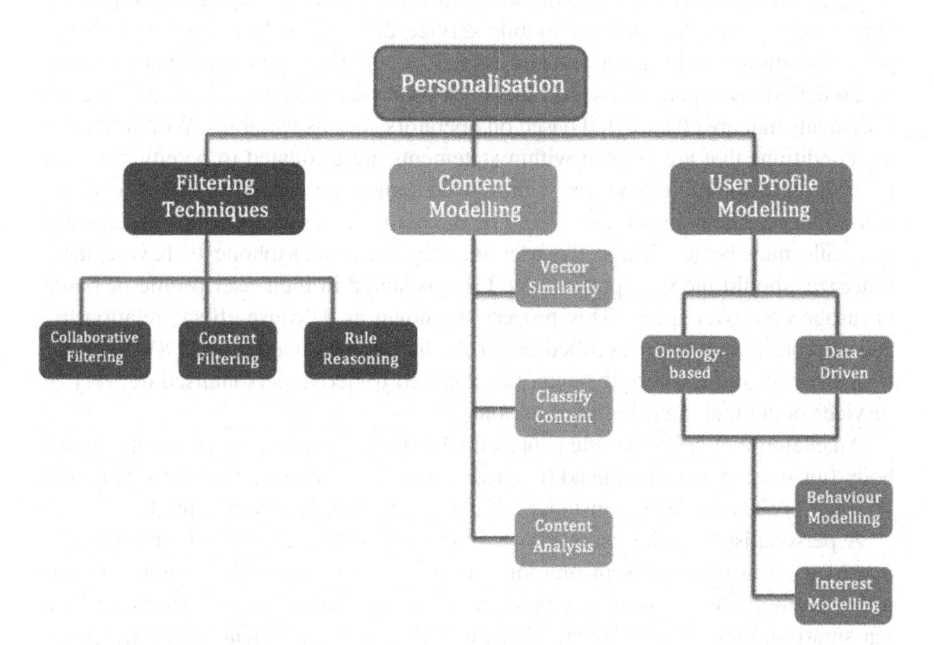

Fig. 4 Extract of the key categories associated with the personalisation of user-based services within pervasive environments

delivery of personalised media content through smart-phones) [3]. Personalisation via rules allows mobile-based systems, in particular, to specify rules based on user profiles (developed initially within ontology models) and apply ontological 'if-then' statements to these profiles to extract relevant information concerning a user action or preference. Rules effectively rely on pre-defined classes of users to determine which content or which service is presented to a user at any time [39]. The overall effectiveness of using a rule-based approach to personalisation relies heavily on the accuracy of the knowledge contained within the rule base (i.e. an ontology model).

Recent research has focused on the use of semantic technologies (via ontological modelling) to enable user representation and personalisation. Advantages of using semantic based approaches include their interoperability, the expressive power of ontology-models and their ability to model complex relationships in different domains of knowledge. Semantics enable applications to infer additional knowledge about core concept. The core underlying components within the Semantic Web include the use of metadata, logical rules and user-ontology models of information. Researched conducted by Razmerita in [40] has focused on the development of an ontological framework for modelling user behaviours in pervasive environments, in particular focusing on modelling for personalisation. As an example, we could model a user's preferences and infer that they have a vision problem, should they always select a media preference of 'Audio' while using a smart-phone application.

The Semantic Web Rule Language (SWRL) has emerged as a core rule-based component within ontology models. SWRL is a combination of a rule mark-up language known as RuleML and the standard Web Ontology Language (OWL) [40]. Particularly within the field of mobile-service delivery in healthcare, rule-based personalisation can be used to define a set of logical expressions that can then be used to deliver personalisation feedback at any time. Rules are simply logical statements that are constructed based on operators such as variables. Within SWRL, the conditions that are created within statements are evaluated to be either true or false. If all of the pre-conditions within the statement are true, it will then lead to a consequence. The consequence would be the output. For example, the consequence of a rule may be to change the user interface of a smartphone to have a large font-size, should the user prefer large text (as stated in their user profile or based on usage data over time). This process is known as a 'cause-effect' relationship and is more commonly described as a rule. Ontologies make use of SWRL rules for personalisation, where they can be exploited to derive personalised delivery of services or content via rule-based reasoning.

An example of a SWRL rule is presented in Fig. 5, where each rule consists of a body (an antecedent) and a head (a consequent). These rules are stored simply as a series of OWL individuals (instances of classes) within an ontology model.

A personalised service would involve a series of logical rules, the information provided via user profiles in ontology models and context information from the user's surroundings. For mobile-based services users may make requests for help via smart-phones or tablet devices within different environments (for example, a user may request travel help when driving, or set up personalised reminders for taking medication at specific times or places). Alongside the rules, reasoning is a

```
rule ::= 'Implies(' { annotation } antecedent consequent ')'
         antecedent ::= 'Antecedent(' { atom } ')'
         consequent ::= 'Consequent(' ( atom } ')'
```

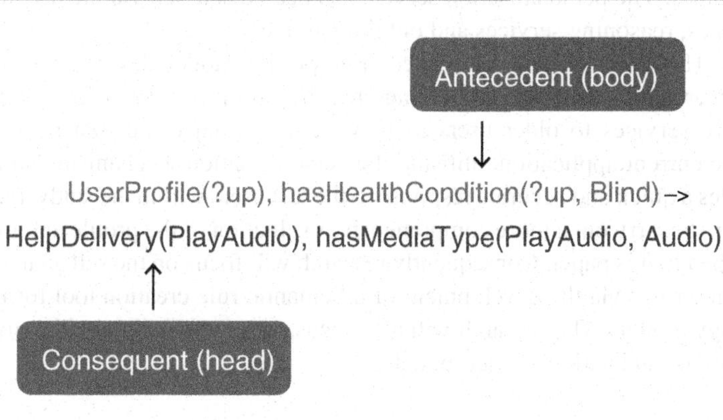

Fig. 5 An example SWRL rule, taken from the user profile ontology in [3], which shows how a user can have a medical health condition of blindness, where media will therefore be displayed in audio format

requirement and is used to infer information based on existing content. Reasoning engines store the user profile information and the rules, and perform forward based changing when a user requests help. If a user makes a request, this is sent to the personalisation services and contextual information is captured at this point. The reasoning engine is then used to check if any antecedent of a rule is met before firing a particular rule. The outcome, or consequence of that rule is used to fire additional rules, if required. Forward changing therefore makes use of user behaviours and context information in order to provide personalised service delivery, which is environment dependent.

8 Conclusions

Within this paper, a comprehensive review of research was undertaken, within the areas of context modelling, user profile modeling and personalisation in context-aware environments. In addition to this, the paper detailed the conceptualisation and development of a novel user profile ontology model and a personalisation component for use in assistive services. We also highlighted existing literature that focuses on the use of semantic rules, reasoning and inference for the personalisation of user-based services. The use of HoD services via MobileSage was discussed, highlighting their research within this domain. Following from this, the paper described a user profile model capable of enabling the delivery of context-aware services through personalised media feedback and context-aware assistance.

The component and model can be adapted to numerous scenarios through the use of personalised semantic rule-based reasoning. To demonstrate the utility of the service, a personalised HoD service is presented through a smart-phone application. The personalisation service has been discussed via the use of semantic web rules, reasoning services and ontology models.

The HoD service and associated user profile model described in this paper have been adopted by the MobileSage research project for providing personalised assistive services to older users as they travel. Initial evaluation results indicate that the current application, utilising the personalisation mechanism and the model provides a quick and accurate response to the test users within the study. Future work will aim to carry out a more comprehensive evaluation of the profile automation tool described in this paper. Consequently research will focus on the self-management of Semantic rules via the development of a Semantic rule creation tool for use within ontology models. The research will also focus on how to maintain identity integrity through the adaptation of user models.

References

1. Cook, D. J., & Das, S. K. (2007). How smart are our environments? An updated look at the state of the art. *Pervasive and Mobile Computing, 3*(2), 53–73.
2. Bharucha, A. J., Anand, V., Forlizzi, J., Dew, M. A., Reynolds III, C. F., Stevens, S., et al. (2009). Intelligent assistive technology applications to dementia care: Current capabilities, limitations, and future challenges. *American Journal of Geriatric Psychiatry, 17*(2), 88–104.
3. Skillen, K.-L., Chen, L., Nugent, C. D., Donnelly, M. P., Burns, W., & Solheim, I. (2013). Ontological user modelling and semantic rule-based reasoning for personalisation of help-on-demand services in pervasive environments. *Future Generation Computer Systems, 34*, 97–109.
4. Zimmermann, A., Specht, M., & Lorenz, A. (2005). Personalization and context management. *User Modeling and User-Adapted Interaction, 15*(3–4), 275–302.
5. Roh, J. H., & Jin, S. (2012). Personalized advertisement recommendation system based on user profile in the smart phone. In *2012 14th International Conference on Advanced Communication Technology (ICACT)* (pp. 1300–1303).
6. Goix, L.-W., Valla, M., Cerami, L., & Falcarin, P. (2007). Situation inference for mobile users: A rule based approach. In *2007 International Conference on. Mobile Data Management* (pp. 299–303).
7. Mikalsen, M., & Kofod-Petersen, A. (2004). Representing and reasoning about context in a mobile environment. In *Proceedings of the First International Workshop on Modeling and Retrieval of Context* (pp. 25–35).
8. Schilit, B., Adams, N., & Want, R. (1994). Context-aware computing applications. In *First Workshop on Mobile Computing Systems and Applications, WMCSA 1994* (Vol. 1994, pp. 85–90).
9. Hong, J., Suh, E., & Kim, S. ((2009). Expert systems with applications context-aware systems? A literature review and classification. *Expert Systems with Applications, 36*(4), 8509–8522.
10. Bettini, C., Brdiczka, O., Henricksen, K., Indulska, J., Nicklas, D., Ranganathan, A., et al. (2010). A survey of context modelling and reasoning techniques. *Pervasive and Mobile Computing, 6*(2), 161–180.
11. Want, R. (2013). Keynote 1: The golden age of pervasive computing. In *2013 IEEE International Conference on Pervasive Computing and Communications (PerCom)* (p. 1).

12. Baldauf, M., Dustdar, S., & Rosenberg, F. A survey on context-aware systems. *International Journal of Ad Hoc and Ubiquitous Computing, 2*(4), 263–277.
13. Hoareau, C., & Satoh, I. (2009). Modeling and processing information for context-aware computing: A survey. *New Generation Computing, 27*(3), 177–196.
14. Chen, L., Nugent, C., & Wang, H. (2011). A knowledge-driven approach to activity recognition in smart homes. *IEEE Transactions on Knowledge and Data Engineering, PP*(99), 1.
15. Rodríguez, J., Bravo, M., & Guzmán, R. (2013). Multi-dimensional ontology model to support context-aware systems. In *Workshops at the Twenty-Seventh AAAI Conference on Artificial Intelligence*, March 2012 (pp. 53–60).
16. Fischer, G. (2001). User modeling in human-computer interaction. *User Modeling and User-Adapted Interaction, 11*(1–2), 65–86.
17. Kuflik, T., Kay, J., & Kummerfeld, B. (2012). Challenges and solutions of ubiquitous user modeling. In *Ubiquitous display environments* (pp. 7–30). New York: Springer.
18. Rojbi, S., & Soui, M. (2011). User modeling and web-based customazation techniques: An examination of the published literature. In *2011 4th International Conference on Logistics (LOGISTIQUA)* (pp. 83–90).
19. Ye, J., Dobson, S., & McKeever, S. (2012). Situation identification techniques in pervasive computing: A review. *Pervasive and Mobile Computing, 8*(1), 36–66.
20. Adomavicius, G., & Tuzhilin, A. (2005). Toward the next generation of recommender systems: A survey of the state-of-the-art and possible extensions. *IEEE Transactions on Knowledge and Data Engineering, 17*(6), 734–749.
21. Held, A., Buchholz, S., & Schill, A. (2002). Modeling of context information for pervasive computing applications. In *Proceedings of SCI*.
22. Carmagnola, F., Cena, F., & Gena, C. (2011). User model interoperability: A survey. *User Modeling and User-Adapted Interaction, 21*(3), 285–331.
23. Liu, C. H., Chang, K. L., Chen, J. J. Y., & Hung, S. C. (2010). Ontology-based context representation and reasoning using owl and swrl. In *2010 Eighth Annual Communication Networks and Services Research Conference (CNSR)* (pp. 215–220).
24. Andrejko, A., Barla, M., & Bieliková, M. (2007,). Ontology-based user modeling for web-based information systems. In *Advances in Information Systems Development* (pp. 457–468). New York, Springer.
25. Zukerman, I., & Albrecht, D. W. (2001). Predictive statistical models for user modeling. *User Modeling and User-Adapted Interaction, 11*(1–2), 5–18.
26. Skillen, K. L., Chen, L., Nugent, C. D., Donnelly, M. P., & Solheim, I. (2012). A user profile ontology based approach for assisting people with dementia in mobile environments. In *Conference Proceedings of IEEE Engineering in Medicine and Biology Society* (Vol. 2012, pp. 6390–6393).
27. Viviani, M., Bennani, N., & Egyed-Zsigmond, E. (2010). A survey on user modeling in multi-application environments. In *2010 Third International Conference on Advances in Human-Oriented and Personalized Mechanisms, Technologies and Services (CENTRIC)* (pp. 111–116).
28. Noy, N. F., & McGuinness, D. L. (2001). *Ontology development 101: A guide to creating your first ontology*. Stanford, CA: Stanford University.
29. S. C. for B. I. Research, "Protégé - A free, open-source ontology editor and framework for building intelligent systems." [Online]. Available: http://protege.stanford.edu/.[Accessed:22-Sep-2014].
30. Eyharabide, V., & Amandi, A. (2012). Ontology-based user profile learning. *Applied Intelligence, 36*(4), 857–869.
31. Kobsa, A. (2001). Generic user modeling systems. *User Modeling and User-Adapted Interaction, 11*(1), 49–63.
32. Pan, J., Zhang, B., Wang, S., Wu, G., & Wei, D. (2007). Ontology based user profiling in personalized information service agent. In *7th IEEE International Conference on Computer and Information Technology, CIT 2007* (pp. 1089–1093)

33. Heckmann, D., Schwartz, T., Brandherm, B., Schmitz, M., & von Wilamowitz-Moellendorff, M. (2005). Gumo-the general user model ontology. In *User modeling* (pp. 428–432). Heidelberg: Springer.
34. Sutterer, M., Droegehorn, O., & David, K. (2008). Upos: User profile ontology with situation-dependent preferences support. In *2008 First International Conference on Advances in Computer-Human Interaction* (pp. 230–235).
35. Von Hessling, A., Kleemann, T., & Sinner, A. (2004). Semantic user profiles and their applications in a mobile environment. *Artificial Intelligence in Mobile Systems (AIMS 2004)* (p. 59).
36. Burns, W., Chen, L., Nugent, C., Donnelly, M., Skillen, K.-L., & Solheim, I. (2012). A conceptual framework for supporting adaptive personalized help-on-demand services. In *Ambient Intelligence* (pp. 427–432). Berlin: Springer.
37. Halbach, T. (2013). The European MobileSage project - situated adaptive guidance for the mobile elderly overview , status , and preliminary results. In *The Sixth International Conference on Advances in Computer-Human Interactions* (pp. 479–482).
38. Gao, M., Liu, K., & Wu, Z. (2010). Personalisation in web computing and informatics: Theories, techniques, applications, and future research. *Information Systems Frontiers, 12*(5), 607–629.
39. Razmerita, L. (2011). An ontology-based framework for modeling user behavior?A case study in knowledge management. *IEEE Transactions on Systems, Man and Cybernetics, Part A: Systems and Humans, 41*(4), 772–783.
40. Horrocks, I., Patel-Schneider, P. F., Boley, H., Tabet, S., Grosof, B., & Dean, M. (2004). SWRL: A semantic web rule language combining OWL and RuleML. *W3C Member Submissions ,21*, 79.

Monitoring and Coaching the Use of Home Medical Devices

Yang Cai, Yi Yang, Alexander Hauptmann, and Howard Wactlar

1 Introduction

Home medical devices (e.g. infusion pumps, inhaler, nebulizers, etc.) which are used by patients at home on their own, are becoming more and more prevalent, due to their cost-saving advantages. However, non-professional users, especially for elderly people with cognitive decline, may sometimes wrongly operate a medical device (e.g. not following required operating procedures). More seriously, the device-use error can lead to fatal results. For example, according to [13], during 2005–2010, 710 deaths linked to the use of one kind of home medical devices, the infusion pumps, which intravenously deliver life-critical drugs, food and other solutions to patients. Therefore, it's critical to have some external mechanisms to supervise patient's use of these devices and keep the use-error from happening. One straightforward solution to this problem would be let a professional person play the supervisor's role. However, because it contradicts to the main objective of home medical devices, i.e. reducing the cost, this solution is obviously infeasible. Instead of using "expensive" human's supervision, in this paper, we propose a cognitive assistive system whose objective is to automatically monitor the use of home medical devices.

The cognitive assistive system has two-fold functionalities: perception and recognition. On one hand, the system should be able to perceive user's operations. Since various advanced sensors been developed in the past decades, we are able to perceive user's operations from many different aspects. For example, the Kinect[1]

[1] http://en.wikipedia.org/wiki/Kinect.

Y. Cai • Y. Yang • A. Hauptmann (✉) • H. Wactlar
Carnegie Mellon University, 5000 Forbes Ave, Pittsburgh, PA 15213, USA
e-mail: yangcai1988@gmail.com; yee.i.yang@gmail.com; alex@cs.cmu.edu; hdw@cs.cmu.edu

A. Briassouli et al. (eds.), *Health Monitoring and Personalized Feedback using Multimedia Data*, DOI 10.1007/978-3-319-17963-6_14

that was invented recently can provide us not only the RGB information but also the depth that cannot be captured by traditional cameras. On the other hand, another important requirement for building a successful cognitive assistive system is to be able to recognize the operations so that use-errors can be identified [9]. However, unlike the well-developed perception module, it is still unknown whether the current techniques are mature enough to fulfill this requirement. Even though many techniques have been proposed to recognize human actions [1, 6, 15, 16, 20], none of them has been applied to recognize operations involved in using home medical devices and therefore, we are not sure if they are adequate in such scenario.

Since the lack of corresponding database is the main reason causing the situation, we construct a database (called PUMP) which was specially designed for studying the use of home medical devices and present it in this paper. Particularly, we take the example of patients using an infusion pump as a typical type of home medical device operation for collecting this database. An infusion pump is a device that infuses fluids, medication or nutrients into a patient's blood stream, generally intravenously. Because they connect directly into a person's circulatory system, infusion pumps are a source of major patient safety concerns. Because of this significant impact, we used an infusion pump as the sample device. To collect the data we first define an operation protocol for correct use of an infusion pump [7]. Then, each user was asked to simulate the use of infusion pump for several times. Their operations were recorded by three Kinect cameras from different views as shown in Fig. 1. Seventeen volunteers participated in data collection and 68 multi-view operation sequences were generated respectively. The operation sequences were then manually annotated by locating the temporal intervals of all operations in each sequence. The database will be released to public for research purpose.

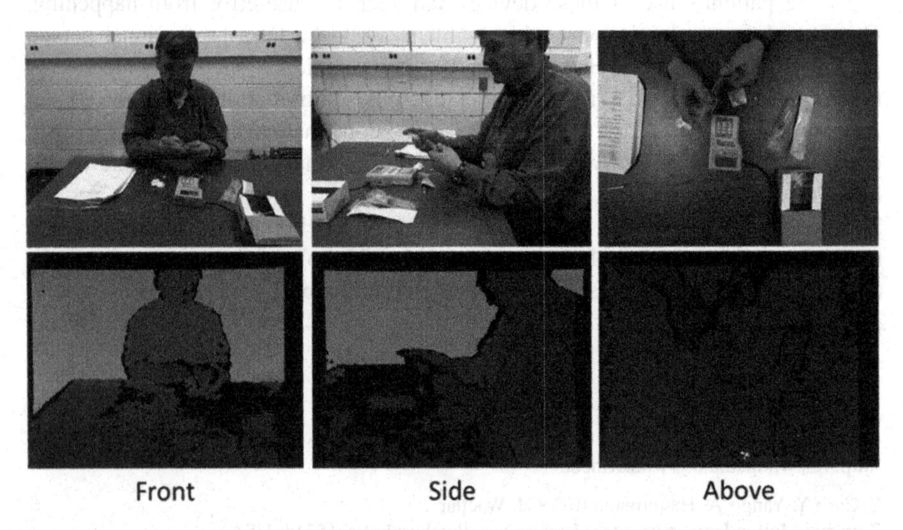

Fig. 1 Examples of data recorded in PUMP database. The *first row* is the RGB data while the *second row* is the corresponding depth data. All user operations were captured by three Kinect cameras from different views

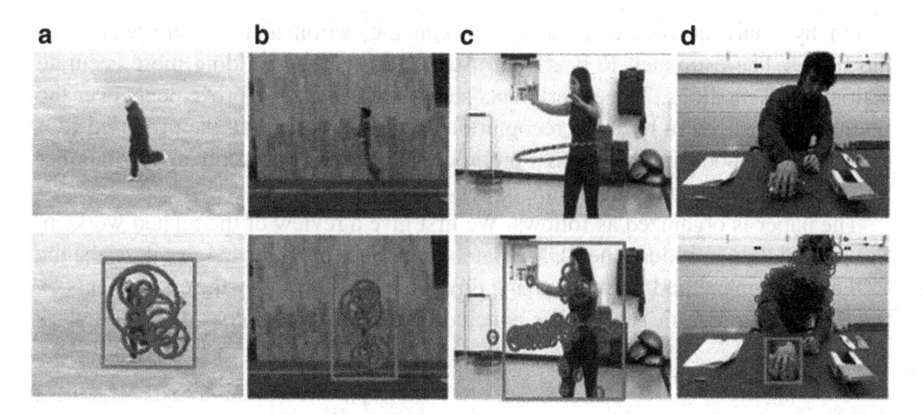

Fig. 2 Comparison between four actions and corresponding extracted MoSIFT features [4]. Only features in *green box* are relevant to actions by definition. (**a**) is "running" from [11], (**b**) is "jumping" from [3], (**c**) is "hula hoop" from [18] and (**d**) is "turning a device on" recorded by ourself (Color figure online)

After building the database, we then evaluate the recognition performance of the existing approaches on the database. Even though using state-of-the-art approach which demonstrates near perfect performance in recognizing general human actions, we observe significant performance drop when applying it to recognize device operations. A subtle and overlooked unique characteristic of actions involved in using devices restrains the performance of the existing action recognition algorithms.

The uniqueness is illustrated in Fig. 2. It shows four actions and the corresponding extracted MoSIFT features [4]. The regions that are relevant to the target actions are indicated by green boxes. Figure 2a–c show the actions of "running", "jumping" and "hula hoop" selected from three popular action recognition datasets [3, 11, 18]. For all three cases, most feature points in the whole frame are inside the green box. Because the "noise" points outside the box are relatively few, it is safe to use all features in the whole frame to model an action. However, as shown in Fig. 2d, for the action "turning a device on", a typical action in using a home medical device, only a very small part of the features lies in the green box compared to all the features extracted from the whole frame. In this case, it is no longer reasonable to use all the features to represent actions, since the representation will be contaminated by substantial amount of essentially random noise. Such differences in feature distributions can be attributed to the fact that the relevant motion of the action in Fig. 2d is non-dominant and with a relatively small area compared to co-occurring non-relevant motion in the frame. We call this type of actions as *tiny actions*. Most of device operations are tiny actions, because we usually operate a device only with a body part, such as hand or foot, instead of the whole body.

To recognize tiny actions, it's critical to focus on the local area where the target action happens, namely the region of interest (ROI). Therefore, in the second part of this paper, we introduce a simple but effective approach to estimating ROI for recognizing tiny actions. Specifically, the method learns the ROI for an action by analyzing the correlation between the action and the sub-regions of the frame. The estimated ROI is then used as a filter for building more accurate action representations. The experiments show performance improvements over the traditional methods in terms of recognition precision. Note that, the proposed ROI estimation method can be used as a preprocessing step before applying any number of existing methods in the literature of action recognition.

The paper is organized as follows. We first give a review of the related works in Sect. 2 and then introduce the PUMP database in Sect. 3. In Sect. 4, we describe the ROI estimation method for recognizing tiny actions, followed by the experiments on the PUMP database in Sect. 5. We conclude our work and discuss the future work at Sect. 7.

2 Related Works

2.1 *Existing Action Databases*

We give a review of current existing action databases and compare them to our proposed one using three taxonomies: (1) RGB videos or RGB+Depth videos, (2) single camera or multiple cameras and (3) significant action or tiny action.

2.1.1 RGB Videos Versus RGB+Depth Videos

The videos of most current databases are RGB videos captured by traditional cameras, such as UCF50 [18], Hollywood2 [12], HMDB [10], KTH [21], Weizmann [3], UT-Interaction [19] and IXMAS [24]. Thanks to the greater availability of RGB+Depth cameras (e.g. Kinect), recently a few 3D databases which contain RGB+Depth videos have been proposed. For example, the MSR3D [23]. Compared to traditional RGB videos, the RGB+Depth videos preserve the additional depth information which could be useful for action analysis. PUMP is a RGB+Depth database.

2.1.2 Single Camera Versus Multiple Cameras

The single camera database refers to those recording actions only use one camera each time, while each action in multi-camera databases was simultaneously recorded by multiple cameras with overlapped views. Again, most of current databases are single camera ones (e.g. UCF50, Hollywood2, HMDB, KTH, Weizmann, UT-Interaction and MSR3D). There are only a few multi-camera action databases have been published, such as IXMAS where each action was captured by five cameras from different views. Since we use three cameras in PUMP, it is therefore a multi-camera database.

2.1.3 Significant Actions Versus Tiny Actions

We classify actions into two types in terms of their motion strength and relative area compared to whole motion region. The motion of significant actions is strong and dominant one compared to other co-occurred motion in a frame. In contrast, tiny actions' motion is weak, non-dominant and with relative small area. By this definition, the actions in KTH, Weizmann, UT-Interaction, MSR3D and IXMAS are mainly significant actions. For UCF50, Hollywood2 and HMDB, they contain videos with both significant actions and tiny actions. As shown in Fig. 4, in PUMP database, all actions are about hands operations, whose movement are weak and only taking a small area compared to co-occurred motion on body, head and etc. Therefore, they are all tiny actions.

2.2 *Recognizing Human-Object Interaction*

Accurately speaking, the action recognized in cognitive assistive system can be further categorized as human-object interaction which is an important group of actions recognition problems [1]. As [1] indicated, existing methods for recognizing interactions between human and object can generally be classified into two classes by judging if recognition of objects and actions are independently or collaboratively. For methods falling into first categories, objects recognition serves the following action recognition. For example, objects are usually recognized first and then actions are recognized by analyzing the object's motion. As for methods in second class, object and action are recognized in a collaborate fashion and the recognition of objects and actions serve each other. This work falls into the first category. However, different from previous works, we novelly use the object (device) as a cue for estimating the ROI.

2.3 *Detecting Salient Region*

Similar to ROI detection, salient region detection also finds a sub-region in an image or video that is considered to be salient. However, despite the similarity, their difference also worths noting. The saliency of a region is defined by its visual uniqueness, unpredictability, rarity and is caused by variations in image attributes, such as color, gradient, edges, and boundaries [5]. In other words, the saliency detection is not task-dependent by relies on the above rules. However, the ROI detection in visual-based coaching system is task-dependent. For example, the region including device operations may not be salient, due to the weak motion, but is of interest. Due to this difference, existing approaches for salient region detection cannot be directly applied to solve our problem.

3 PUMP Database

3.1 Data Collection Methodologies

We selected the Abbot Laboratories Infusion Pump (AIM Plus Ambulatory Infusion Manager) as an example of home medical devices for research. With the help of a medical devices expert, we first defined an operation protocol for correct use of infusion pump, as shown in Table 1. Then, as illustrated in Fig. 3, we set up a "workplace" for data recording. Specifically, we used three Kinect cameras to record user's operation from three views: front, side, above. Before each time of recording, an infusion pump with off-state, two refilled syringes and a box of alcohol pads were prepared on the table.

Each user was asked to perform the operation following certain procedures for several times. In each time, they followed either the exact same procedures as described in Table 1 (correct operation protocol) or the predefined wrong procedures (to simulate the use-errors). Table 2 listed four types of predefined wrong procedures. They were different from the correct operation protocol by including steps disordering and steps missing. During the data recording, there were videos where users unintentionally deviated from the operation protocol they were asked to follow. Since these videos in fact reflected the use error that user made in real-life, we kept them in our database and provided additional error descriptions if the errors they made not belonged to any of the four predefined errors (Due to the limited space, we didn't include the descriptions in our paper but kept them as an independent file in the database).

Table 1 The proposed operation protocol for correct use of infusion pump

1	Turn the pump on
2–5	Press buttons to set up infusion program
6	Uncap pump tube end
7	Clean pump tube end using alcohol pad
8	Open arm port
9	Clean arm port using alcohol pad
10	Flush arm port using syringe
11	Connect arm port and pump tube end
12	Press "START" button to start infusion
13	Press "STOP" button after infusion
14	Disconnect arm port and pump tube end
15	Clean pump tube end using alcohol pad
16	Cap pump tube end
17	Clean arm port using alcohol pad
18	Flush arm port using syringe
19	Clean arm port using alcohol pad
20	Cap arm port
21	Turn the pump off

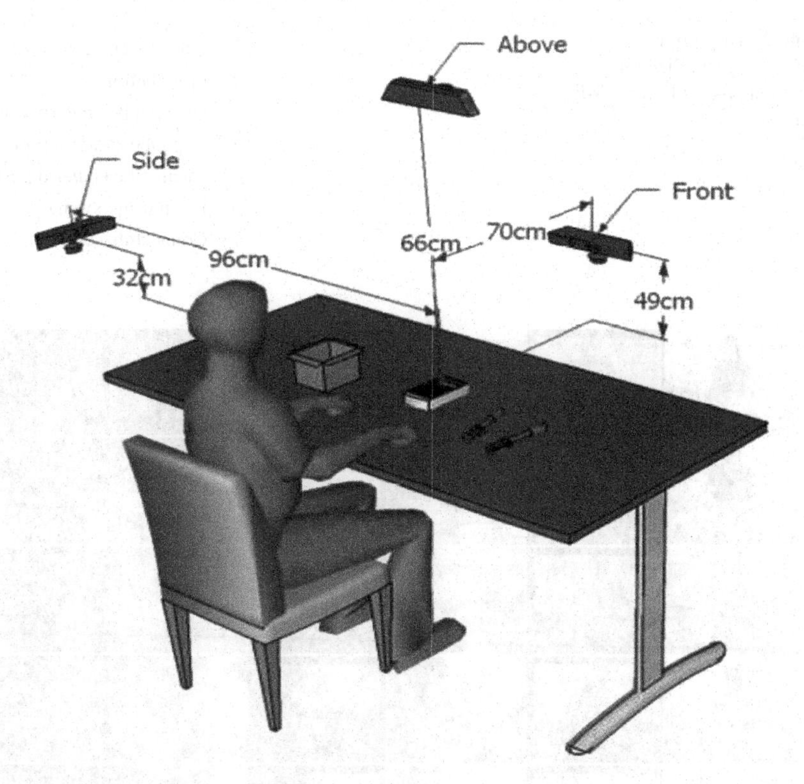

Fig. 3 An illustration of data collection setting for PUMP database [8]

Table 2 The four types of predefined wrong operation protocols

W1	Switch step 6 and 7:
	Disorder flushing using syringe with cleaning arm port
W2	Remove step 4 and 6:
	Forget cleaning arm port and pump tube end
W3	Remove step 13 and 17:
	Forget cap arm port and pump tube end
W4	Remove step 9 and 10:
	Forget press buttons to start and stop infusion

Note that it only listed the differences between the correct protocol and wrong one and non-mentioned parts were same as correct protocol

Since some different steps in operation protocol were in fact the same actions, we therefore categorized them into one action class. We further combined the classes with same actions but operating different devices into one class, which finally leaded to seven action classes. We listed the aggregated classes in Table 3 and gave the snapshots of corresponding actions in Fig. 4.

Table 3 The action categories generalized from operation protocol for PUMP database

1	Turn the pump on/off
2	Press buttons
3	Uncap tube end/arm port
4	Cap tube end/arm port
5	Clean tube end/arm port
6	Flush using syringe
7	Connect/disconnect

Fig. 4 An illustration of seven action classes generalized from operation protocol of PUMP database. (**a**) Turn the pump on/off. (**b**) Press buttons. (**c**) Uncap tube end/arm port. (**d**) Cap tube end/arm port. (**e**) Clean tube end/arm port. (**f**) Flush using syringe. (**g**) Connect/disconnect tube end and arm port

Table 4 The inventory of PUMP database

Video type	OpenNI	RGB	Depth
View count	3		
Participant count	17		
Operation sequences	68		
File count	204		
File format	ONI	AVI	AVI
Frame rate	≈20	20	20
Resolution	640 × 480		
Average duration (m)	4.31		
Total duration (h)	14.64		
Disk storage (GB)	157	21	27

Note that due to the Kinect hardware issue, the frame rate of OpenNI videos was not constant but with little variation

Based on the generalized action categories, we manually annotated all sequences by locating the temporal intervals of all actions in each sequence.

3.2 Database Statistics and Recording Details

There were 17 volunteers participating in the data recording. Each user was asked to operate the infusion pump for four times with different appearances. Specifically, two times were correct operations while the others two came from the wrong operations. We finally constructed a database containing 68 operation sequences where each sequence had three synchronized RGB+Depth videos recorded from different views.

We adopted OpenNI[2] for Kinect recording and stored the raw data in ONI file format. To facilitate the use of this database, we also provided calibrated RGB videos and depth videos extracted from raw OpenNI data in our database. In Table 4, we show detail statistics of the PUMP database. The average video duration was 4.31 in and the total video duration was 14.64 h.

4 ROI Estimation for Recognizing Tiny Actions

4.1 Notations

Let (x, y, z) be the coordinates of the corner index of a cuboid region, A_j be an action of class j, M be the number of action classes. Let $T^{A_j}(x, y, z)$ be a density map that describes the probability for a region at (x, y, z) belonging to the ROI of A_j. Specifically, we call $T^{A_j}(x, y, z)$ as the *ROI template*.

[2]http://openni.org/.

4.2 Action-Region Correlation Estimation

Noticing the ROI for an action can be interpreted as regions that have strong correlation with the action, we estimate the correlation between an action A_j and each region at (x, y, z) in the "3D" frame recorded by Kinect cameras.

To represent each region, we generate sliding windows starting from the origin of "3D" frame and calculate the bag-of-words (BoW) [22] representation for each cuboid window. To reduce the computation cost in following steps, we then apply the principal component analysis (PCA) on the BoW of each window and only keep the dimensions corresponding to the top K largest eigenvalues. The fisher score [2] is used to estimate the correlations between each region and an action.

Let $D_b^{(x,y,z),A_j}$ and $D_w^{(x,y,z),A_j}$ be action class A_j's *between class distance* and *within class distance* [2] respectively for all regions at (x, y, z) among all training data. Then, $D_b^{(x,y,z),A_j}$ and $D_w^{(x,y,z),A_j}$ are defined as:

$$D_b^{(x,y,z),A_j} = \sum_{k=1}^{M} \left(\mu_{\delta(k=j)}^{(x,y,z)} - \mu^{(x,y,z)}\right)^T \left(\mu_{\delta(k=j)}^{(x,y,z)} - \mu^{(x,y,z)}\right)$$

$$D_w^{(x,y,z),A_j} = \sum_{k=1}^{M} \sum_{b \in B_k^{(x,y,z)}} \left(b - \mu_{\delta(k=j)}^{(x,y,z)}\right)^T \left(b - \mu_{\delta(k=j)}^{(x,y,z)}\right)$$

where $\delta(z)$ is an indicator function that outputs 1 if z is true and 0 otherwise, $B_k^{(x,y,z)}$ is a set of BoWs of A_k at region (x, y, z), $\mu^{(x,y,z)}$ is the mean of BoWs at region (x, y, z) for all action classes and $\mu_{\delta(k=j)}^{(x,y,z)}$ is the mean of BoWs at region (x, y, z) for action class A_j or action classes other than A_j(depending on if $k = j$). Then the fisher score $F^{(x,y,z),A_j}$ for an action class A_j and a region at (x, y, z) is simply:

$$F^{(x,y,z),A_j} = \frac{D_b^{(x,y,z),A_j}}{D_w^{(x,y,z),A_j}}.$$

If one region (x, y, z) is highly correlated to action A_j, it will then have relatively small *within class distance* $D_w^{(x,y,z),A_j}$ and large *between class distance* $D_b^{(x,y,z),A_j}$, which gives large fisher score $F^{(x,y,z),A_j}$. Therefore, fisher score can be an indicator of correlation between an action and regions.

By normalizing the fisher score at different regions, we get the representation of the ROI template $T^{A_j}(x, y, z)$:

$$T^{A_j}(x, y, z) = \frac{F^{(x,y,z),A_j}}{\sum_{x,y,z} F^{(x,y,z),A_j}}.$$

4.3 *ROI Adaption and Noise Filtering*

For a given input video sequence, we simply attach the ROI template $T^{A_j}(x, y, z)$ to each frame of the video sequence. Then, all feature points with ROI score lower than threshold λ are removed. In this way, we model the action only based on features in ROI.

5 Experiments

5.1 *Experimental Setting*

The total 68 videos are divided into two-fold and we in turn use one fold as training and the other one as testing data. For each video, we extract the MoSIFT [4] as low-level features and encode them into visual words [22] using a codebook with vocabulary size of 1,000. Then three different methods are used for generating the action representations (see Sect. 5.2 for details). SVM classifier with RBF kernel is adopted for action classification and two-fold cross validation is used for classification model training. Specifically, the training and testing is done independently for each view. To evaluate the effectiveness of different methods, the mean average precision (MAP) [17] which is the average precision (AP) over all actions is computed.

5.2 *Action Representation Methods*

The bag-of-words model (BoW) is adopted for action representation. Each high dimensional local feature point (e.g. MoSIFT) is first mapped to the closest cluster center using the pre-trained codebook and then the cluster's id is assigned to the feature as "visual word". After that, a pooling step is applied to calculate the statistics of all the visual words in the video segment and represent it as vector with same dimension of the codebook. This vector representation is called the BoW of the video segment. In this paper, we experiment with three different pooling methods for action representation as introduced below.

5.2.1 Whole Frame Based Pooling (WF-BoW)

Most of existing BoW-based action recognition approaches [4, 21] use this pooling method. Namely, all visual words in the whole frame are aggregated together first and the frequency for each visual word is calculated then. Finally, the normalized frequency histogram is used as the final representation.

5.2.2 Depth-Layered Multi-Channel Based Pooling (DLMC-BoW)

Since the videos are recorded by Kinect camera, then each feature point extracted from the key frame has not only the x and y coordinates but also the depth z coordinate. Based on this observation, in [14], z axis is first divided into several depth-layered channels, and then features within different channels are pooled independently, resulting in a multiple depth channel histogram representation. In our implementation, we uniformly divide the depth space into five channels.

5.2.3 ROI Based Pooling (ROI-BoW)

In order to estimate the ROI, $200 \times 200 \times 100$ pixels (in the order of x, y and z axis) sliding cuboid windows with moving step of 40 pixels are generated and represented as BoW by aggregating all visual words inside the cuboid window. We then apply the PCA on the BoW of each window and only keep the dimensions corresponding to the top 100 largest eigenvalues. After that, the ROI template is calculated using the method described in Sect. 4.2. Note that, all these process can be done off-line. At online testing stage, all visual words with ROI score lower than 0.5 are filtered out and the final BoW representation is built only based on the left visual words belonging to the ROI.

5.3 *Experimental Results and Analysis*

5.3.1 ROI Visualizations

To qualitatively evaluate the proposed ROI estimation method, in Fig. 5, we visualize estimated ROIs for each action of all three views using density map where higher intensity means high probability of belonging to be the ROI. Because the cameras are put at different positions, the ROI for the same action but different

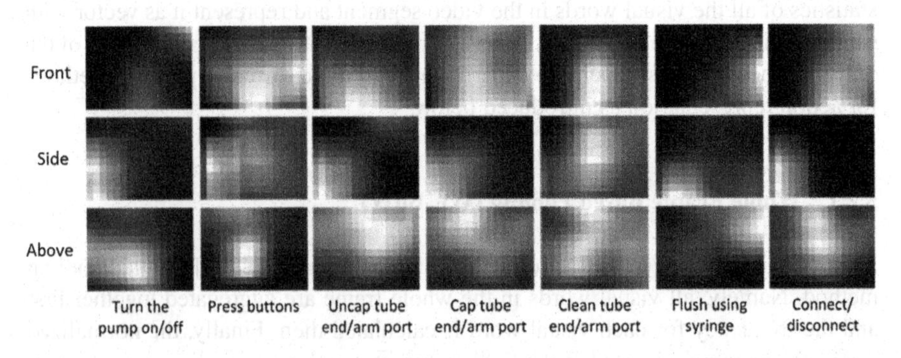

Fig. 5 The visualizations of estimated ROI for each action. Each row corresponds to one of the three views. The action examples of different views can be found in Fig. 3

views can be different. Comparing the action examples shown in Fig. 3 and the corresponding ROIs in Fig. 5, we can find the estimated ROIs make sense intuitively. For example, for the "front" view, the ROI is in the middle for action "press buttons" while is on the right for action "Flush using syringe". This is because the "pressing buttons" always happen in the middle of the frame while for "flush using syringe" users always need to take out the syringe from syringe bag which is placed on the right side of the frame. Also, we can observe some of actions' ROI are dense and sharp (e.g. "Connect/disconnect" in side view) while others are relative sparse and soft (e.g. "Cap tube end/arm port" in front view). This difference can be attributed to the different motion patterns of different actions. For example, "Connect/disconnect" concentrates in a narrow area but the motion of other actions like "Cap tube end/arm port" distributes in relatively larger areas.

5.3.2 Action Recognition Performance

In Table 5, we summarize the experimental results of three action representation methods. For each method, the first three columns correspond to the performance on three different views while the last column is the fusion results given by manually selecting the best performance among the three views. The fusion results can be interpreted as the best performance that the cognitive assistive system achieves using that method. Comparing the average fusion results of three action representation methods, we can see that the *ROI-BoW* achieves the best performance and improves the MAP for 3.33 % compared with *WF-BoW*. *DLMC-BoW* has almost the same performance as *WF-BoW*.

If we further compare three methods' performance on each single view, we observe a different performance changing pattern. For the "front" and "side" views, both *ROI-BoW* and *DLMC-BoW* show significant improvements over the *WF-BoW*. Specifically, on average, *DLMC-BoW* improves for 4.42 and 10.20 % while *ROI-BoW* improves for 8.84 and 19.28 % on those two views. However, for the "above" view, both *ROI-BoW* and *DLMC-BoW* don't show improvement but in fact slightly decrease the performance compared to *WF-BoW*. The reason causing the inconsistent performance changing pattern on different views is illustrated in Fig. 6. It visualizes of extracted MoSIFT features in example frames with the same time stamp but different views for action "Flush using syringe". Again, we use green boxes to indicate the regions that are relevant to the target action. We can see that for "front" and "side" views, only a very small part of the features lies in the green box compared to all the features extracted from the whole frame, while most features are inside the green box for the "above" view. Therefore, due to the difference in camera positions, actions recorded by "front" and "side" cameras are always the most typical tiny actions. Because both *DLMC-BoW* and *ROI-BoW* can be interpreted as feature location based visual word weighting methods (*DLMC-BoW* does it implicitly by using SVM to weight features at different depth differently while *ROI-BoW* does it explicitly by hard weighting features inside ROI 1 and outside 0), they are most effective when actions are typical tiny actions.

Table 5 MAP comparison of three action representation methods on PUMP database

	WF-BoW				*DLMC-BoW*				*ROI-BoW*			
Actions	Front	Side	Above	Fusion	Front	Side	Above	Fusion	Front	Side	Above	Fusion
Turn the pump on/off	0.8356	0.5931	0.7222	0.8356	0.863	0.7761	0.8232	0.863	0.894	0.8348	0.8696	**0.894**
Press buttons	0.8044	0.5841	0.7703	0.8044	0.8312	0.7812	0.782	0.8312	0.8833	0.8058	0.8378	**0.8833**
Uncap tube end/arm port	0.4933	0.2679	0.6335	0.6335	0.6252	0.4034	0.5201	0.6252	0.6541	0.5215	0.5326	**0.6541**
Cap tube end/arm port	0.3466	0.2898	0.4356	0.4356	0.3724	0.3303	0.4352	0.4352	0.4455	0.398	0.3923	**0.4455**
Clean tube end/arm port	0.8684	0.6967	0.8676	0.8684	0.8834	0.7331	0.8445	0.8834	0.9202	0.8438	0.8401	**0.9202**
Flush using syringe	0.831	0.6718	0.8329	0.8329	0.8845	0.7589	0.8038	0.8845	0.948	0.9048	0.7931	**0.948**
Connect/ disconnect	0.451	0.3361	0.6349	**0.6349**	0.48	0.3704	0.5302	0.5302	0.5037	0.4801	0.5335	0.5335
Average	0.6615	0.4914	0.6996	0.7208	0.7057	0.5933	0.677	0.7218	0.7498	0.6841	0.6856	**0.7541**

For each method, the performance is evaluated on each camera independently. The best performance among the three cameras is manually selected and shown in the "fusion" column

Fig. 6 The visualization of extracted MoSIFT features in frames with the same time stamp but of different views

However, even though *ROI-BoW* improves the performance significantly, if looking into the absolute performance of each action, we realize only actions "Turn the pump on/off", "Press buttons", "Clean tube end/arm port" and "Flush using syringe" achieve reasonable high average precision while the performance of left three actions is still low. To build an applicable cognitive assistive system, we have to accurately recognize all the actions. Therefore, it still requires special effort to further boost the performance of the difficult actions.

6 An Interaction Model for Coaching Use of Home Medical Devices

Separate from the detection problem is the interaction model for dealing with the user. The detection module, as described in the above sections, handles the analysis of the camera feeds and reports whether or not a step has been detected as correctly performed, and with what level of confidence. This is reported to an interaction module. The interaction module will have prior knowledge what the typical average confidence of the detection for the particular step is, what the acceptable sequences of steps are, and what the importance is for each step based on the scale which we will describe later. The user may also interact with the system to go through the instructions by using a mouse.

6.1 Importance Levels

We assign four different levels of importance to each step in the procedure. The interaction module functions differently for each step, depending on the level of importance of that step. An harmless or obvious step would be one where is unnecessary to warn the user if not performed, such as turning the system on. This would be categorized as level 0 importance. A very important step would be placed at Level 3 importance, which would require the system to make it hard for a user

Table 6 Example of warnings and reminders at different levels of error importance and confidence

	Importance			
Confidence	Level 0	Level 1	Level 2	Level 3
High	No action, green light	Audio-visual correction information	Flashing audio-visual correction information	Red, flashing visual correction with loud audio and pop-up warning
Moderate	Yellow light	Visual reminder with soft audio	Flashing audio-visual reminder with soft audio	Red, flashing visual reminder with soft audio and pop-up warning
Low	Yellow light	Visual reminder only	Flashing visual reminder with beep	Red, flashing visual reminder with soft audio and pop-up warning

to ignore the error warnings issued by the system, e.g. by combining loud, flashing audio-visual notification. Table 6 shows the different actions by the system given their importance levels.

6.2 *Event Detection Certainty*

One of the ways to prevent user frustration with system miss-recognitions is to have the system communicate what the certainty of its detection is, with appropriately adapted feedback. The quickest way for a user to begin doubting and ultimately ignoring the system, is to report an event as detected or not detected when (to the user) it is clearly the opposite. To combat this, the system incorporates an algorithm to decide whether an action is detected as incorrect or not detected. The warning for a user error with a high enough confidence rating will be shown to the user. Furthermore, the system adapts the presentation of the message to convey the confidence level.

- Generally, moderate to high confidence detection of a correct step is only indicated with a green light, without otherwise disrupting the user.
- Less confident detections of the correct step might be indicated in yellow.
- Moderate to high confidence detection of an incorrect action triggers a strong warning
- Low confidence detection of an incorrect action would trigger a suggestion or reminder, without commitment by the system that an actual error was made.

6.3 *Warning Levels*

Instead of confusing the user with percentages of confidence in the event detection, warnings will be presented appropriately to denote both error severity and confidence levels. This will prevent the system from making blatant statements that might contradict reality such as noting an event as not detected when to the user it clearly happened. These confidence and severity appropriate warnings allow the system to function usefully if it occasional events are misclassified.

7 Conclusion and Future Work

In this paper, we proposed a cognitive assistive system to monitor the use of home medical device. To build such a system, accurately recognition of user actions is one of the most essential problems. However, since few research has been done in this specific direction, it is still unknown if current techniques are adequate to solve the problem. In order to facilitate the research in this area, we made three contributions in this paper. First of all, we constructed a database where users were asked to simulate the use of infusion pump following predefined procedures. The operations were recorded by three Kinect cameras from different views. All the data was manually labeled for experimental purpose. Secondly, we performed a formal evaluation of some existing approaches on the proposed database. Because we realized current methods can hardly deal with tiny actions involved in using home medical devices, we made our third contribution by introducing an ROI estimation method and applying the ROI for building more accurate action representations. The experiments show significant performance improvements over the traditional methods by using the proposed methods.

Finally, we outline an interaction model how to handle different levels of errors recognized with varying certainty and appropriately provide feedback to the user reflecting the severity/importance of the mistake as well as the confidence of the system in the correctness of the recognition.

Currently, we treat all the actions in operating home medical devices as independent ones and recognize them separatively. However, it obvious they are mutually related because they are operations in a procedure. Therefore, in the future, we will focus on leveraging the inner relations between actions to further improve the recognition performance.

Acknowledgements This work was supported in part by the National Science Foundation under Grant No. IIS-0917072. Any opinions, findings, and conclusions expressed in this material are those of the author(s) and do not reflect the views of the National Science Foundation.

References

1. Aggarwal, J. K., & Ryoo, M. S. (2011). Human activity analysis: A review. *ACM Computing Surveys (CSUR)*, 43(3), 1–43.
2. Belhumeur, P. N., Hespanha, J. P., & Kriegman, D. J. (1997). Eigenfaces vs. fisherfaces: Recognition using class specific linear projection. *IEEE Transactions on Pattern Analysis and Machine Intelligence*, 19(7), 711–720.
3. Blank, M., Gorelick, L., Shechtman, E., Irani, M., & Basri, R. (2005). Actions as space-time shapes. In *International Conference on Computer Vision*.
4. Chen, M. Y., & Hauptmann, A. (2009). MoSIFT: Reocgnizing human actions in surveillance videos. In *CMU-CS-09-161*.
5. Cheng, M. M., Zhang, G. X., Mitra, N. J., Huang, X., & Hu, S. M. (2011). Global contrast based salient region detection. In *IEEE Conference on Computer Vision and Pattern Recognition*.
6. Dollar, P., Rabaud, V., Cottrell, G., & Belongie, S. (2005). Behavior recognition via sparse spatio-temporal features. In *Joint IEEE International Workshop on Visual Surveillance and Performance Evaluation of Tracking and Surveillance*.
7. Gao, Z., Chen, M. Y., Detyniecki, M., Wu, W., Hauptmann, A., Wactlar, H., et al. (2010) Multi-camera monitoring of infusion pump use. In *IEEE International Conference on Semantic Computing*.
8. Gao, Z., Detyniecki, M., Chen, M. Y., Hauptmann, A. G., Wactlar, H. D., & Cai, A. (2010). The application of spatio-temporal feature and multi-sensor in home medical devices. *International Journal of Digital Content Technology and Its Applications*, 4(6), 69–78.
9. Gao, Z., Detyniecki, M., Chen, M. Y., Wu, W., Hauptmann, A. G., & Wactlar, H. D. (2010). Towards automated assistance for operating home medical devices. In *International Conference of Engineering in Medicine and Biology Society*.
10. Kuehne, H., Jhuang, H., Garrote, E., Poggio, T., & Serre, T. (2011). HMDB: A large video database for human motion recognition. In *International Conference on Computer Vision*.
11. Laptev, I. (2005). On space-time interest points. *International Journal of Computer Vision*, 64(2/3), 107–123.
12. Marszałek, M., Laptev, I., & Schmid, C. (2009). Actions in context. In *IEEE Conference on Computer Vision and Pattern Recognition*.
13. Meier, B. (2010). F.D.A. Steps up oversight of infusion pumps. In *New York Times*.
14. Ni, B., Wang, G., & Moulin, P. (2011). RGBD-HuDaAct: A color-depth video database for human daily activity recognition. In *International Conference on Computer Vision Workshops*.
15. Niebles, J. C., Wang, H., & Fei-Fei, L. (2008). Unsupervised learning of human action categories using spatial-temporal words. *International Journal of Computer Vision*, 79(3), 299–318.
16. Parameswaran, V., & Chellappa, R. (2006). View invariance for human action recognition. *International Journal of Computer Vision*, 66(1), 83–101.
17. Philbin, J., Chum, O., Isard, M., Sivic, J., & Zisserman, A. (2007). Object retrieval with large vocabularies and fast spatial matching. In *IEEE Conference on Computer Vision and Pattern Recognition*.
18. Reddy, K., & Shah, M. (2012). Recognizing 50 human action categories of web videos. *Machine Vision and Applications Journal*, 25(5), 97–81.
19. Ryoo, M. S., Aggarwal, J. K. (2010). UT-Interaction dataset. In *ICPR Contest on Semantic Description of Human Activities (SDHA)*.
20. Ryoo, M. S., & Aggarwal, J. K. (2011). Spatio-temporal relationship match: Video structure comparison for recognition of complex human activities. In *International Conference on Computer Vision*.
21. Schuldt, C., Laptev, I., & Caputo B (2004) Recognizing human actions: A local sVM approach. In *International Conference on Pattern Recognition*.
22. Sivic, J., & Zisserman, A. (2003). Video google: A text retrieval approach to object matching in videos. In *International Conference on Computer Vision*.

23. Wang, J., Liu, Z., Wu, Y., & Yuan, J. (2012). Mining actionlet ensemble for action recognition with depth cameras. In *IEEE Conference on Computer Vision and Pattern Recognition*.
24. Weinland, D., Özuysal, M., & Fua, P. (2010). Making action recognition robust to occlusions and viewpoint changes. In *European Conference on Computer Vision*.

Printed in the United States
By Bookmasters